Cognitive Radio Engineering

Mario Boella Series on Electromagnetism Information and Communication

Piergiorgio L. E. Uslenghi, PhD – Series Editor

The Mario Boella series offers textbooks and monographs in all areas of radio science, with a special emphasis on the applications of electromagnetism to information and communication technologies. The series is scientifically and financially sponsored by the Istituto Superiore Mario Boella affiliated with the Politecnico di Torino, Italy, and is scientifically cosponsored by the International Union of Radio Science (URSI). It is named to honor the memory of Professor Mario Boella of the Politecnico di Torino, who was a pioneer in the development of electronics and telecommunications in Italy for half a century, and a Vice-President of URSI from 1966 to 1969.

Published Titles in the Series

Fundamentals of Wave Phenomena, 2nd Edition
by Akira Hirose and Karl Lonngren (2010)

Scattering of Waves by Wedges and Cones with Impedance Boundary Conditions
by Mikhail Lyalinov and Ning Yan Zhu (2012)

Complex Space Source Theory of Spatially Localized Electromagnetic Waves
by S.R. Seshadri (2013)

The Wiener-Hopf Method in Electromagnetics
by Vito Daniele and Rodolfo Zich (2014)

Higher-Order Techniques in Computational Electromagnetics
by Roberto Graglia and Andrew Peterson (2015)

Forthcoming Titles

Slotted Waveguide Array Antennas
by Sembiam Rengarajan and Lars Josefsson

Cognitive Radio Engineering

ISMB Series

Charles W. Bostian

Bradley Department of Electrical and Computer Engineering,
Virginia Tech

Nicholas J. Kaminski

Trinity College Dublin

Almohanad S. Fayez

Intel

SciTECH
PUBLISHING
an imprint of the IET

scitechpub.com

SciTech PUBLISHING
an imprint of the IET

Published by SciTech Publishing, an imprint of the IET
www.scitechpub.com
www.theiet.org

ISBN 978-1-61353-211-9 (hardback)
ISBN 978-1-61353-212-6 (PDF)

Typeset in India by MPS Limited
Printed in the UK by CPI Antony Rowe Ltd

Contents

Foreword   x
Acknowledgments   xi

1 Introduction   **1**

1.1   What Is a Cognitive Radio and Why Is It Needed?   1

1.2   Book Coverage and Philosophy   2

1.3   The Origin of Cognitive Radio   3

1.4   Overview of Cognitive Radio Operation   5

1.5   An Illustrative Example of Cognitive Radio Application:
      Dynamic Spectrum Access in the Broadcast Television Bands   11

    1.5.1   Introduction   11

    1.5.2   Identifying Frequencies for Cognitive Radio Operation:
        TV Channel Occupancy in the United States   12

    1.5.3   FCC Rules and Commercial Standards for Unlicensed Television
        Band Devices (TVBDs)   13

    1.5.4   FCC Rule Implementation by the IEEE 802.11af Standard   16

    1.5.5   TV White Space Databases   16

    1.5.6   Standard ECMA-392   16

    1.5.7   Cognitive Radio Design for U.S. TV White Space   17

1.6   Can a Radio Really Be Cognitive?   18

2 Cognitive Engine Design   **23**

2.1   Introduction   23

2.2   The Basic Function of a Cognitive Engine   24

2.3   Cognitive Engine Organization   25

    2.3.1   Optimizer   27

    2.3.2   Objective Analyzer   27

    2.3.3   Ranker   28

	2.3.4	Knowledge Base	28
	2.3.5	Radio Interface	28
	2.3.6	Sensor	28
	2.3.7	User Interface	29
	2.3.8	Controller	30
2.4	Tools and Techniques for CE Component Design		30
	2.4.1	Machine Learning	30
	2.4.2	Optimizers	35
	2.4.3	Estimation	38
	2.4.4	Sensing	39
2.5	Cognitive Engine Architecture		42
	2.5.1	Broad Considerations	42
	2.5.2	Monolithic Versus Distributed	44
	2.5.3	Standards	45
	2.5.4	Original CE Architecture	49
	2.5.5	CSERE Architecture	53
2.6	Information Flow in Cognitive Engines		55
	2.6.1	Example Use of Uncertainty Coefficient	57
2.7	Conclusion		60
3	**RF Platforms for Cognitive Radio**		**65**
3.1	Introduction		65
3.2	Preliminary Considerations in Choosing an RF Platform		66
3.3	RF Architectures		67
	3.3.1	Receivers	67
	3.3.2	Transmitter	68
3.4	Receiver RF Specifications		71
	3.4.1	Introduction	71
	3.4.2	Noise, Noise Performance, and Weak Signal Behavior	72
	3.4.3	Strong Signal Behavior	77
3.5	Transmitter RF Specifications		85
3.6	MAC and Performance Considerations		88
3.7	Radio Frequency Integrated Circuits		90
	3.7.1	Introduction	90
	3.7.2	Example: RFM69CW	90
	3.7.3	Computational Support for RFICs	92
3.8	Platforms for Software Defined Radio		94
	3.8.1	Introduction	94
	3.8.2	Packaged RF Front Ends and All-in-one Platforms	94
3.9	Conclusion		99

**4   Cognitive Radio Computation
      and Computational Platforms**   **103**

4.1   The Role of Computing and Cognitive Radio
       Architecture   103

4.2   Control Flow and Data Flow Computer Architectures   103

       4.2.1   Control Flow Computing   103
       4.2.2   Data Flow Computing   106

4.3   Overview of Computational Devices (GPP, DSP, FPGA)   107

       4.3.1   Digital Signal Processors   107
       4.3.2   General Purpose Processors   108
       4.3.3   Field-programmable Gate Arrays   108
       4.3.4   Alternative Computational Devices   109
       4.3.5   Computational Heterogeneity   109

4.4   Models of Computation   110

       4.4.1   Reactive and Real-time Systems   111
       4.4.2   Data Flow Models of Computation   112
       4.4.3   Process Algebra   119
       4.4.4   Calculus of Communicating Systems and Π-calculus   122

4.5   Models-of-Computation Use   123

4.6   Conclusion   123

**5   Integrating and Programming RF and Computational
      Platforms for Cognitive Radio**   **127**

5.1   SDR Platforms   127

5.2   Choosing a Platform   128

       5.2.1   Choosing Between RF Alternatives   128
       5.2.2   Processor Choices   129
       5.2.3   Benchmarks   130
       5.2.4   Processor Interconnect   130
       5.2.5   Other Considerations   131

5.3   Programming   131

       5.3.1   Classic Approach   132
       5.3.2   Model-Based Design   134
       5.3.3   Application of Models-of-Computation   134

5.4   Concluding Remarks   142

6   Cognitive Radio Evaluation   **147**

6.1   Introduction   147

6.2   Performance Evaluation Principles   148

6.3   Metrics and Factors for Cognitive Radio Evaluation   149
       6.3.1   Purpose   150
       6.3.2   Language   151
       6.3.3   Actions   154
6.4   Practical Evaluation Methods   154
       6.4.1   Setup   155
       6.4.2   Logging   155
       6.4.3   Encoding   155
       6.4.4   Interpolation   157
       6.4.5   Alternative Approaches to Evaluation   159
6.5   Example Evaluation   159
       6.5.1   Setup Phase   159
       6.5.2   Logging Phase   160
       6.5.3   Encoding Phase   161
       6.5.4   Interpolation   163
6.6   Example Code   167
       6.6.1   Free FEC Cognitive Radio   167
       6.6.2   Fixed FEC Cognitive Radio   172
       6.6.3   Interpolation Code   176
6.7   Conclusion   178

7   **Cognitive Radio Design for Networking**   **179**
7.1   Networks of Cognitive Radios Versus Cognitive Networks   179
7.2   Cognitive Network Goals   180
7.3   Interaction Methods for Cognitive Radios   182
       7.3.1   Social Language   183
7.4   Components of Interaction   187
       7.4.1   Observability   187
       7.4.2   Understanding   189
7.5   Analyzing Interactions   190
       7.5.1   An Example Analysis   191
       7.5.2   Analysis Results   192
7.6   Group Learning   194
7.7   Building a Cognitive Network with Social Language   196
       7.7.1   MAC Layer Considerations   196
       7.7.2   Behavior-based Design and Social Language   197
       7.7.3   Tasks and Behaviors   197
       7.7.4   Hardware Considerations and Implementation   198

    7.7.5   Implementing Behaviors in Software and Hardware   200

    7.7.6   Network Evaluation   207

    7.7.7   The Entry Scenario   207

    7.7.8   Social Learning   210

    7.7.9   Total System Behavior   211

8   Cognitive Radio Applications   215

   8.1   Introduction   215

   8.2   Zoned Dynamic Spectrum Access   215

   8.3   Cognitive WiFi and LTE Operation in TV White Space Spectrum   216

    8.3.1   WiFi Frequency Translators   216

    8.3.2   LTE Frequency Converters   221

   8.4   LTE Cognitive Repeaters for Indoor Applications   221

   8.5   Cognitive Radio and Cognitive Radar: Communications and Radar System Coexistence   222

    8.5.1   Cognitive Radar   222

    8.5.2   Legacy Radar and Communications System Coexistence   224

   8.6   Ka Band Geostationary Satellite Applications   228

    8.6.1   Introduction   228

   8.7   Public Safety and Emergency First Responder Communication   232

    8.7.1   Introduction   232

    8.7.2   The Virginia Tech Public Safety Cognitive Radio   233

    8.7.3   Current (2016) Situation   235

   8.8   Cognitive Radio and Autonomous Vehicles   236

   8.9   Smart Grids   239

Index   245

Foreword

The Mario Boella series contains textbooks and research monographs in all areas of Radio Science, with a special emphasis on the applications of electromagnetism to information and communications technologies. The series is scientifically and financially sponsored by the Istituto Superiore Mario Boella affiliated with the Politecnico di Torino, Italy, and is scientifically co-sponsored by the International Union of Radio Science (URSI). It is named to honor the memory of Professor Mario Boella of the Politecnico di Torino, who was a pioneer in the development of electronics and telecommunications in Italy for half a century and was a Vice-President of URSI from 1966 to 1969.

This book on *Cognitive Radio Engineering* is the seventh volume published in the ISMB series.

It addresses a topic that has received much attention in recent years, and is co-authored by three well known experts in the field. After explaining what cognitive radio is and why it is important, it proceeds to the design, evaluation and applications of cognitive radio. This volume may be used both as a textbook and as a reference monograph.

Piergiorgio L. E. Uslenghi
ISMB Series Editor
Chicago, May 2016

Acknowledgments

This book began as a keynote address for the 2013 North American Radio Science Meeting and an invited paper for *Radio Science Bulletin*. We are grateful to Amir Zaghloul and W. Ross Stone for those invitations and to Piergiorgio L.E. Uslenghi for encouraging us to turn that material into a book.

Alexander R. Young played an important role in the initial planning for this book and assisted with early drafts. Although he did not continue as a co-author, we are grateful for his contributions and friendship.

The faculty, students, and staff of the Virginia Tech Center for Wireless Telecommunications (CWT) contributed greatly to the development of cognitive radio, and we have referenced much of their work. The authors were very fortunate to have them as colleagues and friends. Their work was funded by the National Science Foundation, the National Institute of Justice, the Defense Advanced Research Projects Agency, and the United States Air Force. We are grateful for their support and for the advice and encouragement provided by an outstanding group of program managers, including but not limited to Preston Marshall, Joseph Evans, Joseph Heaps, and Bruce Fette.

Joseph Mitola started it all, and we are grateful for his many contributions and his willingness to share ideas with newcomers to the field. We note with thanks that then student Christian Rieser (now with MITRE) first brought Mitola's pioneering dissertation on cognitive radio to Charles Bostian's attention. Christian adamantly would not accept Charles' initial refusal to let the laboratory pursue cognitive radio research. Charles changed his mind, and many fruitful years followed.

We are grateful to many people whose support and assistance made this book what it is. Those whose comments improved early drafts include Ahmed Eltawil (U.C. Irvine) and Stephen J. Shellhammer (Qualcomm). Brad Brannon (Analog Devices) and Damian Anzaldo (Maxim Integrated) generously shared their knowledge and discussed their design techniques with the authors and helped us secure needed graphical material. Matt Ettus, Neel Pandeya, and Erik Luther (all of Ettus Research), Patrick Murphy (Mango), and Iyappan Subbiah (RWTH Aachen University) provided valuable information and advice. Douglas Sicker (Carnegie Melon University) and Thomas Rondeau (DARPA) generously reviewed the draft manuscript and offered many excellent suggestions which we gratefully adopted.

Our wives Frieda Bostian, Jennifer Kaminski, and Sarah Feda deserve a special and heartfelt word of thanks for their continuing support and encouragement.

CHAPTER 1

Introduction[1]

1.1 What Is a Cognitive Radio and Why Is It Needed?

In this book, we define a cognitive radio as a transceiver which is (a) aware of its environment, its own technical capabilities and limitations and those of the radios with which it may communicate, the rules governing its operation, and its user's needs, priorities, and limitations; (b) capable of acting on that awareness and past experience to configure itself in a way that optimizes its performance subject to some set of constraints; and (c) capable of learning from experience. In a real sense, a cognitive radio is an intelligent communications system that designs and redesigns itself in real time.

This definition intentionally says nothing about software-defined radio (SDR) or dynamic spectrum access (DSA), a technology where radios detect open channels and operate in them, subject to a priority system and to limitations on the interference they can cause to other users. Figure 1.1 presents a simple example of a cognitive radio demonstrating DSA in the 400 MHz band [1]. DSA is an important application for cognitive radio, but radios can implement some forms of adaptive DSA without needing or being able to learn anything. Cognitive radio is much more than DSA. While cognitive radio grew out of the SDR community, cognitive radios can be and are built using radio frequency integrated circuits (RFICs) and analog hardware radio frequency (RF) platforms.

An ideal cognitive radio would respond to its user's communication needs – or even anticipate those needs – in a way that provides acceptable performance subject to a set of constraints. These constraints may include legal requirements, mission-related operational procedures, resource availability, and equipment limitations.

Here are a few examples of the practical problems that cognitive radio can solve.

Spectrum Shortage: Cognitive radios allow spectrum to be shared much more efficiently and used dynamically and intelligently.

Equipment Cost: Cognitive radios can intelligently perform the complicated tweaking that low-cost complementary metal-oxide-semiconductor (CMOS) RF devices require to accommodate changes in frequency and operating mode.

---

[1]Much of the historical and descriptive material in this chapter has previously appeared in *Radio Science Bulletin* and is reproduced here with permission.

(a)

(b)

(c)

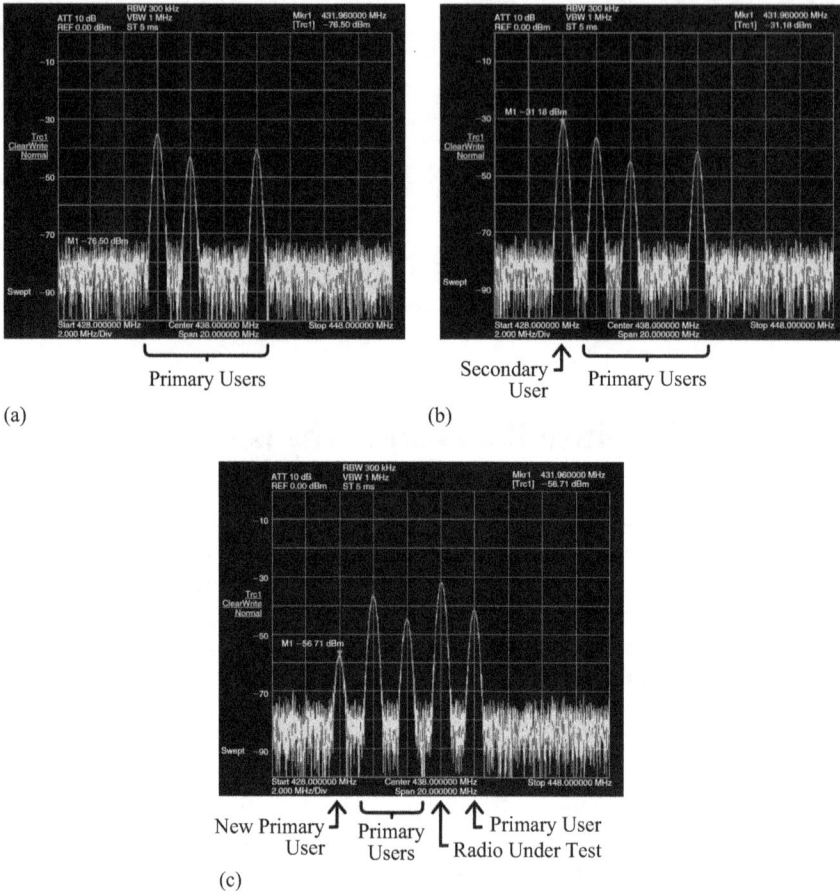

Figure 1.1   Time sequence of spectrum analyzer displays illustrating cognitive radio dynamic spectrum access in the 400 MHz band. (a) Initial situation: primary users occupy three channels. (b) The radio under test finds an open channel and occupies it as a secondary user. (c) A new primary user appears in the channel occupied by the radio under test, and the radio under test finds a new open channel and moves into it. © 2013 IEEE. Reprinted, with permission, from Reference 1.

Energy Conservation: Cognitive radios can recognize legacy radios and configure themselves to interoperate with them, solving what can be a major problem in public safety communications for disaster response.

Network Self-Organization: Cognitive radios can organize themselves to maintain desired levels of network performance without human intervention.

1.2 Book Coverage and Philosophy

Cognitive Radio Engineering is both a text and a reference book about cognitive radio architecture and implementation, intended for readers who want to design and build working cognitive radios. We intend it as a complement to the available literature which tends to focus on algorithmic issues and ideal cognitive radios with relatively little attention to practical devices. With it, we hope to take the reader from conceptual block diagrams through the design and evaluation of illustrative prototypes. An important goal is to bridge the divide between radio engineers, who often have little experience with the computational resource and timing issues inherent in cognitive radios, and computer engineers who often are unaware of RF issues like dynamic range, intermodulation products, and acquisition time. In writing, this book we emphasized the cognitive radio architecture and design processes developed in the authors' laboratory at Virginia Tech (VT). We acknowledge and appreciate that many others have worked or are working in this space. A 2015 search of the *IEEE Xplore* digital library for the term "cognitive radio" returns 16, 215 entries!

Following a brief overview of cognitive radio history and a high-level look at cognitive radio operation, we present a detailed study of cognitive engine design and analysis. A cognitive engine is an intelligent software package that controls a frequency-agile and mode-agile RF platform, which may be an SDR or an RFIC. After treating RF subsystems, we turn our attention to computational platforms for and computation issues in cognitive radios. We follow this with a chapter on system integration and then discuss evaluation methods for cognitive radio. The next chapter focuses on cognitive radio design for networking emphasizing our efforts to build intelligent networks of autonomous radios, avoiding what we term the "faculty meeting problem" where the radios use up their computational and bandwidth resources telling each other everything they know and how important they are. We conclude with a chapter on cognitive radio applications in communications and in autonomous vehicles.

Cognitive radio regulation and security are two related issues which are still evolving at the time of writing. Rather than treating these matters in separate chapters, we have chosen to consider them when they arise in conjunction with other topics.

1.3 The Origin of Cognitive Radio

Joseph Mitola originated the idea of a cognitive radio in the late 1990s [2]. He provided the perfect name and a coherent and most influential body of ideas for an emerging trend that had been developing for a few years but lacked clear focus. (An inspiring visionary, Mitola is also widely credited with originating the term "SDR" to replace names like "Modular Multifunction Information Transfer System." See Reference 3 for an interesting history of the field and a discussion on the contributions of Mitola and others.) He presented cognitive radio as a kind of super-intelligent personal data assistant (PDA) operating at the application level that would communicate conversationally with its user and set up whatever radio links were necessary to satisfy the user's needs.[2] While Mitola's thesis focused on the application layer, his work soon inspired others to look at cognitive radio as a solution to physical layer problems [4].

The beginnings of cognitive radio coincided with the recognition at the Defense Advanced Research Project Agency (DARPA), the Federal Communications Commission (FCC), and other U.S. government agencies that a different approach to spectrum access was badly needed [5–7]. The demand for spectrum for broadband mobile devices was outrunning the supply.

---

[2]It may be argued that a portion of Mitola's vision was realized by iPhone's Siri, but here the intelligence is in a remote server rather than in the radio.

Top-down frequency management, based on exclusive spectrum occupancy, was seen to be both inadequate and extremely inefficient. Policy makers envisioned a future in which cognitive radios could opportunistically find and use vacant spectrum, sharing it based on a cooperative- and priority-based system that allowed decentralized intelligent management of interference. Funding soon followed with programs like DARPA's XG (Next Generation) in 2003 [5] and WANN (Wireless Adaptable Network Node) in 2006 [8] and the National Science Foundation (NSF) Network Technology and Systems (NeTS) program in 2004 [9].

At the same time, RF technology was changing rapidly. Complementary metal-oxide-semiconductor (CMOS) technology made low-cost (RF) front ends available, but these typically required frequent adjustment of hardware parameters in response to changes in frequency, bandwidth, or output power. Concurrently, it became possible for microprocessors to perform many radio functions, and the SDR was born [10]. Since a cognitive radio is aware of its own operational capabilities and needs, researchers anticipated that it could intelligently adjust its own internal hardware and software parameters as necessary to meet its instantaneous physical layer performance requirements.

These factors all blended into the idea of cognitive radios as transceivers that would dynamically access the spectrum, share it with other users, and optimizing their performance across multiple objectives and constraints. They would do this intelligently and learn from experience, all with minimal user involvement. Cognitive radios would deal intelligently with a wide variety of channel- and equipment-related imperfections, ranging from propagation effects to RF component nonlinearities. By the mid 2000s, cognitive radio was a hot research topic whose time had come, both as a major extension of ongoing work in adaptive SDR (see, e.g., Reference 11) and as an independent field in its own right.

The first cognitive radio publications following Mitola's influential dissertation [2] were somewhat conjectural in nature and focused on the promise of cognitive radio [12], theoretical architectures and mathematical techniques for implementation [13], and game theoretic analysis of possible cognitive radio behavior [14]. The first successful prototype cognitive radio architecture consisted of an intelligent software package called a cognitive engine controlling an electronically managed mode-agile and frequency-agile RF platform (called, in context, "the radio"). This was commonly, but not necessarily, a SDR. Figure 1.2 provides an early view of this concept. In terms originated by Christian Rieser, the cognitive engine "turns the radio's knobs" and "reads the radio's meters," functioning very much as a human operator would. In later versions of this architecture, a central cognitive controller accesses a variety of modules encompassing sensors, optimizers, policy verifiers, and radio platforms. See Figure 1.3. The first prototype cognitive radios, employing the VT cognitive engine and genetic algorithms, were built by Rieser *et al.*, in 2004 [4, 16]. The RF unit was a 5.8-GHz Proxim *Tsunami* radio with the following electronically settable knobs: transmitter power, modulation type and index, forward error correction (FEC), uplink/downlink time slot ratio ("fibs"), and center frequency. The test radios established a video link on a fixed frequency. A jammer was then turned on. The radios were not allowed to change frequency but cooperatively adjusted all of the other knobs to minimize the effect of the jammer. If the jammer went away and subsequently returned, the radios remembered their earlier settings and returned immediately to them. Figure 1.4 shows the radio setup, the video images, before and after the cognitive engine eliminated the jammer interference, and the before and after knob settings as evolved by the cognitive engine's Wireless System Genetic Algorithm.

The initial prototype was soon followed by an SDR-based cognitive radio using an Ettus Research Universal Software Radio Peripheral (USRP). In a demonstration at DySPAN 2007,

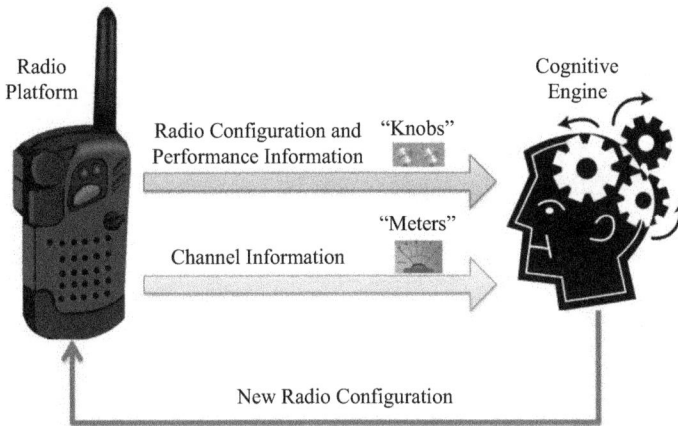

Figure 1.2   Original Virginia Tech cognitive engine concept. ©2012 Radio Science Press, Belgium. Reproduced, with permission, from Reference 15.

a pair of these radios operating in an interference filled environment found an open frequency and established and optimized a data link. At this point, the learning and optimization parts of the cognitive engine were fully operational, and the user could assign relative weights to the parameters that were to be optimized. See Figure 1.5 [17].

1.4 Overview of Cognitive Radio Operation

In this section, we present the Virginia Tech (VT) approach to cognitive radio engineering as developed in the authors' laboratory since that is the focus of our book. We recognize that researchers at many university, corporate, and government agencies have made significant contributions and developed their own architectures and methodologies.

Mitola visualized the cognition process in terms of the radio executing a cognition loop [2] as shown in Figure 1.6. The VT cognitive engine implemented Mitola's loop and added a second (inner) loop within it (see Figure 1.7). It introduced *predict* and *compare* functions to guide the cognitive engine in choosing parameter values and to ensure that the cognitive engine understood the situation with which it was dealing.

The cognition process starts with the radio platform, which reports its status and observations to the cognitive engine (CE). The CE combines this RF data with the output of other sensors and with its knowledge of the user's needs and the policies and limitations governing its operation. It then synthesizes a scenario (e.g., "I am dealing with co-channel interference from a QPSK signal."). The CE queries its knowledge base to see if it has encountered this scenario (case) before and, if so, what action it took. It uses the results to determine starting values for radio parameters and passes these to a genetic algorithm-based multi-objective optimizer that attempts to optimize the radio's performance (optimum performance is defined as complete satisfaction of the user's needs). If the CE is unable to synthesize a scenario, it may initialize the system with a set of random parameters whose values are within ranges consistent with acceptable radio operation and see what happens. When this process is finished, the optimizer generates an array of knob settings (settable radio parameters) that will be passed to the radio

Cognitive Radio Architecture

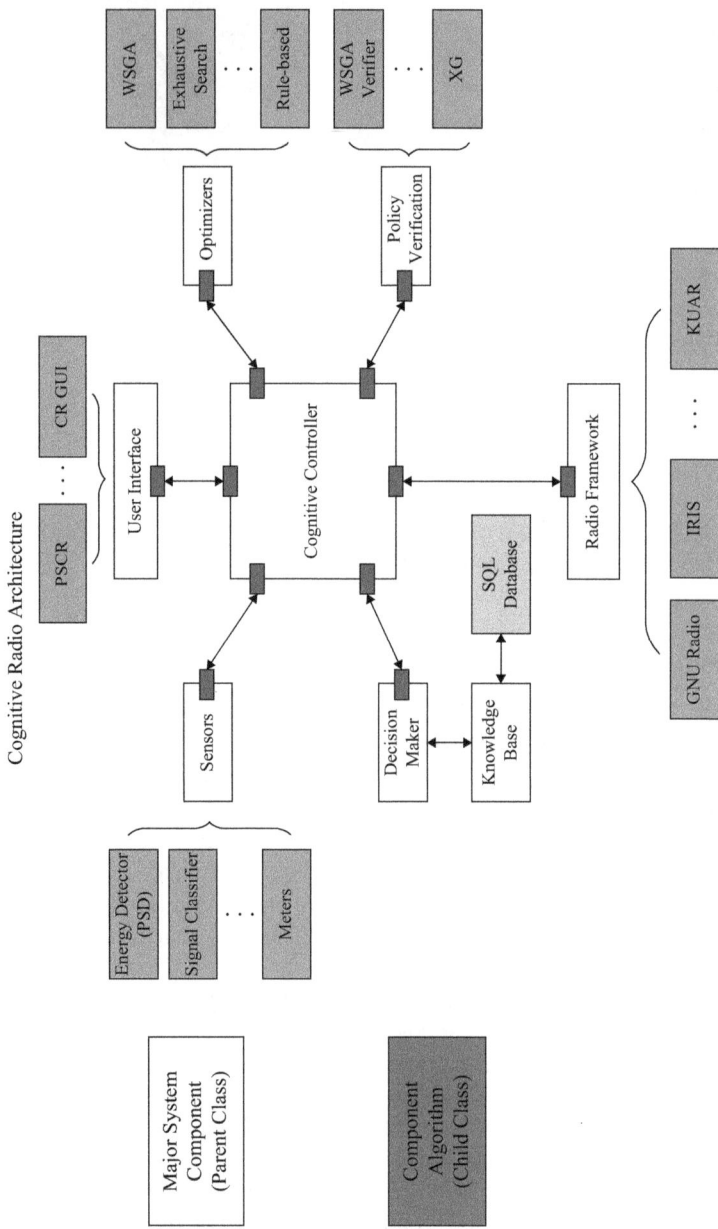

The cognitive engine coordinates components to realize: sensing, learning, and optimization. It enables development of new components for testing and comparison.

Figure 1.3 Modular cognitive radio architecture. ©2012 Radio Science Press, Belgium. Reproduced, with permission, from Reference 15.

- Cognitive
  Radio
  Testbed Link

- Interferer
  Degrades
  Broadband
  Wireless Link

Interferer   Broadband Wireless Link

- WSGA Evolves
  Radio Operation

Interferer	Link $\longrightarrow$	Link
Tx = 11 dBm	Tx = 6 dBm	Tx = 16 dBm
QAM 16	QAM 16	QPSK 4
Fibs 12	Fibs 22	Fibs 4
3/4 FEC	3/4 FEC	1/2 FEC
Freq = b−5	Freq = a−5	Freq = a−5

- Link Quality of Service
  (QOS) is restored

IEEE MILCOM 2004 - Rieser

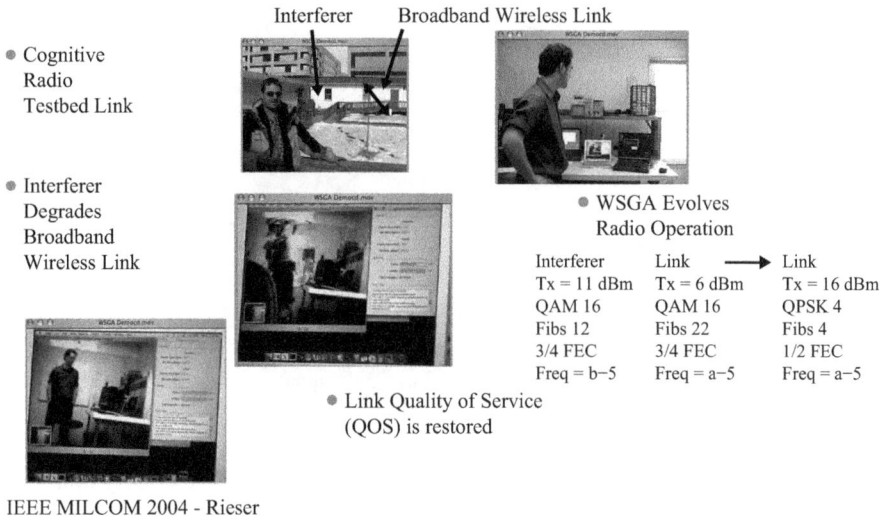

Figure 1.4   First demonstration of a cognitive radio. ©2012 Radio Science Press, Belgium. Reproduced, with permission, from Reference 15.

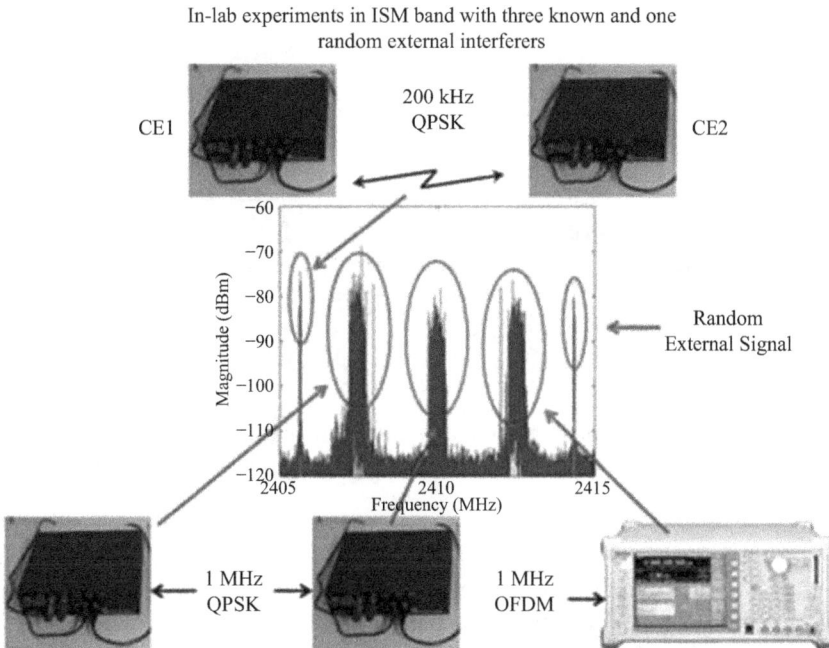

In-lab experiments in ISM band with three known and one random external interferers

CE1

200 kHz
QPSK

CE2

Random
External Signal

Magnitude (dBm)

−60
−70
−80
−90
−100
−110
−120

2405   2410   2415
Frequency (MHz)

1 MHz
QPSK

1 MHz
OFDM

Figure 1.5   Cognitive radio demonstration at DySPAN 2007. ©2012 Radio Science Press, Belgium. Reproduced, with permission, from Reference 15.

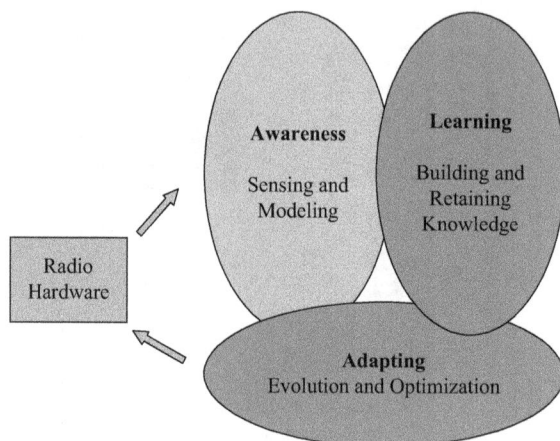

Figure 1.6   Original cognition loop concept. ©2012 Radio Science Press, Belgium. Reproduced, with permission, from Reference 15.

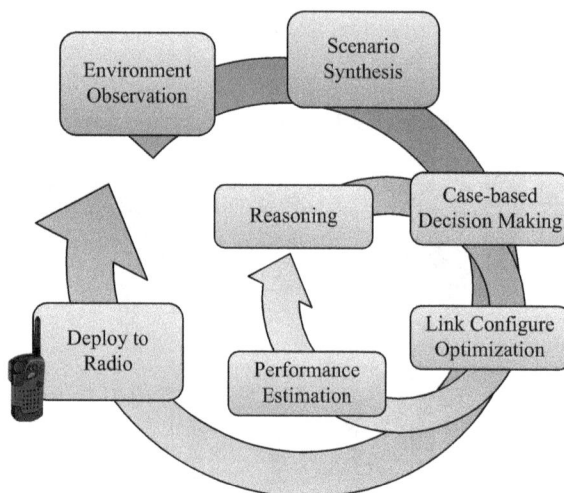

Figure 1.7   Virginia Tech implementation of a nested cognition loop. ©2012 Radio Science Press, Belgium. Reproduced, with permission, from Reference 15.

platform. As part of the optimization process, the CE has predicted what the radio performance will be after each trial set of radio parameters is applied, and it will later compare the final set of these predictions with the actual radio performance to determine whether or not it took the correct action. The predictions are based on standard formulas – for example, calculating bit error rate from signal-to-noise ratio (SNR). The new knob settings are loaded into the radio and the radio's observations are reported to the cognitive engine, the actual and predicted performances are compared and the results stored, and the cycle begins again. An important reason for making this comparison is to determine whether or not the cognitive engine correctly synthesized the scenario during the previous cycle. Clearly, it did not if the knob settings calculated in the previous cycle made the performance worse, and the CE will remember this and revise its assumed scenario and try something different.

The first cognitive engine version that our laboratory developed is a multi-threaded system written in C++ and running under Linux, shown schematically in Figure 1.8. We will describe it first here because the sequence of operations more closely corresponds to those undertaken by the nested cognition loop of Figure 1.7. A thorough treatment of both this architecture and the more recent Cognitive System Enabling Radio Evolution (CSERE) architecture appears in Chapter 2.

The radio platform used with the cognitive engine of Figure 1.8 is an SDR (often in development projects a USRP) and appears at the top of the figure. It communicates with the cognitive engine through a CE-Radio Interface which interprets between the radio software layer and the cognitive engine software layer. The radio describes itself (specifies its capabilities) to the cognitive engine's Radio Resource Monitor through software that provides (a) Hardware/Platform-specific application programming interfaces (APIs) and (b) translating radio configuration commands generated by the cognitive engine to the needed radio instructions. The Performance API allows access to information about the RF environment and the communications link performance to the cognitive engine's Wireless Modeling System and Radio Performance Monitor. The Wireless Monitoring System, the Radio Performance Monitor, and the Radio Resource Monitor all provide input for the Cognitive System Controller. At this point in the cycle, the controller has an updated awareness of (a) the *RF environment*, (b) the *channel performance* that the radio is currently delivering in that environment, and (c) the *radio resources* that currently are being used to deliver that performance and which could be called upon for improved performance.

We describe the process of monitoring the channel performance, RF environment, and radio operation as *reading meters*. We characterize these as *actual meters*, when the parameters measured are directly available from the radio hardware (signal and noise levels in dBm, for example), and as *derived or computed meters* when values are not directly measurable by the hardware and must be calculated (bit error rate, for example.)

The user communicates with the cognitive engine through the CE-User Interface which consolidates information about the user's needs and priorities and the policies which the user wishes to govern them. These data pass through a set of policy and security checkers to the controller. At this point, the controller has sufficient information about the RF environment, the radio performance, the user's needs and priorities, and the governing policies to query the Knowledge Base for guidance about whether or not this scenario has been encountered before. The Knowledge Base's response and the other data available to the Cognitive Controller provide the needed input to the Decision Maker, which provides starting values (these might be the set of chromosome values that the knowledge base remembers were used the last time the radio encountered the current situation), weighting factors (e.g. what relative weights to give the optimizing transmitter power, bit error rate, and occupied bandwidth), etc., to the Wireless System Genetic Algorithm (WSGA). The WSGA determines the radio configuration for the next cognition cycle and passes the necessary commands back up through the hardware/platform API.

The developers of the VT Cognitive Engine chose genetic algorithms as an optimization process because these provide both a computationally efficient trial-and-error process, and a way to implement learning, patterned on nature's way of remembering successful adaptations. While the technique can be criticized for its tendency to get close to an optimum solution quickly and then either taking long times to reach the exact optimum or even oscillate between

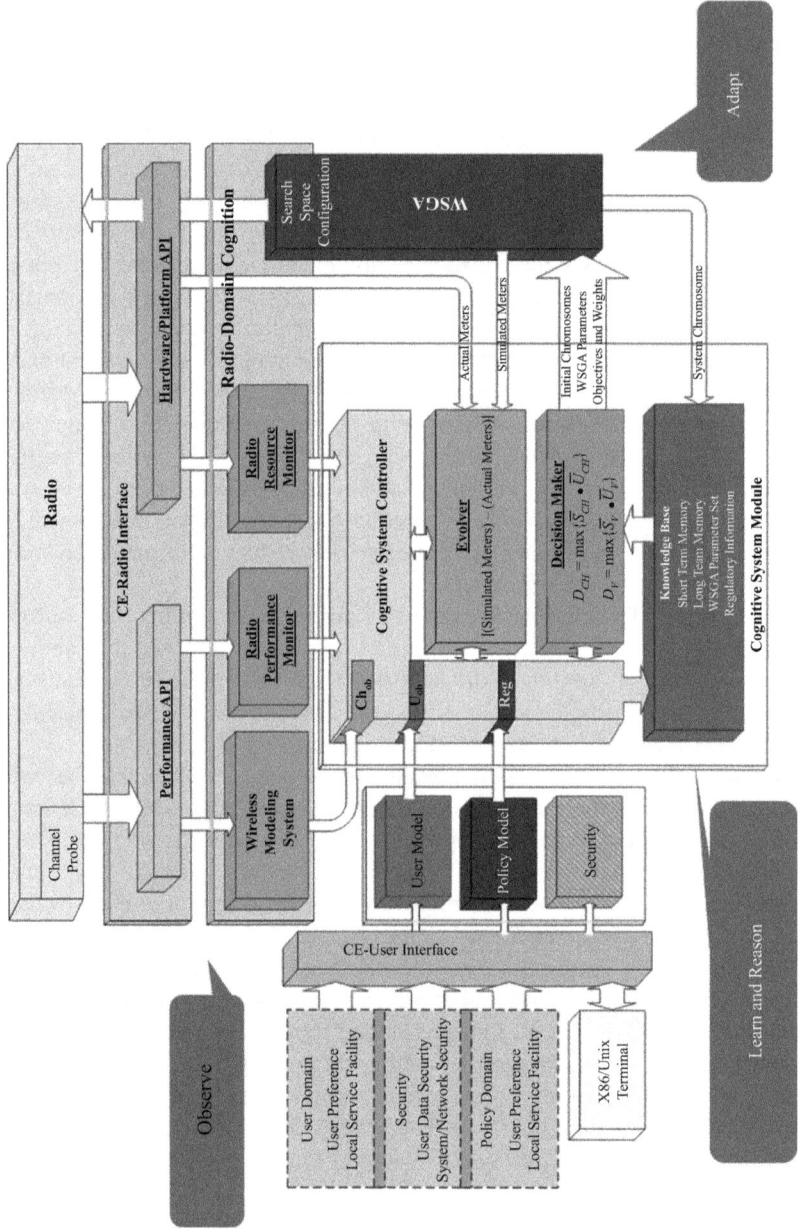

Figure 1.8 Architecture of the first Virginia Tech cognitive engine. ©2012 Radio Science Press, Belgium. Reproduced, with permission, from Reference 15.

suboptimal solutions, for purposes of practical radio operation, getting close to the optimum is sufficient.

1.5 An Illustrative Example of Cognitive Radio Application: Dynamic Spectrum Access in the Broadcast Television Bands

1.5.1 Introduction

For cognitive radios like Mitola originally proposed, DSA is only the first step of a continuing series of actions where radios find an open channel, organize a network, and carry out a mission, constantly evaluating and attempting to improve their performance. But this first step is critical, particularly from the point of view of regulatory authorities who are used to allowing licensed radios exclusive access to a radio channel. Users will not readily give up their exclusive access to a channel so that someone else may share it. Shared use of a channel – whether the sharing is sequential in time with no interference or simultaneous in time with interference held below some acceptable limit – is a revolutionary concept which must be approached cautiously. The other cognitive radio processes offer significantly lower regulatory hurdles.

Just defining an "open" channel is complicated. Except perhaps when spectrum is being reallocated and its ownership might be ambiguous, there are few if any channels without assigned users. A channel could be said to be open if, at that time, no user is operating within effective range of the cognitive radio that wants to use that channel. "Effective range" could mean that the cognitive radio's operation cannot be detected at all by the channel's assigned users when they are active, or it could mean that interference caused by the cognitive radio will be at an acceptably low level. Then there are the time questions: is a channel open because there are no assigned users physically located within effective range, or is it because users within range are temporarily inactive? How long will that inactivity or out-of-range positioning last? How quickly must the cognitive radio give up a shared channel when it determines that the channel is no longer available?

Propagation effects are often overlooked in discussions of what constitutes an open channel. This is probably because many researchers may only be familiar with the UHF spectrum assigned to cellular telephony and television broadcasting. In this frequency range, radio propagation is not influenced significantly by atmospheric conditions. In some frequency bands, particularly below about 60 MHz and above 10 GHz, propagation conditions can change drastically with weather, time of day, solar activity, etc. A channel may be open because current propagation conditions make the whole band useless and "closed," or (at say 50 MHz) it may be occupied by a strong signal from a station hundreds of miles away that is normally too distant for interference. Obviously, there is no point in occupying an "open" channel in a "closed band." The importance of propagation issues is pointed out and discussed further in Reference 18.

Two obvious ways for identifying an open channel are (a) consulting a database (see Section 1.5.5) and (b) "listening" (checking for the presence of a received signal) on that channel. The risks of (a) include that the database may not be up to date, or that unusual propagation conditions will make signals appear at locations where they should not. The risks of (b) include "hidden transmitters," failure to recognize channels occupied only by receivers used for radio astronomy or other sensing applications, and systems like satellite downlinks where the incoming signals are too weak to detect without specialized equipment but where a cognitive radio's transmission could seriously disrupt the intended receiver. Requiring the cognitive radio first to consult a database and then listen would seem to be the safest approach.

The first large-scale commercial application of cognitive radio applications in DSA will probably be that proposed for the television broadcast bands. Note that this is more a test of cognitive radio techniques than of full cognitive radio operation, because autonomous radio behavior will be restricted by regulatory concerns. With the possible exception of *sensing only* devices (see below), the radios will not be fully autonomous, finding open frequencies and optimizing link and network parameters subject to their users' priorities. While their operation will thus fall short of Mitola's vision for cognitive radio, the radios will implement some of the underlying technologies and are consistent with the common perception of cognitive radio as dynamic spectrum access. In this section, we provide a brief overview of cognitive radio operational requirements under current (as of 2014, last revised May 17, 2012) U.S. regulations. Our goal is to illustrate one set of approaches to solving the problems of finding and using open channels, and to expose the reader to some of the regulatory requirements that a cognitive radio designer may expect to encounter. We will not attempt here a comprehensive discussion of all the policy, economic, and technical issues associated with dynamic spectrum access in the TV broadcast bands.

1.5.2   Identifying Frequencies for Cognitive Radio Operation: TV Channel Occupancy in the United States

Broadcast TV spectrum in the United States was first allocated in 6 MHz VHF band channels based on the original analog black and white National Television System Committee (NTSC) standard of the 1940s. Over the years, some numbered channels were deleted (Channel 1, for example) and many others were added in the UHF band, but backward-compatibility require-ments kept the same channel designations through the advent of color TV and the demise of analog TV broadcasting in 2009. In the U.S. Advanced Television System Committee (ATSC) digital TV (DTV) standard, the channel number shown on a user's set is no longer tied to the transmitter frequency. The ATSC Program and System Information Protocol (PSIP) include a virtual channel designation set by the broadcaster or cable system that the receiver displays, and several DTV transmissions may occupy the 6 MHz bandwidth of an old analog channel [19]. When broadcast TV transitioned from analog to digital, stations were allowed to keep their historical channel designations. Thus, a station known in analog days as Channel 7 may now broadcast digital content identified as Channels 7.1, 7.2, 7.3, etc. (These are all contained within the station's assigned 6 MHz.) Nevertheless, the old channel numbers are still used by the FCC to designate allocatable 6 MHz portions of the spectrum. With this notation, the subset of current U.S. TV channels available for unlicensed DSA operation is listed in Table 1.1. The 108 MHz bandwidth once occupied by channels 52–69 was reallocated for other uses.

Because of both the limitations of 1940s analog technology and of the desire to pro-mote the design of low-cost television sets, analog television receivers had poor selectivity and linearity. Consequently, transmitters in a given geographical area were not permitted to operate in that area's *taboo channels*. These channels were either adjacent in frequency to already assigned channels or else their frequencies were related in ways that could allow unwanted intermodulation products to appear in a receiver's passband. (See Reference 20 for an

Table 1.1 TV channels available for unlicensed television band devices.

Channel designation	Frequency range (MHz)
2	50–54
5–6	76–88
7–13	174–216
14–36	470–608
38–51	614–698

interesting discussion of the "taboo channel" issues.) This practice resulted in a significant amount of unused or inefficiently used spectrum. For the same performance, digital TV requires less transmitter power, uses spectrum more efficiently, and is less sensitive to interference than analog TV, and these factors are what allowed channels 52–69 to be reallocated, freeing 108 MHz of spectrum between 698 and 806 MHz [21]. At the time of writing, only parts of this 108 MHz have been allocated. In the United States, it is not available to unlicensed TV band devices, although some authors refer to it as if it were and specify the available spectrum as 54–806 MHz instead of 54–698 MHz.

Even with the more efficient allocation of spectrum for DTV broadcasting, TV channels unoccupied by broadcast TV signals are available at most locations. The U.S. FCC and its counterparts in other countries are moving to make these available for shared use via cognitive radio techniques for what are called unlicensed Television Band Devices (TVBDs), provided that these provide sufficient protection both for television signals and for wireless microphones and similar devices (variously called Low Power Auxiliary Devices (LPAD) and Program Making Special Event (PMSE) devices) that already operate in unused TV channels. This last term is common in Europe [22].

Now that the set of available channels is defined, we next must consider how open channels are to be defined, found, and used.

1.5.3 FCC Rules and Commercial Standards for Unlicensed Television Band Devices (TVBDs)

Almost all non-Federal-government radio transmitter operated in the United States must comply with rules established by the FCC. These rules typically specify maximum physical layer parameters values (transmitter output power and spectral mask, for example) but do not dictate how these values are to be achieved or the details of the network protocols that radios must use. Standards bodies develop and adopt documents that dictate the radios' specifications in great detail, ensuring that radios designed and manufactured in compliance with the standards will both be interoperable with each other and in compliance with the rules. The rules themselves do not require compliance with any particular standard, and historically in the United States this has led to competing and mutually incompatible standards for AM radio stereo broadcasting and cellular telephone systems, to give just two examples. Under ideal circumstances, worldwide standards insure that radio systems can be operated anywhere in the world, in full compatibility with each other and in full compliance with local regulatory rules (which may vary from country to country and even from region to region within a country). An example where this works well is in the worldwide use of WiFi systems, all compliant with the IEEE 802.11 series of standards.

Television band devices (TVBDs), prototypical cognitive radios, are defined by the FCC as "unlicensed intentional radiators that operate on the available TV channels in the broadcast frequency bands" as listed in Table 1.1. They are governed by applicable parts of the FCC rules, primarily Part 15, Subpart H [23], and will probably be designed and manufactured in compliance with several standards. We will focus on IEEE 802.11af since that covers the radio issues addressed here, but the reader should be aware that the Internet Engineering Task Force Protocol to Access White Space (IETF PAWS) database will govern database access, and that there are other applicable standards in existence (IEEE 802.22) or under development (IEEE 802.15.4m).

At the time of writing, the proposed IEEE 802.11af standard is still going through the approval process. Our source for information about it is [24]. In the description to follow, we attempt to combine the terminology used in both the rules and the standard. See Table 1.2. Combining the terminology is complicated because the FCC rules are less restrictive and include fewer technical details than does a standard which compiles with them.

The FCC rules are subject to revision, and we advise the reader to seek the most up-to-date version when referring to them. In the online version, found in the *Electronic Code of Federal Regulations* (www.ecfr.gov), the date of rule publication in the Federal Register appears at the end of each numbered paragraph. The TV white space rules were initially issued on November 17, 2006, and appear in 71 FR 66876 and 71 FRR 66897 as the FCC's *First Report and Order and Further Notice of Proposed Rule Making*. These were revised in the *Second Memorandum Opinion and Order*, 75 FR 75814, December 6, 2010. The most recent update at the time we wrote this book is the *Third Memorandum Opinion and Order*, 77 FR 29245, May 17, 2012.

For an excellent and comprehensive discussion of the FCC rules as they existed in 2009 and the technical issues associated with cognitive radio operation in TV white space, see Reference [25], remembering that details of the FCC rules have changed twice since that work was published. We will reference it frequently in the following material. Policy, standards, and

Table 1.2   FCC and IEEE 802.11af terminology.

Informal description	FCC term	IEEE 802.11af term
Cell	(none)	Basic Service Set (BSS)
Base Station	Fixed Device	Registered Location Secure Server (RLSS)
Mobile terminal with geolocation and base station capabilities	Mode 2 Device	Geolocation-Database-Dependent Enabling Station (GDD-Enabling STA)
Mobile user terminal without geolocation and with access to database information	Mode 1 Device	Geolocation-Database-Dependent Station (GDD-Dependent STA)
Mobile user terminal using sensing rather than data base information for channel access	Sensing Only Device	(none)

system level considerations are discussed in detail in Reference 26, which predates the 2012 revision of the FCC rules.

The FCC rules recognize a hierarchy of *fixed devices* and three types of *personal/portable device* called *mode 2, mode 1*, and *sensing only*. All except the sensing only devices must obtain an operating channel from a TV bands database, either by themselves or through an intermediate device above them in the hierarchy. The fixed devices are somewhat analogous to base stations, while the mode 1 devices are like tablets, personal computers, or phones. Mode 2 devices have similar capabilities to fixed devices but with the added feature of mobility and might be used, for example, as access points. Fixed devices must operate at a specified fixed location and are authorized to select and use channels provided to them by a TV bands database which they access via the Internet. Fixed devices may initiate and operate networks by assigning channels to and otherwise supervising the operation of other fixed devices or personal/portable devices. A mode 2 personal/portable device has an internal geo-location capability and the ability to access a TV bands database either directly via the Internet or through an intervening fixed or mode 2 device. A mode 2 device may select an available channel from the database and initiate and operate a network, and it may provide a list of available channels to mode 1 devices for their use. Personal/portable devices are allowed to operate on available channels only in TV channels 21–36 and 38–51. This includes open channels adjacent to occupied TV channels; fixed devices are not allowed to operate in these adjacent channels.

The group of stations to which a fixed device assigns channels is somewhat analogous to a geographic cell controlled by a base station and is called a Basic Service Set (BSS) in IEEE 802.11af. The process of obtaining channels for operation is as follows. Starting at the top of the hierarchy, fixed devices and mode 2 devices must access a TV bands database to determine the TV channels that are available for operation at their location. The list is valid for 48 hours, but a mode 2 device must contact a database every time it is powered up or moves more than 100 meters from the location of the last query. Mode 2 devices may obtain data on available channels for a bounded geographic area and use this information to govern mobile operation within that area. A mode 1 device originates contact with a fixed or mode 2 device by transmitting either on a channel which it hears the fixed or mode 2 device using or on a channel which the higher level devices have announced as available for this purpose. A mode 1 device may operate only after it has received a list of available channels from a fixed or mode 2 device that has contacted a database and verified that the mode 1 device has a valid FCC identifier.

A fixed or mode 2 device must broadcast an encoded Contact Verification Signal (CVS) to allow any mode 1 device that has received a list of available channels from the higher level device to insure that it is still within receiving range of the higher level device. Encoding ensures that a received signal actually originates from the device that distributed the channel authorizations to the mode 1 device rather than from an impostor. A mode 1 device must receive a contact verification signal at least once every 60 seconds while it is active. If it fails to do so, it must immediately re-establish a list of available channels or else cease operation. Note that mode 1, mode 2, and fixed devices are not required to monitor their assigned channels for the presence of a primary user. (A primary user is an incumbent radio service that has first claim to a channel.) They depend on the database for permission to operate. Thus, whether or not they are really cognitive radios is debatable.

Sensing only devices operate differently from the above, and are subject to a strict FCC certification process. In the words of the FCC rules, "Devices authorized under this section must demonstrate with an extremely high degree of confidence that they will not cause harmful interference to incumbent radio services. . .The application [for certification] must include a full explanation of how the device will protect incumbent authorized services against interference." Required detection thresholds measured at the output of an omnidirectional antenna are -114 dBm, averaged over a 6-MHz bandwidth, for ATSC digital TV signals; -114 dBm for NTSC analog TV signals (some of which are still authorized for special applications); and -107 dBm averaged over a 200-kHz bandwidth for wireless microphone and similar signals. A sensing only device must verify that no signal above threshold is detected for a minimum observation time interval of 30 seconds and must recheck the channel once every 60 seconds. Once the initial check is completed, there is no minimum observation time interval specified for the 60 second checks. Note that the sensing only devices can function as true cognitive radios, but their designers must convince the FCC that they will operate without causing harmful interference. This is a task that we hope our readers will accomplish!

Even if a channel is sensed as open at a particular location, legally it may not be available. In particular, operation is not permitted in channels 3 and 4 to avoid interference with external devices like DVD players that use these channels over wired connections to TV receivers. Channels 14–20 may be used by public safety agencies in some metropolitan areas, and Channel 37 is protected for radio astronomy. The reader should check the current FCC rules for up-to-date information [23].

1.5.4 FCC Rule Implementation by the IEEE 802.11af Standard

While the FCC rules require devices at the top of the TVBD hierarchy to interact frequently with a TV bands database operated by an official TV bands database administrator using approved protocols and procedures to guarantee the security and accuracy of these interactions, the rules do not prescribe implementation details. These are addressed in the IEEE 802.11af standard. They include the provision of Registered Location Secure Server (RLSS), which act as local databases and which distribute operational parameters to devices authorized to receive them. The protocols by which this information is requested and distributed include Channel Availability Query (CAQ) and Channel Schedule Management (CSM) procedures, within which there are specified message sequences and formats. See [24] for details.

1.5.5 TV White Space Databases

The FCC authorizes entities called TV White Space Administrators to operate and maintain databases that unlicensed radios may query. One publicly available example at the time of writing is the Google Spectrum Database (https://www.google.com/get/spectrumdatabase/channel/). Figure 1.9 shows the result of querying this database for a hypothetical portable device at a location in Blacksburg, VA. Twenty-one channels are available. For details about the developing protocol that white spaces device will use to communicate with authorized databases, the reader should consult the Internet Engineering Task Force (IETF) web site. At the time of writing, the current draft (July 2014) was available online at http://tools.ietf.org/pdf/draft-ietf-paws-protocol-14.pdf

1.5.6 Standard ECMA-392

In June 2012, the European Computer Manufacturers Association (ECMA-International)® adopted Standard ECMA-392, [27]. This document provides very detailed waveform and

Available Spectrum: 21 Channels (126 MHz)

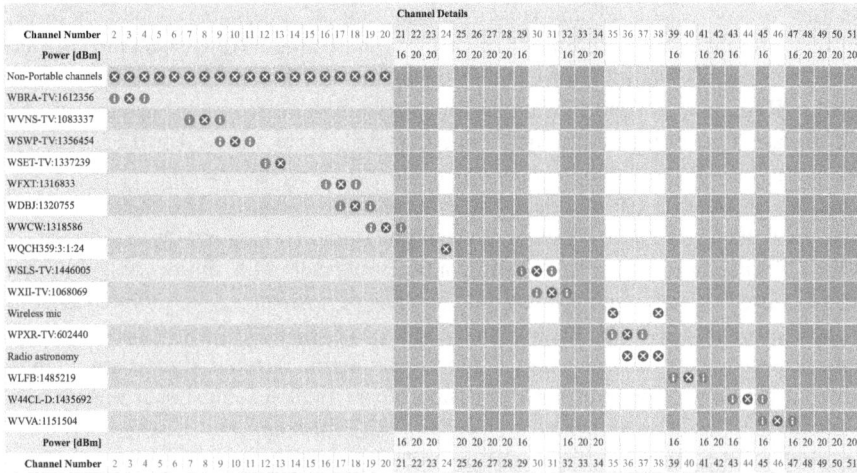

	Channel Details																																															
Channel Number	2	3	4	5	6	7	8	9	10	11	12	13	14	15	16	17	18	19	20	21	22	23	24	25	26	27	28	29	30	31	32	33	34	35	36	37	38	39	40	41	42	43	44	45	46	47	48	49	50	51
Power [dBm]																				16	20	20		20	20	20	20	16			16	20	20					16		16	20	16		16		16	20	20	20	20

Non-Portable channels — rows listed:
WBRA-TV:1612356, WVNS-TV:1083337, WSWP-TV:1356454, WSET-TV:1337239, WFXT:1316833, WDBJ:1320755, WWCW:1318586, WQCH359:3:1:24, WSLS-TV:1446005, WXII-TV:1068069, Wireless mic, WPXR-TV:602440, Radio astronomy, WLFB:1485219, W44CL-D:1435692, WVVA:1151504

| Power [dBm] | | | | | | | | | | | | | | | | | | | | 16 | 20 | 20 | | 20 | 20 | 20 | 20 | 16 | | | 16 | 20 | 20 | | | | | 16 | | 16 | 20 | 16 | | 16 | | 16 | 20 | 20 | 20 | 20 |
| Channel Number | 2 | 3 | 4 | 5 | 6 | 7 | 8 | 9 | 10 | 11 | 12 | 13 | 14 | 15 | 16 | 17 | 18 | 19 | 20 | 21 | 22 | 23 | 24 | 25 | 26 | 27 | 28 | 29 | 30 | 31 | 32 | 33 | 34 | 35 | 36 | 37 | 38 | 39 | 40 | 41 | 42 | 43 | 44 | 45 | 46 | 47 | 48 | 49 | 50 | 51 |

Figure 1.9 Example of TV White Space information. Generated by Google Spectrum Database, https://www.google.com/get/spectrumdatabase/ind [Accessed 4 Aug 2014], and reproduced under terms of Creative Commons.

protocol-level specifications for a beacon-based system for operation in the TV bands. At the time of writing, it seems to be less familiar to the U.S. cognitive radio community than the FCC rules and IEEE 802.11af and to have been developed independently of them. See [27] for further discussion of ECMA-392.

1.5.7 Cognitive Radio Design for U.S. TV White Space

Designing a cognitive radio operating in TV white space that satisfies the FCC rules (and, preferably, IEEE 802.11af) involves a number of difficult RF issues which are commonly overlooked in discussions of the signal processing aspects of the problem (i.e., in articles presenting theoretical methods for and simulations of detecting signals from TV stations, wireless microphones, or other licensed users). These are discussed in detail in [25]. See Chapter 3 for a general discussion of RF specifications and terminology.

The first RF problem is covering the required frequency range (54–698 MHz) with an independently tunable transmitter and receiver while meeting the FCC's stringent restrictions on harmonic and intermodulation production radiation. The nearly $13\times$ frequency range $(698/54 = 12.9)$ means that harmonics up to the 13th must be considered, along with local oscillator (LO) leakage and all of the intermodulation products that result. This requires highly sophisticated synthesizer, mixer, and filter designs and integration. The calculations necessary to convert the field strength requirements found in the rules to spectral masks are complicated and are derived and explained in Reference 25. They yield required spurious signal suppression values in the 55–69 dB range, depending on frequency separation from the transmitter carrier, and as large as 95 dB to protect radio astronomy receivers in Channel 37. The reader should be aware that these numbers reflect the FCC rules as they were in 2009 and that a designer must redo the calculations for the current rules. European maximum power limits for out-of-band emissions fall in the 50–63 dB range [28]; here protection for U.S. TV channels is not an issue. Architectural and circuit-based approaches to designing radios capable of meeting these kinds of specifications appear in References [29, 30]. An interesting FPGA-based proof-of-concept prototype transceiver with limited frequency coverage (432.0–434.2 MHz in four channels) is presented in Reference 31; spurious emission issues are not discussed. The authors of Reference 32 discuss a novel technique that converts signals from a 2.4-GHz IEEE 802.11 b/g/n access point to the TV band with promising out-of-band emission results.

A second RF issue is the practical implementation of sensing. The required -114 dBm threshold for detecting a TV or wireless microphone signal corresponds to an SNR of -15 dB in a 6-MHz TV bandwidth at best, assuming a 6-dB receiver noise figure, and on the order of -20 dB for more realistic assumptions about antenna performance. The problem is simplified by the presence of characteristic features in the spectra of TV and wireless microphone signals and complicated by the practical reality that narrowband interference signals are often present from unknown sources [25]. Further discussion of practical sensing issues and results from prototype sensors based on earlier versions of the FCC rules appears in Reference 33.

If a suitable RF platform (i.e., an SDR or an RFIC) was available that meets the frequency coverage and emission specifications described above and provides an electronic interface to control, the other major pieces of the design problem are developing a cognitive engine, obtaining the necessary networking software and hardware, and selecting a computational platform on which it all can run. See Chapters 3 and 4. Likely cognitive engine candidates will most probably implement the rule-based approach described in IEEE 802.11af, with the sensing-only purely cognitive radio approach pushed farther into the future.

1.6 Can a Radio Really Be Cognitive?

Cognitive radio is an application of artificial intelligence (AI), and that carries a lot of baggage, ranging from difficulties of definition to the observation that, while everyone may agree that the proposed solution to an unsolved problem is clearly AI, it is not recognized as AI once the problem is solved.[3] This applies to cognitive radio in the *cognitive* versus *adaptive* quandary, in which a working radio that solves a new problem is often described as adaptive rather than cognitive. For us, the key distinction is that a "smart" radio is cognitive if it is capable of learning and adaptive if it is not. We feel that a truly cognitive radio is capable of surprising its designers by arriving at unexpected solutions to problems and remembering these for future applications, while an adaptive radio is limited to a finite set of predefined outcomes. But this distinction does not keep researchers (ourselves included) from calling their products "cognitive" even when some obvious learning-enabling features are not implemented in order to allow the operator rather than the radio to make key decisions, as in our Public Safety Cognitive Radio [34].

The idea that a cognitive radio could surprise its designers may require further elaboration. In the simplest case, the radio's knobs and meters are specified at design, and the radio is limited to some finite range of values or set of discrete settings for each physical layer parameter. These ranges and sets are certainly known to the designer; the surprise may be the particular

---

[3]While this observation may well have originated earlier, we thank Karen Zita Haigh for first bringing it to our attention. See reference AFRL 11 in Reference 35.

combination of values chosen by the radio in a given circumstance. This is certainly true in most practical cases observed to date, where the cognitive engine selects software modules from pre-loaded libraries and uses one of them, for example, to implement a particular modulation format. But, conceptually at least, we can imagine a cognitive engine that is able to modify its own software and/or switch in new hardware on the fly, thus, in a sense, "growing" new knobs and meters. Based on our work with cognitive engines discussed in the chapters to follow, we feel that such developments are probable and that cognitive radios will do surprising things.

Surprises also come about through emergent behavior of networks of cognitive radios. Here, the network may behave in unexpected ways based on information encoded and stored in the behavior of individuals nodes. This is analogous to the intelligent collective behavior of a swarm of bees, and we will discuss it further in Chapter 7.

This philosophical, theoretical, and algorithmic issues underlying cognitive radio and, more broadly, cognitive dynamic systems have been explored by Haykin in a series of influential papers, beginning with Reference 36. He identified a set of fundamental cognitive tasks (radio scene analysis, channel identification, and transmit-power control and dynamic spectrum management) and placed these in a cognition loop as a framework for implementation. In later work (see, e.g., Reference 37), he and his co-author expanded their analysis to general systems embodying cognitive perception and control, including cognitive radio and cognitive radar.

Bibliography

[1] A. R. Young and C. W. Bostian, "Simple and Low-cost Platforms for Cognitive Radio Experiments [Application Notes]," *IEEE Microwave Magazine*, vol. 14, no. 1, January 2013, pp. 146–157. doi: 10.1109/mmm.2012.2226543

[2] J. Mitola, "Cognitive Radio: An Integrated Agent Architecture for Software Defined Radio," Ph.D. Dissertation, Royal Institute of Technology (KTH), Stockholm, Sweden, May 2000.

[3] (2015) Software-defined Radio. Wikipedia. Accessed September 30, 2015. [Online]. Available: http://en.wikipedia.org/wiki/Software-defined_radio

[4] C. Rieser, T. Rondeau, C. Bostian, and T. Gallagher, "Cognitive Radio Testbed: Further Details and Testing of a Distributed Genetic Algorithm Based Cognitive Engine for Programmable Radios," in *IEEE Military Communications Conference (MILCOM 2004)*, 2004.

[5] "The XG Vision Request for Comments," BBN Technologies XG Working Group, 2003.

[6] P. Kolodzy, "Spectrum Policy Task Force Findings and Recommendations," in *International Symposium on Advanced Radio Technologies (ISART 2003)*, 2003.

[7] M. McHenry, "Frequency Agile Spectrum Access Technologies," in *FCC Workshop on Cognitive Radios*, 2003.

[8] Wirless Adaptable Network Node WANN Proposers Day Announcement. Federal Grants. Accessed September 30, 2015. [Online]. Available: http://www.federalgrants.com/Wireless-Adaptable-Network-Node-WANN-Proposers-Day-Annoucement-5588.html

[9] J. B. Evans. (2004, Feb) The NeTS Program. Accessed September 30, 2015. [Online]. Available: http://archive.cra.org/Activities/workshops/nsf.wireless/Evans_ProWiN_NeTS_Program2.pdf

[10] J. Mitola, "Software Radios: Survey, Critical Evaluation and Future Directions," in *IEEE National Telesystems Conference*, 1992.

[11] T. Newman and G. J. Minden, "A Software Defined Radio Architecture Model to Develop Radio Modem Component Classifications," in *First IEEE International Symposium on New Frontiers in Dynamic Spectrum Access Networks, 2005. DySPAN 2005*. Institute of Electrical & Electronics Engineers (IEEE), 2005. doi: 10.1109/dyspan.2005.1542674

[12] T. Costlow, "Cognitive Radios Will Adapt to Users," *IEEE Intelligent Systems*, vol. 18, no. 3, p. 7, May 2003. doi: 10.1109/mis.2003.1200720

[13] R. E. Ramos and K. Madani, "A Novel Generic Distributed Intelligent Re-configurable Mobile Network Architecture," in *IEEE VTS 53rd Vehicular Technology Conference, Spring 2001. Proceedings (Cat. No.01CH37202)*. Institute of Electrical & Electronics Engineers (IEEE), 2001. doi: 10.1109/vetecs.2001.945031

[14] J. Neel, R. M. Buehrer, B. H. Reed, and R. P. Gilles, "Game Theoretic Analysis of a Network of Cognitive Radios," in *The 2002 45th Midwest Symposium on Circuits and Systems, 2002. MWSCAS-2002*. Institute of Electrical & Electronics Engineers (IEEE), 2002. doi: 10.1109/mwscas.2002.1187060

[15] C. W. Bostian and A. R. Young, "Cognitive Radio: A Practical Review for the Radio Science Community," *Radio Science Bulletin*, vol. 342, 2012, pp. 15–25.

[16] T. Rondeau, B. Le, C. Rieser, and C. Bostian, "Cognitive Radios with Genetic Algorithms: Intelligent Control of Software Defined Radios," in *SDR Forum Technical Conference SNR2004*, 2004.

[17] K. Nolan, P. Sutton, L. Doyle, T. Rondeau, B. Le, and C. Bostian, "Dynamic Spectrum Access and Coexistence Experiences Involving Two Independently Developed Cognitive Radio Testbeds," in *IEEE DySPAN 2007, Dublin, Ireland*, 2007.

[18] J. Lunden, V. Koivunen, and H. V. Poor, "Spectrum Exploration and Exploitation for Cognitive Radio: Recent Advances," *IEEE Signal Processing Magazine*, vol. 32, no. 3, May 2015, pp. 123–140. doi: 10.1109/msp.2014.2338894

[19] B. Lechner, R. Chernock, M. Eyer, A. Goldberg, and M. Goldman, "The ATSC Transport Layer, Including Program and System Information Protocol (PSIP)," *Proceedings of the IEEE*, vol. 94, no. 1, January 2006, pp. 77–101.

[20] C. Rhodes, "Interference Between Television Signals due to Intermodulation in Receiver Front-Ends," *IEEE Transactions on Broadcasting*, vol. 51, no. 1, March 2005, pp. 31–37.

[21] R. Rast, "The Dawn of Digital TV," in *IEEE Spectrum*, Oct 2005. Accessed November 20, 2013. [Online]. Available: http://spectrum.ieee.org/consumer-electronics/audiovideo/the-dawn-of-digital-tv

[22] FAQs for Unlicensed LPAD (Including Wireless Microphone) Registrations. Federal Communications Commission. Accessed November 22, 2013. [Online]. Available: http://www.fcc.gov/help/faqs-unlicensed-wireless-microphone-registrations

[23] (2013, November) Electronic Code of Federal Regulations. Title 47. Part 15. Subpart H – Television Band Devices. U.S. Government Printing Office. Accessed November 22, 2013. [Online]. Available: http://www.ecfr.gov/cgi-bin/text-idx?SID=b1aceff0f2270193cad0c9a1995bda9b&node=47:1.0.1.1.16.8&rgn=div6

[24] A. Flores, R. Guerra, E. Knightly, P. Ecclesine, and S. Pandey, "IEEE 802.11af: A Standard for TV White Space Spectrum Sharing," in *IEEE Communications Magazine*, pp. 92–100, October 2013.

[25] S. Shellhammer, A. Sadek, and W. Zhang, "Technical Challenges for Cognitive Radio in the TV White Space Spectrum," in *Information Theory and Applications Workshop, 2009*, pp. 323–333, February 2009.

[26] R. Saeed and S. Shellhammer, *TV White Space Spectrum Technologies*. Boca Raton, FL: CRC Press, 2012.

[27] MAC and PHY for Operation in TV White Space. ECMA-International. Accessed May 31, 2014. [Online]. Available: http://www.ecma-international.org/publications/files/ECMA-ST/ECMA-392.pdf

[28] M. Schuhler, A. Jaschke, M. Tessema, and C. Kelm, "Flexible RF Front-End for Communication in TV White Spaces," in *Microwave Conference (EuMC), 2013 European*, pp. 1087–1090, October 2013.

[29] J. Kim and H. Shin, "Design Considerations for Cognitive Radio Based CMOS TV White Space Transceivers," in *SoC Design Conference (ISOCC), 2011 International*, pp. 238–241, November 2011.

[30] K.-F. Un, P.-I. Mak, and R. Martins, "A 53-to-75-mW, 59.3-dB HRR, TV-Band White-Space Transmitter Using a Low-Frequency Reference LO in 65-nm CMOS," *Solid-State Circuits, IEEE Journal of*, vol. 48, no. 9, 2013, pp. 2078–2089.

[31] A. Prata, A. Oliveira, and N. Carvalho, "An Agile Digital Radio System for UHF White Spaces," *Microwave Magazine, IEEE*, vol. 15, no. 1, January 2014, pp. 92–97.

[32] T. Matsumura and H. Harada, "Prototype of UHF Converter for TV White-Space Utilization," in *Wireless Personal Multimedia Communications (WPMC), 2012 15th International Symposium on*, September 2012, pp. 123–127.

[33] R. Balamurthi, H. Joshi, C. Nguyen, A. Sadek, S. Shellhammer, and C. Shen, "A TV White Space Spectrum Sensing Prototype," in *New Frontiers in Dynamic Spectrum Access Networks (DySPAN), 2011 IEEE Symposium on*, May 2011, pp. 297–307.

[34] B. Le, F. Rodriguez, Q. Chen, B. Li, M. ElNainay, T. Rondeau, and C. Bostian, "A Public Safety Cognitive Radio Node," in *Software Defined Radio Forum Technical Conference, Denver, CO*, 2007.

[35] K. Haigh. Selected Publications by Karen Zita Haigh. [Online]. Available: http://www.cs.cmu.edu/ khaigh/papers.html

[36] S. Haykin, "Cognitive Radio: Brain-Empowered Wireless Communications," *IEEE Journal on Selected Areas in Communications*, vol. 23, no. 2, 2005, pp. 201–220.

[37] S. Haykin and J. Fuster, "On Cognitive Dynamic Systems: Cognitive Neuroscience and Engineering Learning from Each Other," *Proceedings of the IEEE*, vol. 102, no. 4, 2014, pp. 608–628.

CHAPTER 2

Cognitive Engine Design

2.1 Introduction

A cognitive engine (CE) is an intelligent package that turns the knobs and reads the meters of a controllable radio system. As discussed in Chapter 1, this concept was initially introduced as a software entity that interacts with an electronically configurable radio transceiver [1]. Early examples exhibited limited functionality and software portability, often focusing on fairly specific problems or approaches that were very dependent on particular aspects of the support platform. Subsequent work on the topic has developed many of the trends pioneered by these early examples and expanded the domain of a CE. However, it is important to consider that not much more than a decade has passed since the inception of the concept of a CE for radio application; there are a great many challenges left unsolved and capabilities left undiscovered. In fact, the full scope of CE is only beginning to develop and there is a vast territory left to explore.

While a CE is typically considered as operating in conjunction with a single radio, a CE need not be tied to a single radio. In fact, the CEs that currently see the most widespread use operate over entire networks under the guise of network resource managers.[1] The fundamental purpose and the logical organization of a CE remain invariant with regard to the number of radios over which the system operates; only the goals and specific techniques employed change. Further, it is important to note that these concepts are natural evolutions of wireless and computational technology and their existence alone does not necessarily represent a groundbreaking advancement. Rather, the key factor that defines a CE is its abstraction of reasoning from the core of radio operation and this factor does not depend on the number of radios involved.

This chapter aims to provide readers with the equipment necessary for successful expeditions into the unknown of CEs. Thus, we place more emphasis on the explanation of fundamental CE principles and framework than on the examination of any particular machine learning or artificial intelligence (AI) technique. We provide practical and specific examples where possible, but as the CE concept spans several fields, a complete discussion of all topics is simply impossible. Furthermore, this chapter will focus on CE concepts through the lens of radio applications. With this chapter in hand, readers should be equipped to begin their exploration of a concept that could fill several books.

---

[1]While typically not referred to as CEs, the elements that implement self-organizing functionality in cellular networks fit within our definition. See Reference 2 for more on self-organizing networks.

2.2 The Basic Function of a Cognitive Engine

Fundamentally, cognitive radio is a new tool for designing radio systems. That is, in the cognitive radio paradigm, the designer specifies *techniques* for realizing a particular radio system for a particular situation. Note that this is distinct from directly specifying a radio system for a particular situation; rather, an additional layer of abstraction has been added between the designer and the radio system. This additional layer is anthropomorphized as being cognitive because it mirrors the actions of an engineer in designing wireless systems (although we will return to this cognitive designation to discuss some of the subtle nuances). This additional cognitive layer is the cognitive engine and its fundamental purpose is to design a radio systems on demand.

As a result of this abstraction, the designer is conceptually removed from the problem of designing a specific radio system. In anthropomorphic terms, the designer's role shifts from doing radio design to teaching radio design; that is, the cognitive engine learns how to design a radio from the designer. Note that learning occurs during the construction of the CE. A designer encodes her knowledge and experience into an entity that goes on to approach a particular problem of radio system design (perhaps frequency and modulation selection) in a particular way (perhaps through the application of an optimization algorithm). This is exactly analogous to an engineering student learning a particular algorithm from a professor. Both the CE and the student are initially limited by the methods they have been taught. The distinction between the two (and ultimately the challenge for CE designers) is that students come to the professor already possessing the ability to develop concepts on their own (We hope!), whereas cognitive engines begin their existence with the designer. That is, a CE only does what it is told and never develops a concept or approach unless the capability to do so is explicitly constructed by its designer. Therefore, the challenge for a CE designer is not to teach a machine the design of a particular radio system, but rather to teach a machine how to design a radio system.

While such anthropomorphic language is useful for communicating the general purpose of a CE, it naturally leads to several debates and arguments. In an effort to avoid becoming bogged down in these, we will adopt a slightly less vague model for a CE than a student learning to design radio systems. Rather consider a CE to be an agent with particular inputs (meter values) and outputs (knob settings). This agent must apply some process, determined by its designer, to arrive at good settings for the knobs, where the definition of good is determined by the designer. For this purpose, it is often useful to consider a set of knobs collectively as a waveform.[2] Several measures of goodness maybe defined for a waveform, and for convenience the term waveform is often used to refer to union of a particular set of knobs and their associated goodness measures. Thus, a CE may be considered as an agent that has been designed to determine the best waveform for a given set of meter values.

Note that the agent model of a CE presented above is a very inclusive definition for a cognitive engine in that many communication systems may be cast in this form. Consider, for example, the Modulation Coding Scheme (MCS) selection in 802.11. In this scheme, the module coordinating WiFi transmission determines a MCS based on whether or not it has recently received an acknowledgment (of course this is a gross simplification, but it captures

---

[2]Our definition of waveform is similar to, but distinct from, that used in the context of Software Communications Architecture (SCA) [3]. Rather than capturing all the transformations applied to information, we are using the term to simply refer to the knob settings that eventually control these transformations.

the general concept). In essence, the WiFi controller determines a MCS waveform based on the goodness metrics of robustness, in terms of whether the packet is received and subsequently acknowledged, and user mission, in terms of high throughput. This selection is made on the basis of the algorithm taught to the controller by its designer.

While we hope that the above example has clarified our concept of a CE, it has likely raised questions about what distinguishes CEs as a new concept. This is, in fact, the point of the proceeding discussion. More than a particular methodology or technique a CE is a framework for advanced radio system design that allows for the incorporation of a wide variety of techniques. This concept has certainly existed prior to the inception of the term, but has not necessarily been recognized as such. The key advancement in the recognition of the term is in the beginning of a field of study focused solely on the development of radio design algorithms. While the WiFi MCS selection scheme certainly represents a CE approach, it is much within the sphere of standard communication approaches to the problem. That is, WiFi represents a natural evolution of communication techniques, whereas the identification of the CE concept allows for more radical leaps. While the distinction initially seems subtle, the impact is significant. As a brief, immediate example, consider that CE researchers regularly discuss constructing systems that learn a good approach during the course of regular operation. This approach seems to be an anathema in the world of WiFi, where every aspect of a radio system is specified in lengthy standards documents and even minor variations of MCS require several pages of rigid definition. Therefore, CE is conceptual advancement, in which a new paradigm for design enables new features more than it is a new approach in and of itself.

This new paradigm pushes beyond prior techniques, focusing as it does on developing automatic radio system design rather than simply refining parameters. These two notions of automatic system design and parameter refinement certainly overlap, but they differ in scope and approach. A cognitive engine represents the encoding of a designer's knowledge and experience into a software entity, and as such it focuses on the processing of information related to radio design. The cognition loop of Chapter 1 is entirely based on this concept of effective use of available information. During the *awareness* stage, the CE gathers environmental information and builds a model of its current situation, i.e., processing environmental information into a useful form. In the *learning* stage, this information is stored for later use. Finally, the *adaptation* transitions from the situational information to knob settings (called action information in Section 2.6) based on the methods made available by the designer and the goals given by the user. This action information is then encoded as a radio on the available hardware. Cast in this light, a CE processes information, based on the experience shared by the designer, with the objective of creating a radio system to serve the user's goals.

2.3 Cognitive Engine Organization

The information processing that must be completed by a CE is readily categorized into several broad tasks. While determining a set of knob values for use is the final goal of a CE, much of the work done in a CE involves gathering information and manipulating it into a useful form. Characterizing these tasks according to how information is used and shaped provides a natural organization for a CE.

The individual information processing tasks of a CE are handled within components. Each component is responsible for advancing the CE toward its goal through some operation on

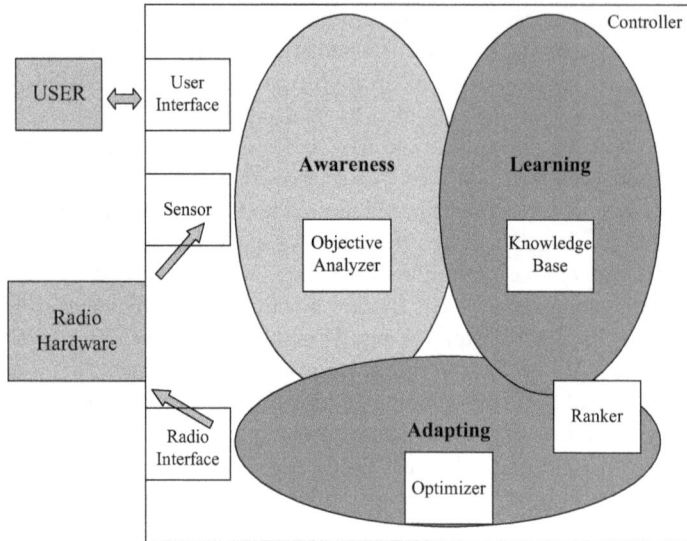

Figure 2.1   Cognitive engine components in the cognition loop. ©2012 Radio Science
Press, Belgium. Adapted, with permission, from Reference 4.

information. Recall that the fundamental purpose of a CE is to design a radio system automatically. At the core of a successful design process is the application of relevant, usefully
formulated information to the problem of setting parameters within an applicable model.
The definition of individual information processing tasks decomposes the question of setting parameters into more tractable pieces (even if these pieces encompass large fields of
study) and the encapsulation of these tasks into components allows the selection of a particular
model through the combination of applicable techniques.

Figure 2.1 provides a reference of how the components of a cognitive engine fit into the
cognition loop discussed in chapter one. Each component encapsulates a specific task within the
cognition loop, framing duties that may be accomplished by a range of techniques. For example,
the optimizer encapsulates the central task of adaptation in the cognition loop, defining the
responsibilities that the chosen algorithm must address. Two components, the controller and
user interface, serve to augment the cognition loop to provide a complete cognitive engine. The
controller coordinates the actions of the rest of the components, bounding the space in which
the loop operates. The user interface provides connectivity to an external user, allowing her to
direct the loop to meet her needs. All together the cognition loop provides a framework for the
individual components of a cognitive engine.

In this section, the individual components are discussed in terms of their role in information processing and how they affect the assumed model for the radio system under design. The
concept of the waveform provides a useful handle for tracking how information is manipulated
within the CE through the involvement of each component. However, note that the components discussed here represent a logical organization for a CE and while this organization is
occasionally reflected in the software architecture of the CE (see Section 2.5.5), this need not
be the case. Furthermore, every component discussed here is not necessarily compulsory and

a particular CE may not include all or even any aspects of a particular component. Rather the components discussed below provide a view of the breadth of information processing applied by a CE.

2.3.1 Optimizer

The optimizer is the single component most commonly associated with the CE. Fundamentally the optimizer provides the adaption capabilities of a CE by selecting a good waveform for use by searching through a parameter space. This search may be accomplished in a number of ways, from mathematical approaches to random number generation; yet fundamentally the optimizer performs a search. Defined as a search through a parameter space, the operation of optimizer exemplifies many of the core concepts of a CE. More than any other single component, the optimizer is responsible for determining a collection of knobs for use. However, this selection is made on the basis of supporting information in the form of calculated metrics and indications of waveform goodness. Furthermore, the searching operation, and indeed the parameter space itself, depends heavily on the radio model that is applied in the design process. Therefore, while the optimizer functionality is certainly core to a CE, this functionality requires both the support of several other components to provide supporting information and the collection of all components to provide a complete model.

A cognitive engine is often constructed completely around the optimizer; the most straightforward CEs often consist of little more than this component. Optimization techniques tend to be best suited to particular radio models, often depending on specific assumptions or the availability of information in certain forms. Thus, the technique selected for optimization typically dictates the requirements for other components. See Section 2.4.2 for a discussion of several optimization techniques and the requirements they impose.

2.3.2 Objective Analyzer

The objective analyzer supports the operation of other components by calculating objective (non-subjective) metrics for use within the CE. Calculation of such metrics represents the CE's development of situational awareness by refining raw information into a more useful form. These metrics are well defined, repeatedly calculable, and objective in that they have the same meaning in any situation. Such metrics may include expected throughput, or estimated additive white Gaussian noise (AWGN) bit error rate (BER). These metrics are not knobs to be applied; rather, they are a distillation of some particular information useful within the CE. These metrics represent a direct manipulation of information into a form more useful within the CE.

Note that the objective analyzer need not be realized as a separate entity in software. Rather this logical component is often realized within other software entities; this component encapsulates the informational translation that takes place throughout the CE. By defining the metrics available to the optimizer, this component largely defines the radio model employed by the CE. Furthermore, this component encapsulates the approach to determining a given set of metrics, whether through assuming a particular noise environment or through the use of various statistical techniques. Thus, this component provides a logical encapsulation of both the key metrics used in the CE and the method by which they are determined.

2.3.3 Ranker

The ranker serves a parallel purpose to the objective analyzer. This component is responsible for calculating the goodness of a waveform to support the selection process. In contrast to the objective analyzer, the metrics calculated by this component are subjective measures of goodness. Furthermore, while the objective analyzer defines the model used in design, the ranker defines the measure of goodness. This component operates closely with the optimizer to determine waveforms for use, but is separate from the optimizer component to provide an encapsulation of goodness determination. The ranker sits between the adaption and learning by supporting the actions of the optimizer and providing a point at which learned information may be applied to guide future options. This encapsulation allows designers to align a CE operation with the goals of a user regardless of which techniques are applied.

2.3.4 Knowledge Base

The knowledge base provides information storage. This component is more than simply a database, however, as the storage of information within a CE includes the distillation of that information into trends through artificial learning. Thus, the knowledge base is responsible for handling the learning duties of the cognitive engine. Note that this learning may be the form of correlation calculations, machine learning, or case-based reasoning. This learning typically takes place over the knobs and their resulting performance as a means to bypass or aid the optimizer in the selection of a waveform. Since the knowledge base handles information about the limitations of available hardware, this learning can take place over the use of the hardware as well. The knowledge base component provides an encapsulation of the learning that takes place in a CE.

2.3.5 Radio Interface

The radio interface component provides the interface between the radio and the CE, linking the cognitive operation to the actual radio hardware. This interface chiefly provides the conduit allowing the CE to control radio operation and receive statistics about communication, such as packet loss information. As such, this component defines the set of knobs available to the CE and the time scale on which they may be changed. Note that this interface may connect the CE to radio hardware as discussed in Chapter 3 or to a simulation thereof. Typically, this component is fixed from the perspective of the CE, presenting some collection of knobs and method for setting them. Selection of an appropriate radio platform and its corresponding interface is further discussed in Chapter 5.

2.3.6 Sensor

The sensor component provides the interface between any sensors and the CE. Similar to the radio component, this component links the cognition to the radio hardware. However, in contrast to the radio component, the sensor is more focused on providing external information to the cognitive engine. Thus, the radio component can be thought of as providing links to hardware internal awareness and control within the cognitive radio (CR) and the sensor component can be thought as providing external awareness. Sensor information is made available for repacking by the objective analyzer and subsequent use in the optimizer. Additionally, this component

CSERE

Main | Spectrum Sensor | Radio | Knowledge Base | Objectives | Rank | GA Optimizer

Knobs

Name	# of Bits	Min. Value	Max. Value
Modulation	2	0	2
Tx Power	4	0	1
Bit Rate	6	125000	10000000
Roll off	3	0	1
Frequency	10	400000000	500000000

Population Sample

Modulation	Tx Power	Bit Rate	Roll off	Frequency

Seed

Modulation	Tx Power	Bit Rate	Roll off	Frequency

Population Size	10	Population Size
Safe Generations	5	Safe Generations
Crossover Probability (%)	20	Crossover Probability
Mutation Probability (%)	5	Mutation Probability
Max Children (% of Population)	10	Max Children

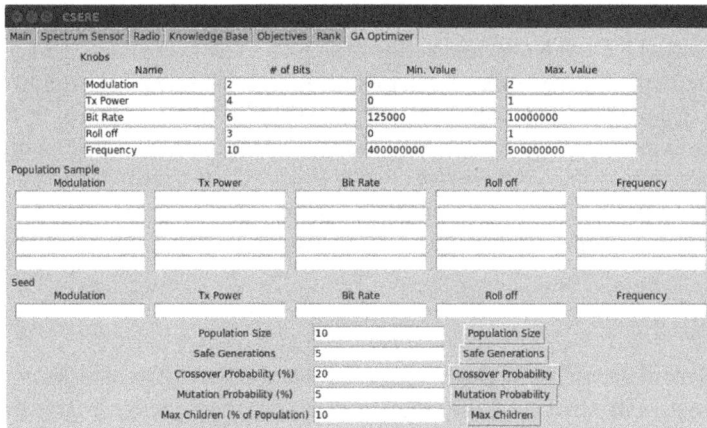

Figure 2.2   Example CE GUI.

may determine a separate category of knobs from those associated with the radio component in that they do not directly affect the completion of a CE's goal; rather, the knobs associated with the sensor component simply affect the information that is gathered into the CE. These sensor control knobs allow the CE to adjust its own situational awareness instead of establishing communication for a user. The component encapsulates the input of situational information to the CE.

2.3.7 User Interface

As the final interface component, the user interface provides the interface between the CE and its user. Ultimately, a CE exists to serve the needs of some user and this component provides a means for a user to guide the CE. This guidance will naturally take different forms for different users. For example, human users may want to maintain some direct control over radio actions or machine users may require highly structured interactions. This component provides a way for external users to oversee the operation of a CE typically through setting radio models or through guiding the determination of goodness.

Figure 2.2 presents one possibility for a CE graphical user interface (GUI) from the Cognitive System Enabling Radio Evolution (CSERE) CE. The research focus of this CE, which will be discussed further in Section 2.5.5, is reflected in the GUI through exposing several settings to the user. Thus, each CE component includes its own GUI panel which displays different information or settings associated with that component; Figure 2.2 displays the panel for a Genetic Algorithm (GA)-based optimizer component. These displays allow the user to monitor and modify the behavior of individual components through a convenient graphical interface. While this level of sophistication is useful in the case of CSERE, other CEs may not require the same degree of complexity in the user interface. Indeed many CEs will require only a means to input the data to be communicated, a method to indicate the type of service required, and an activation button. More sophisticated CEs may only require the user as a source of data and learn her preferences over time. Finally, there is certainly no requirement for a graphical interface, which may be infeasible on several platforms. In these cases, a console-based text interface is typically used.

Several applications of CEs may not directly interact with human users at all. In the context of machine-to-machine communications, for example, a CE may simply be activated by its designer and left to serve the needs of an artificial user. Such interaction would certainly not be conducted in a manner intelligible to humans, whether that be graphical or text based. In this case, the user interface may resemble the radio interface more than the graphical examples presented above. A machine user interface would likely take the form of some self-contained communication system that passes a predefined protocol. Perhaps even this could be provided through a cognitive system, although this would certainly lead to a highly complex system.

2.3.8 Controller

The controller is the component that coordinates the actions of the other components. This component can be realized in a range of ways, from simply a static action loop to a sophisticated control center. In the most static case, the design specifies an order of operations that is maintained throughout the lifetime of the CE. In more complex realizations, the controller is empowered to swap components at will, employing the encapsulation provided by components to alter any aspect of the CE's operation. As such the controller itself can be a higher level CE designing a system of lower level cognitive engine components to design radios through the selection of appropriate components and setting of their parameters.

2.4 Tools and Techniques for CE Component Design

The tasks of a CE, as encapsulated by components, cover a wide range of topics and potential approaches. Several methods are available for fulfilling the needs of nearly any CE component. This section provides an introduction to some of the more commonly applied methods in an effort to provide some clarity in the face of such a breadth of alternatives. This section does not attempt to offer a complete guide for techniques applicable to CEs, but rather a starting point for more in-depth exploration.

2.4.1 Machine Learning

Machine learning is a topic that has captured the imagination of humans for centuries. More recently this concept has coalesced into a field of study focused on realizing machines capable of learning. Herein, we avoid the interesting (and often heated) discussion of what constitutes learning and instead accept any acquisition and later use of information as sufficient demonstration that learning has occurred. This definition provides a broad basis for the goal of machine learning. In the context of CEs, clearly these operations are the domain of the knowledge base component. Recall that this component has the loosely defined task of managing the information relevant to a CE's operation. Machine learning is typically applied to allow a CE to refine its approach to designing communication systems based on its own past experience. We will discuss more specific examples of how this may occur during our discussion of significant machine learning techniques.

As belabored above, machine learning is a broad field that occasionally inspires arguments about which techniques should or should not be included. In an effort to find a path through this swamp of ideas and concepts, we will characterize some of the machine learning techniques that are particularly significant to the operation of CE. These techniques are by no means the only options but rather are commonly seen approaches because they tend to offer clear advantages to CE operation. Rather than limiting the reader to these techniques, it is our hope that this discussion demonstrates the methods of applying machine learning to CE such that readers are inspired to explore alternative approaches or new applications of machine learning.

2.4.1.1 Reinforcement Learning

Reinforcement learning applies the theories of behaviorist psychology [5] to the creation of artificial agents that learn on the basis of rewarding desired behavior or punishing undesired behavior. Perhaps the most clear (and iconic) foundation for this technique is exemplified by Pavlov's work, in which dog's were conditioned to drool at the sound of a bell [6]. The simple principles of tallying a reward value for various actions based on their results have clear application to artificial entities which excel at such book-keeping exercises. In fact, the success of artificial implementations of this technique is, ironically, one of the factors that drove the transition away from behaviorist psychology toward cognitive psychology [7].

The fundamental operation of reinforcement learning is book-keeping. In the most basic form, a CE would select some action and adjust a goodness value for that action based on its result. The designer might select one value for a good result and some other value for a bad result, representing how severely actions should be rewarded or punished. The CE would then be able to select actions based on their goodness score. Clearly, this would result in the CE always selecting the first action that was rewarded, which may not be desirable. To combat such fixations with the first-rewarded choice, the selection process is made stochastic, forcing the CE to explore other options with a given probability. This alternative may be the option that has been selected least, an option chosen at random, or an option selected through some other means. Some approaches also simulate forgetting by incrementally resetting reward and punishment values. Designing a successful reinforcement learning system then comes down to a pool of options, reward/punishment values, an exploration probability, an exploration strategy, and forgetting time. For a more general discussion, see Reference 8 and for a CR discussion, see Reference 9.

Reinforcement learning focuses on applying experience to make decisions during operation; that is, there is no pre-training phase necessary. This attribute frees the CE designer from the need to collect training samples and use them to train the CE prior to its deployment. However, in a manner mirroring human beings, this focus on online learning can result in the CE initially making several non-optimal choices. Therefore, while reinforcement learning allows CEs to adapt to a changing environment, it is only suitable for situations where mistakes are acceptable.

Overall, reinforcement is well suited to many CR problems. The fundamental model of reinforcement learning is straightforward, typically resulting in systems that are straightforward to debug. Furthermore, since the fundamental operation of reinforcement learning is simply keeping track of the results of an action, integration within a CE is typically simple. Rather than deploying a reinforcement centric system, it is possible to use only some of the elements of reinforcement learning to add learning capabilities to systems focused on other techniques. (Some aspects of this approach are discussed in terms of case-based learning in Section 2.4.1.4.) Finally, reinforcement systems can be applied to learn (and track) a good choice among any pool of options as long as a reward or punishment may be defined.

On the other hand, reinforcement learning is not a perfect solution. This technique trades the need of *a priori* knowledge for a number of parameters (reward values, exploration probability, forgetting time, etc.) that often take on arbitrary values. The meaning of these values (and the process of setting them) is further clouded when integrating elements of reinforcement learning with other techniques. It is certainly possible to determine appropriate values for a particular situation, often through trial and error, but there is no universal method for determining the values. Refinements of the basic approach address many of these problems, at the cost of increased complexity. Q-learning is a successful model-free refinement of the basic reinforcement approach initially proposed by Watkins [10]. The application of a more modern variant of Q-learning called *probably approximately correct (PAC)* learning to CR is presented in Reference 11.

2.4.1.2 Classification

Classification provides an alternative approach to machine learning that is more structured than reinforcement learning. The fundamental operation of classification is the assignment of points to appropriate groups. These points may represent a range of things, but in the context of CE they are typically situations, as indicated by meter readings. A typical application for classification algorithms is the determination of like scenarios that call for a particular action. In fact, the concept of determining appropriate groups is core to classification techniques. The point that varies between individual techniques is the definition of *appropriate*. Various nuances in this definition allow the individual techniques of clustering to be applied in a variety of ways.

The single most widely known classification technique is most probably artificial neural networks (ANNs) or simply neural networks (NNs). Initially conceived of as a model for the human brain [12], NNs provide a powerful tool for realizing artificial reasoning. This technique relies on training a network of artificial neurons based on a collection of sample inputs and corresponding desired outputs. Each neuron within the network expresses an output based on the aggregation of its inputs as filtered by a nonlinear (typically sigmoidal or similar) activation function. Training proceeds by adjusting the weights on inputs according to one of several algorithms designed to bring the output of the network into agreement with the desired outputs, while maintaining the ability to generalize beyond the training data by not conforming too closely to the input data. The result of training is then a network topology and set of input weights that allow the NN to infer the proper cluster for inputs it has not yet seen. Hinton is considered to be a pioneer in the area of NNs, providing some early work in the area [13] and continuing his work today. Haykin (who also investigates CR design) provides excellent and detailed coverage of these operations in Reference 14.

Given their general ability to learn the connection between inputs and outputs, NNs provide many options to a CE designer. Neural networks are readily applied to the task of developing situational awareness by connecting sensor data to the transmitter's modulation and coding scheme [15]. Alternatively, NNs may be used directly for the selection of an operating point in the spectrum [16]. Even the entire operation of a CE has been claimed to be within the domain of NN application [17, 18].

Such power and wide applicability comes at the dual price of training and inflexibility. NNs are largely restricted to use in situations for which the designer already has relevant samples (consisting of input and output pairs) to train the network. While there are certainly highly complex scenarios that involve harvesting samples using some alternative method and automating the training process, these have not (as of yet) been successfully applied to CEs. Furthermore, even when suitable data are available for training, a NN represents a largely fixed box within the CE. The operation of NNs does not lead itself to integration with other techniques, as reinforcement learning does. This casts an NN as a black boxes that performs a set task for CE.

Based on the requirements for NN use, it is more useful to consider NNs as a replacement for some CE functionality rather than as a tool for learning. In this context, NN's learning occurs prior to run-time and the NN learns directly from the designer. This use of neural networks is the most common application to CEs. For example, rather than designing some complicated method for determining the modulation and coding scheme of a sensed signal, the designers in Reference 15 simply trained a NN to accomplish this task. In this way, NNs perfectly exemplify the concept of a CE as an entity that learns how to design communications systems from their creator. While this technique offers a powerful method for accomplishing tasks that would otherwise be extremely difficult, the inflexibility resulting from the need of training prevents this approach from being a panacea for CEs.

Note that the NN approach is not the only option in clustering for CEs. Rather there are several techniques, each with their own specific abilities and requirements; however, the general themes discussed here in terms of neural networks apply to all of these clustering techniques. The interested reader is directed to References 11 and 19 for exploration of these other techniques.

2.4.1.3 Markov Models

Markov models are a family of stochastic modeling methods based on finite state machines (FSMs). Specifically, Markov models capture the behavior of a system by focusing on the probability of transferring from one state to another. Herein lies the defining feature of Markov models: transition probabilities are assumed to depend only on the current state and not on any prior states [20]. Indeed this assumption is often required to achieve a tractable FSM model for a real system.

Typically, Markov models are applied to CEs as some variant of a Markov decision process (MDP). An MDP extends basic Markov models by adding the concept of an action, or decision, that may be taken within a state, triggering a transition to the next state. The probabilities associated with the action may initially be known or learned over time, often using reinforcement learning as briefly discussed in Section 2.4.1.1. Once transition probabilities are know, the MDP provides a map indicating the sequence of actions most likely to result in the transition from any initial state to any desired state. Given this knowledge and knowledge of a goal state, a CE may easily select actions. The authors of Reference 21 provide an example application of this technique in analyzing a CR system.

Standard MDPs tend to only be applied to analysis of CR systems due to uncertainty regarding the current state in CRs. This uncertainty results from a CR not being able to directly access information regarding other nodes or limitations in the ability of sensors. Thankfully, partially observable Markov decision processes (POMDPs) were developed to handle situations in which the current state may not be directly observable and can therefore cope with uncertainty. In this variant MDP, a CR must maintain a probability distribution over possible states representing the CR's belief about the identity of the current state. This framework provides a flexible system that is better suited to handle uncertainty inherent in CR problems. An example application of POMDPs to CR is given in Reference 22.

A slightly different formulation of a Markov model, the so-called *multi-armed bandit (MAB)* approach, is also becoming popular for cognitive radios. This formulation is based on a theoretical gambler selecting which of several *one-armed bandits*, or slot machines, he should play next. The gambler knows that each slot machine follows a Markov model with each state representing a different winning amount and would like to develop a strategy that maximizes his own money. As discussed in Section 2.4.1.1 for reinforcement learning, the gambler needs to achieve a balance between exploiting machines he believes will pay out and exploring to find even larger payouts. Thus, the formulation of the MAB approach centers on developing an optimal sequence of slot machines to play in order to maximize the winnings of the imaginary gambler. Happily a method to determine such a strategy exists in closed form [23].

In the typical application, a CR plays the part of the gambler and channels represent the slot machines. Here instead of money, the CR is attempting to win uninhibited access to the spectrum. The Markov process for each channel usually models the behavior of a primary user whose use of a given channel would block the access of the CR. Several CR algorithms have been developed based on this approach with References 24 and 25 providing relatively straightforward examples. Furthermore, Christophe Moy [26] is considered leader in the use of MAB approaches for cognitive radio.

2.4.1.4 Case-based Learning

Case-based learning represents a departure from the traditional ML reaping grounds of mathematics and computer science. This approach has its roots in case-based decision theory (CBDT) from economics [27]. CBDT provides a framework for the application of past knowledge to current actions with the goal of advancing future economic positions. To accomplish this, the theory defines the concept of a case as the collection of a particular problem, action, and result. During operation, an entity would collect a memory of cases, which constitutes its past experience. This experience can be applied to any present problem by retrieving the case with the problem most similar to the current one and the most favorable result. The action of the matching case may either be applied or serve as the starting point for the determination of a new action. In this way, the entity in question can apply past experience for future gain.

On the face of it, this approach is not dissimilar to reinforcement learning, as discussed in Section 2.4.1.1. However, there is an important distinction in terms of the manner of storing information. In a basic reinforcement learning approach, only a goodness value is stored for each action. This stems from the typical use case of reinforcement learning as guiding a machine to learn a single task automatically. The result is that reinforcement provides a lightweight solution, requiring storage of only a single value per action, at the cost of handling only a single task (or small group of closely related tasks). Contrast this to the situation of CBDT which focuses on a human being (or economic firm) handling several varying problems in its pursuit of wealth. This yields an approach that requires the storage of much more information, but extends to a wider variety of approaches. The approach suggested by CBDT can be considered as an augmented variation of reinforcement learning.

The rich historical utilization of case-based learning is well suited to CE application. Rondeau, who coined the term case-based learning, discusses the application of the approach to CEs in Reference 1. The inherent ability to track good responses to several nuanced situations is especially well suited to operation within often rapidly changes spectral situations. Furthermore, case-based learning shares reinforcement learning's prowess for smooth integration into the operation of a CE. In evidence of this, Rondeau also displayed the advantage such integration offers to optimization. Even when an exact match for the current situation is not in the memory, a similar situation can be used as a seed for optimizer, reducing the time needed to find a good action. These benefits typically offset the cost of the need for increased memory capacity in CE.

2.4.2 Optimizers

Optimization focuses on finding the best (optimal) value. Before proceeding, consider just how vague this concept really is: the term *best* is extremely subjective; no definition of *value* is even provided. Typically, this is removed by defining a well-formed model, including fitness functions and constraints to outline the concept of goodness and acceptable values. This approach results in a well-defined branch of mathematics with a long history of benefit to numerous fields of study. However, this approach is not the only path available. Alternative efforts focus on modeling biological methods of problem solving, instead of building complete frameworks for developing closed-form solutions. Still other methods model successful humans directly, attempting to apply the experience of human experts. Each of these approaches, and many more not covered here, have different attributes to offer CEs.

In the context of application to CE, we will consider optimization to be the process of searching an action space for the best action. This definition provides more solid ground than the intentionally vague offering above, but does not exclude any of particular family of approaches. For our purposes, it will suffice to define an action space as the collection of actions available to a CE. It may be helpful to relate this concept to a k-dimensional geometric space, where k is the number of knobs a CE has to control, but it is important to remember that this space need not have any particular property, continuity, or attribute commonly defined for geometric spaces. Rather this space is simply a convenient way to refer to the collection of actions. Note that the concept of *best* remains in our updated definition. This allows the CR (or perhaps the CE itself) the latitude to determine what should be accomplished by a given action. This determination is logically encapsulated by the ranker component (Section 2.3.3) and typically connected to the goals of the user. We will discuss several optimization techniques in terms of our definition below.

There are several considerations that a CE designer should consider when selecting or implementing an optimization technique. In practice, the most important of these tend to be how the technique handles uncertainty, sensitivity to parameter setting, and speed of operation. The degree to which these factors matter vary with the particular goals of a CE, but these three tend to be among the top factors that degrade optimizer performance.

Finally, it is important to consider that a CE optimizer operates on multi-objective problems. These objectives tend to be semi-conflicting. For example, consider the maximization of throughput and minimization of interference. Shannon tells us that capacity increases with transmission power, so increasing power is an obvious method to increase throughput. However, increasing power also increases the interference inflicted upon other users, so the goal of minimizing interference is served by decreasing power. In these multi-objective situations *best* can be especially difficult to specify. Thankfully, the concept of Pareto optimality [28,29] from economics provides a solution. A solution is Pareto optimal if no alternative exists that improves all individual objectives. That is, Pareto optimal solutions are those for which improvement in any single objective causes degradation in some other objective. Typical CE optimization problems have a set of such solutions, which is referred to as a Pareto frontier. All solutions on this frontier are generally considered to be equally good. The second most common approach to handling multiple objectives is the linear combination of objectives [30]. This method weights objectives according to the preference with which they are considered before combining the multiple objectives into a single goodness score. Linearization in this fashion can be more simple to implement, but relies on complementary scaling of objectives which can be difficult to achieve.

2.4.2.1 Mathematical Approaches

Mathematical optimization develops formal models, based on closed-form formulas that express the fitness of actions within the action space and constraints on the space. This family of techniques is typically categorized as either linear [31] or nonlinear [32] depending on the nature of the formulas required. This category of techniques is likely the best studied of all those discussed here.

These approaches center on developing a particular mathematical model. This model needs to capture the details of the CR situation in a tractable form and provides a closed-form description of how the situation progresses. This description often takes the form of a utility function over available actions. Given this model, an optimal set of actions can then be found to maximize utility. The authors of Reference 33 provide an example application of this concept.

In the particular situations for which they apply, mathematical approaches offer highly optimal results relatively quickly. The major problem with these techniques is their inability to handle large amounts of uncertainty. Applications of mathematical techniques typically requires a closed-form expression that relates the impact of any possible selection. Additionally, any discontinuities in the action space tend to complicate the process greatly. The result is that in most situations, mathematical optimization in a CE is infeasible due to the lack of information or the complexity of the action space.

2.4.2.2 Evolutionary Algorithms

Evolutionary algorithms are a very popular class of algorithms inspired by the evolution of biological species. This class of algorithms tend to feature several signature mechanisms each modeling a different aspect of the evolution process, including reproduction, genetic mutation, genetic recombination, and natural selection. These processes are used to evolve a population of candidate actions toward a high *fitness*, or utility. The population represents the region of *action shape* under consideration and the various mechanisms of the algorithm shape this region over time. The inherent stochastic nature of several of the mechanisms and region of consideration, as opposed to only considering a single point, combine to avoid being trapped in local extrema. Thus, this class of algorithms does not tend to get stuck in low fitness solutions and can typically proceed very close to optimal solution quickly. However, the techniques work against the algorithm once a nearly optimal action is found because the stochastic nature of the evolution can prevent convergence to a single optimum. Thankfully, in practice these algorithms tend to quickly find actions that are good enough [1].

In the context of CE, the most famous evolutionary algorithm is the GA [1, 34, 35]. AGA follows the standard approach of evolutionary algorithms in the evolution of a population of potential actions, or *chromosomes*, through a number of generations. Each of the typical mechanisms can occur in each generation of a standard GA. Reproduction of high fitness chromosomes transpires through seeding the population for a new generation based on the highest fitness individual, or individuals, of the prior generation. Genetic mutation is represented by randomly changing some aspects of a subset of the population. Recombination takes place as a crossover between two chromosomes where a portion of the action are traded between the two. Finally, natural selection is embodied in the calculation of fitness for each candidate and use of the best for seeding or output. Generational evolution continues until either a generation limit or some fitness threshold is reached.

In order to understand the impact of the various aspects of GAs, let us consider a simple example algorithm. For the sake of this example, we will consider a GA that is attempting

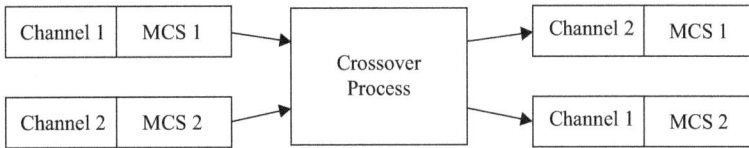

Figure 2.3   Crossover in a GA.

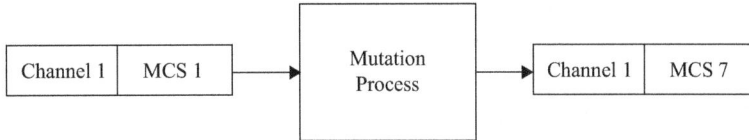

Figure 2.4   Mutation in a GA.

to select a channel and MCS from a given pool of options. The evolution process is then attempting to determine the best pair of channel and MCS for operation as denoted by a particular fitness function. When the optimizer of this CE is called to determine a good pair it begins by initializing a population to some set size, randomly selecting a channel and MCS for each individual. These chromosomes provide the first of several generations to be evolved. Within each generation, each pair of chromosomes is selected for crossover with a probability given by a designer-specified parameter. Pairs selected for crossover trade channel selections, as illustrated by Figure 2.3. Next chromosomes are selected for mutation with a probability specified by the designer. This mechanism randomly selects which subcomponent, or *gene*, of the chromosome to alter and then randomly alters it. For example, a chromosome representing the use of MCS 1 on channel 1 might be mutated to use MCS 7 on channel 1 instead, as shown in Figure 2.4. Once these genetic modifications are complete the fitness function is used to determine a fitness for each chromosome and the best individual is selected. If the fitness is above some threshold, often given by a percentage of the maximum possible fitness, or the generation limit has been reached evolution ends the top performer is returned by the optimizer. Otherwise, the top chromosome is used to seed the next generation. In this example, that might occur by giving half of the new population the channel selection of the seeding individual and the other half the MCS of the seed. Other values can then be filled randomly.

We hope that this example has clarified the basic operation of an evolutionary algorithm, in the form of a GA, and has raised some important points about the class of algorithms. First among these is the issue of sensing. The operation of a GA does not necessarily have an obvious place for sensing; however, environmental information is required to maintain a link with the current situation of the CR. Thus, sensing information is typically used to shape the fitness values assigned by the fitness function, allowing the environmental information to shape the fitness landscape in the action space that is being searched. Next one must consider the impact of the probabilities for crossover and mutation. The crossover probability controls the degree to which the region in parameter space containing the current population of chromosomes is explored, and the mutation probability affects the degree to which the region of parameter space under review is expanded. Appropriate tuning of these two parameters can tailor the behavior of the optimizer as desired. These slightly more suitable aspects of evolutionary algorithms provide a wealth of options for use in CEs.

Evolutionary algorithms represent a point in between a completely random, uncoordinated approach and the more structured mathematical scheme. Only solutions that happen to be represented within a population of chromosomes are considered; yet through promoting the genes of high fitness individuals the strongest attributes are maintained. The fitness function utility provides the core of an evolutionary algorithm, guiding its operation. There are several methods for developing a good fitness function, including the use of machine learning, as discussed in Section 2.4.1. Furthermore, Goldberg has shown that GAs in particular will obtain an optimal solution for a given fitness function, albeit in unknown time [36].

Ultimately, this class of algorithms provides a fairly robust approach that is able to determine rather good, but not perfect, actions in highly uncertain situations [1].

2.4.2.3 Expert Systems

Expert systems take an entirely different approach to optimization. Rather than consider some fitness value that varies over the action space, an expert system attempts to encode the experience of a human designer directly in a CE. To accomplish this, expert systems define a set of situation–action pairs. These pairs define a particular situation in which a given action should be taken, often taking the form of some threshold. For example, a fall back action might be used if the packet loss rate climbs above a certain value. These situation-action pairs are occasionally referred to as *rules* and the expert system implementing them would then be a *rule-based system*.

These systems have several advantages over other options. First, expert systems are often very simple to implement, since only a lookup table is required. Additionally, optimization with an expert system only requires matching a situation and reading a rule, which can be accomplished very quickly. More complex systems may trigger multiple rules simultaneously, which would then require some sort of combination, but this process still tends to be faster than running the operations necessary for a GA, for example. Finally, the determinism of expert systems provides a significant benefit over many other techniques. This is especially true for systems in which regulation and policy are important concerns, as tends to be the case for industrial applications. The nature of expert systems is such that its behavior is completely predictable, given knowledge of the environment, which is often highly desirable or necessary.

These benefits come at the cost of flexibility. Once implemented an expert system is typically a static entity, unable to keep pace with a changing environment. Naturally, a basic rule-based approach could be augmented with machine learning to prevent this, but doing so dilutes the benefits engendered by the expert system. The result is that expert systems only really provide the above benefits to situations in which the designer can foresee all of the challenges that a CE will face and formulate rules for overcoming these challenges. Expert systems shift the responsibility of radio system design heavily toward the human designer and therefore dilute the nature of the CE.

2.4.3 Estimation

Estimation is the distillation of knowledge into a particular figure. This is extremely useful in CE for digesting information for use in reasoning, as is done by the objective analyzer. However, estimation as a field has a greater deal of overlap with machine learning. Several techniques, especially Markov modeling (Section 2.4.1.3), straddle the line between these two areas. Thus, estimation is the sort of topic that pervades several aspects of CE operation. Furthermore, estimation is a CE's single most potent tool in the face of uncertainty. Kumar and Varaiya provide a good discussion of the application of estimation techniques to reduce uncertainty [37]. Hilbe and Robinson outline the use of estimation tool kits that may be integrated into CE functionality [38]. Estimation may be employed nearly anywhere within a CE to distill available information in an effort to combat the lack of certainty.

One common application of estimation in CR systems is channel state estimation. In most real systems, sensing only serves to reduce uncertainty, rather than completely remove it. Thus, the useful information must be distilled out of raw sensing data, typically over the course of several samples, to allow for reasonable decisions to be made. For example, when attempting to find open channels, individual sensor reading tends to exhibit a high degree of variation. Often examination of a series of readings is required to obtain a stable ruling. Accomplishing this typically takes the form of defining an estimator for channel availability based on a statistical measure of sensor information, as exemplified in References 39 and 40. Such an application of estimation to refine raw sensor data illustrates the most typical use of estimation techniques within a CE.

2.4.4 Sensing

Sensing is perhaps the best studied aspect of CR[3] This largely stems from the role of spectrum sensing in providing situational awareness. Ultimately, this situational awareness underpins the operation of the CE by providing the information necessary to make appropriate decisions. This environmental information is the primary purpose of spectrum sensing and thus we will discuss sensing techniques in terms of the information each provides to the CE. Note that there are more aspects to this information than the accuracy of environmental awareness. Specifically, both the timeliness and the reliability of the information must be considered. Additionally, the information required to utilize a particular sensing technique must be considered. We will focus on these aspects in the following discussion.

As another dimension in sensing, the object of the sensing impacts the usefulness of the technique employed. There are several potential reasons to employ spectrum sensing in CR, including finding open channels, discerning the identity of a particular transmitter, or modeling the fading of a particular environment. Each of these purposes represents a different type of information that may be made available through sensing and impose different requirements on sensing techniques. These differing requirements along with variations in terms of the timeliness and reliability of sensor information make spectrum sensing for CR a complex field in its own right. Note that occasionally, trading off among multiple sensing techniques in a single CR may provide a method for taming some of this complexity.

2.4.4.1 Energy Detection

Energy detection sensing provides the most basic form of spectrum sensing. In its simplest form, this method measures the received power level on a given channel. This power is then compared to a noise floor threshold to determine whether or not the received energy is a result of an active transmitter or simply background noise. Naturally, an appropriate setting for the noise floor threshold is vital to the performance of this technique and signals received at a

---

[3]Several surveys of sensing for cognitive provide an overview of the available techniques from perspectives different from that presented here [41, 42].

power very close to this threshold will often be incorrectly categorized as noise [41]. Options for improving the performance of this technique include monitoring a channel over some period of time and then examining the statistical properties of the received energy to determine whether or not it is just noise [43].

The primary benefit of energy detection-based techniques is their simplicity. The only prior information required is the noise floor threshold; no information about other transmitters is required to determine whether there is a transmitter operating. The technique also does not require any specialized hardware. Furthermore, the computational complexity of energy detection algorithms tends to be low. While this is less true for methods that calculate covariance or other statistical measures, energy detection typically provides a very light weight solution. This low complexity enables fairly rapid sensing. Moreover, the small computational footprint of energy detection sensing techniques makes them suitable for use as a "secondary" sensing technique to be used with a more complex technique.

The simplicity that is fundamental to energy detection limits the depth of information that these techniques yield. They are largely limited to determining channel occupancy and little more than that. Statistical processing of the history of received energy may allow for estimating more nuanced information, but other techniques more focused on providing such abilities exist. Fundamentally, energy detection techniques are light weight, low footprint approaches suitable for quick surveys of channel occupancy.

2.4.4.2 Feature Detection

Feature detection sensing is any technique that detects a specific element of a received signal. For example, detecting a digital TV (DTV) pilot, as done in Reference 44, is a common instantiation of feature detection sensing. However, this category is not limited to determining the presence of a specific element of the spectral footprint of a signal. Rather, this approach to sensing includes any method that focuses on detecting the presence of a distinct feature, from waveform elements such as preambles, to cyclostationary features of the received energy. While this broad definition seemingly includes several techniques previous ascribed to energy detection, the distinction lies more in the object of sensing rather than the method. While energy detection focuses on channel occupancy, feature detection represents a refinement that focuses on determining the presence of a specific signal.

In fact, specificity provides the central theme of feature detection sensing. Rather than simply determining if any given signal has been received, the methods in the feature detection category determine if the signal of interest is present. Naturally, this requires prior knowledge of the signal of interest that matches the specific method employed; the signal preamble must be known if that is the object of sensing. Also signal processing is required to separate the wheat, signals of interest, from the chaff, any other elements in the spectrum, which increases the computational cost and can introduce additional delay. The result of these increased burdens is comparatively reliable knowledge that a particular signal is present. Furthermore, the transients of a signal can be used to fingerprint a transmitter, allowing determination of the presence of a specific signal generated by a specific transmitter [45].

Specific knowledge of a signal, or indeed a transmitter, present within a given portion of the spectrum is certainly useful in a CR context. This knowledge is most readily obtainable and applicable in dynamic spectrum access (DSA) scenarios, where primary users are likely willing to provide some prior information to improve the avoidance capabilities of secondary users. Moreover, the cyclostationary techniques, in particular, provide a wealth of possibilities, because signals can be reliability detected without the need of decoding the signal. Combined with the capabilities offered by a CE, cyclostationary signatures of signals can be learned and tracked to determine the behavior of other communications systems in the environment. Adding the ability to fingerprint specific transmitters extends this behavior monitoring to individual users. Learning such behavior would allow for several new applications of CR, including efficient spectral use but also extending well beyond this.

2.4.4.3 Cooperative Sensing

Cooperative sensing is based on the operation of several sensing elements working together to build an understanding of the spectral environment. Typically, this involves distributed sensing entities that pool their knowledge. Each individual sensor employs one of the many available techniques; however, simple energy detection is a common choice for distributed sensing elements due to its simplicity. The defining feature of cooperative sensing is that the information from all of the distributed sensing elements is combined. Such a combination usually occurs within a single aggregation entity, which can then be queried by other entities as needed. Cooperative sensing systems can take the form of a number of CRs working together to pool knowledge, a single CR, employing a network of sensors, or even simply a single pair of communicating elements sharing sensor information. In any case, the defining element of cooperative sensing systems is the combination of knowledge from distributed sensing elements.

Pooling of sensing knowledge provides several benefits. One primary benefit is overcoming the hidden node problem, in which a node is visible from one radio in a communication system, but not visible from all others. As a simple example consider the situation in which radio A is transmitting information to radio B in the presence of a third radio, radio C. If radio B falls within the coverage area of radio C, but radio A does not, radio C is considered a hidden node from the perspective of radio A. Identifying the existence of a hidden radio C using only sensing information from radio A is extremely difficult. However, if radio B shares sensing information, the problem disappears. The same concept applies to simply characterizing the spectral environment. Cooperative sensing allows better understanding in non-reciprocal fading environments. Furthermore, cooperative sensing improves sensing reliability by combining independent samples.

The primary cost of cooperative sensing is the complexity of combining sensing information. Consider that any useful CR system will involve at least two radios, a transmitter and a receiver. Thus, distributed sensing element tend to be readily available and the challenge is simply combining information from these elements in a useful manner. To begin with a mundane problem, the information must be transferred to some common point in order to be combined and then that combined knowledge must be transferred to the points at which it is useful. Since the entire purpose of CRs is designing systems to transfer information, this issue becomes a balancing problem, between transferring sensing information and user information. However, the challenges do not end with determining a balance in information transfer, rather they proliferate. Questions of reliability, or indeed of trust, naturally arise for any situation in which individual sensors may fail or malicious users may exist. Additionally, protocols to allow for time and format synchronization are required but also lead to overhead in processing and delay. Finally, the central issue of combining trusted, synchronized data is non-trivial in and of itself. Thankfully, several methods for accomplishing this exist [46–50]. Cooperative sensing has the potential to improve nearly any sensing information required, at the cost of the overhead necessary to combine said information.

2.4.4.4 Radio Environment Map

Radio Environment Map (REM) sensing constructs a temporal–spatial representation of the spectral environment. In truth, REMs are usually a result of cooperative sensing as discussed above. The defining feature of this particular method is the construction of a map of the spectrum which typically includes a history. This method can be thought of an extension of energy detection to a particular region in space. An image of the energy at all points within this region is produced and updated continuously with a small delay. The resulting map supports frequency selection in a spatially intelligent manner; REMs reveal the locations at which channels are free. Moreover, since the map tracks the environment over time it enables several use-cases, including transmitter localization, propagation modeling, and spatial spectral monitoring.

This method is extremely significant for CR systems due to its potential for use in a database-oriented DSA system. In such a system, a group of sensors would feed information into a central entity, which would construct a REM and make it available in a database. This REM would then be used to determine whether a cognitive radio is able to access a channel based on its location. Furthermore, when the channel is no longer open, all radios in the effected region can be altered, without distributing the operation of those in areas where the channel remains available. The basis for such a system is discussed further in Reference 51.

Since REMs are based on cooperative sensing, they share most of the associated cost. REM approaches to sensing require the additional machinery necessary to construct and transfer REMs. Additionally, some localization system must be available, both for the construction of the REM and its use. This localization system may be based on an external system, such as global positioning system (GPS), or included with the framework that supports the REM. In either case, localization imposes additional requirements.

2.5 Cognitive Engine Architecture

Ultimately, the final form of a CR depends on the goals and constraints of its intended use. These topics are discussed more broadly in Chapter 5, but this section discusses how these factors relate specifically to the CE. Here we provide an overview of the issues that must be considered when composing the techniques discussed above into a CE along with some examples of CE organization.

2.5.1 Broad Considerations

The integration of components raises issues that are distinct from the factors considered when constructing the individual components. Such integration issues will not be unfamiliar to any designer of large scale software systems, but a few of them have special implications in the space of CEs. We provide a discussion of these here.

2.5.1.1 Flow of Control

However sophisticated, the controller of a CE is, it will ultimately be based on one of two possible control flow models: the event-driven model or the polling model. Each of these models have their own benefits and limitations which will shape the core operation of the CE and therefore impact its ultimate operation. We will provide a brief discussion of impact of the trade-offs between these options on CEs here and reserve a more complete discussion until Section 4.4.

The event-driven model of control should be extremely familiar to anyone with experience in embedded or user-interface programming. In this model, the controller waits for an event to occur before activating an appropriate response. The arrival of a primary user or the failure to receive some number of acknowledgments might be events that trigger the action of a CE. This model is typically very responsive and efficient with computational resources, but it can be much more complex to implement, than the alternative.

The key to success in the event-driven model is fast event handling. If an event is handled very quickly the controller will be ready for the next event. Alternatively, if the time necessary for handling a single event exceeds the time between events, the controller will miss potentially important events. It is certainly possible to buffer events, and even sort them according to urgency, but this adds to the complexity of implementation. Ideally implementation of this model involves running each module in parallel so events may be simply directed by the controller and handled entirely within components; however this is not always feasible. Furthermore, the stochastic nature of many CE techniques can make the operation of event-driven systems especially hard to track. This contributes to the complexity of implementation as well as complexity of regulation, as predicting the behavior of an event-driven CE can be difficult. Event-driven CE are akin to complex systems in that small variations in an input situation can lead to large differences in behavior. Finally, event-driven systems typically employ asynchronous time, meaning that the operation of any given module does not necessarily occur concurrently with that of any other module. That is, different asynchronous components can run under different clock domains, complicating interactions between modules. This can result in several problems, such as delaying the handling of events or eroding the concurrency of data. There are several methods to address these issues. For example, Real-time Operating Systems (RTOSs) provide mechanisms to prevent these problems or hardware-driven systems, such as Field Programmable Gate Array (FPGA), offer direct control over clock domains in terms of clock cycles. These, and other methods, have complexity cost which represents one of the primary trade-offs inherent in implementing event-driven CEs.

The polling control flow approach provides an alternative to the event-driven scheme based on looping through operations. In this paradigm, operations are considered for activation in a fixed sequence. Perhaps the controller may first collect data from the sensors, then pass this to the objective analyzer, and then consider whether the optimizer is needed to find new knob settings. The polling control method is directly analogous to the cognition loop approach commonly used to discuss CRs and CEs, which reveals its primary benefit of simplicity. This scheme is very straightforward to discuss, implement, and regulate. Furthermore, since all operations occur in lockstep with a central clock, timing considerations are greatly simplified. The cost of this is, of course, a reduction in the quick reactions that would be offered by a well-implemented event-driven approach. Each operation must wait its turn in line. It is certainly possible to implement very fast control loops to reduce the time each operation must wait (again parallelization proves useful here), but the sequence of events is still fixed. For example, if a CE must always be alert for a primary user, sensor information may be desired between (or indeed during) the operation of all other components, but doing this while maintaining communications this may introduce unwanted overhead. In short, the polling control scheme favors simplicity over reactiveness.

Finally, hybrid solutions of these two basic schemes are certainly possible. Such approaches blend the benefits and limitations of each approach. Combining both polling and event-driven methods can often provide a CE, the exact mix of simplicity and reactiveness required for its operation. The initial conception of the cognition loop fits into this hybrid approach very naturally in that the operation of the CE as a whole often follows a polling structure, periodically progressing through the awareness, learning, and adapting stages, but the operation of an individual component may be organized in a more event-driven fashion. However, note that the cognition loop may also be implemented as a collection of parallel event-handling entities, each running a separate polling loop internally. If the complexities of event-driven operation can be overcome, this organization allows the awareness, learning, and adapting operations to occur simultaneously, with entities only updating each other, through the generation and handling of events, when situations change.

In practice, most CR systems follow this hybrid approach, where even the radio aspect of a CR may exhibit elements of both polling and event-driven methods. Considering the CE within this larger system, interactions with the radio aspect of a CR may require examination of event-driven challenges even if the CE is implemented purely as a polling structure. Thus, a CE designer must be aware of both approaches and the potential challenges that may arise.

2.5.2 Monolithic Versus Distributed

The elements of a CE can be deployed either as a monolithic process on a single computational platform or distributed across several processes, possibly on different platforms. While the reasons for distributing CE elements may not initially be obvious, consider that a CR is typically distributed across hardware elements, as is discussed further in Chapters 4 and 5. This distribution allows the interaction between the radio elements and computational elements of a CR. Additionally, such heterogeneous computation approaches can be implemented in a manner that improves performance. Such performance enhancements drive the demand consideration within CEs.

A monolithic design vastly simplifies the operation of a cognitive engine, since all elements of the CE are rooted to a single process. While this certainly has impact on the development process, simplicity has a much greater impact on the operation of the CR. Recall that CEs are complex sub-systems of complex systems; CEs exist entirely to support the operation of CRs. Having a highly intricate CE typically serves to make the CR behavior unpredictable in real implementations with real noise. Furthermore, complex realizations of CEs are unsuitable for low-power devices, like user handsets, embedded sensors, or any other small-scale wireless devices. On these devices, computational power and battery limits suggest a monolithic design.

Distributed CEs have the ability to apply more computational resources to a situation. Spreading the execution of modules across several computational platforms has great benefits in enabling parallelization. Additionally, specialized computational platforms may benefit some calculations or operations, but hinder other operations. However, all of these benefits come at the cost of complexity, as discussed previously. A significant portion of this complexity arises from the need to coordinate distributed elements and provide communication between them. Coordination and communication issues depend heavily on the hardware architecture employed but typically provide the primary overhead for distributed solutions. Thus, if a distributed CE is necessary, examination of the communication between elements may provide the most direct path toward an operational system. Once operational, a distributed CE may provide suitable computational power for optimizing performance over a large-scale system or an entire network.

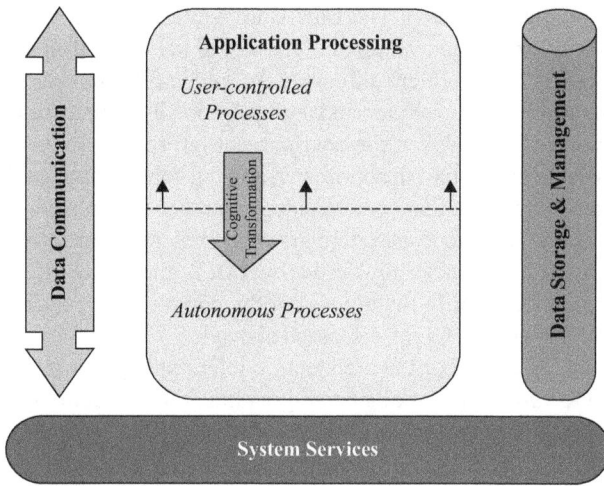

Figure 2.5   General CE architecture. ©2010 Wireless Innovation Forum. Reprinted, with permission, from Reference 52.

2.5.3 Standards

The issue of standards is useful starting point for our examination of CE architectures that allows us to provide some key points about CEs. First, it is important to note that no universally accepted standard exists. In fact, at the time of writing very few standards for CRs or CEs exist at all. However, there has been some effort in this area which attempts to solidify the often fluid topic of cognitive radio in order to drive industrial adoption. However, the standards must find a balance between stable definition and allowing the inherent flexibility of CR.

2.5.3.1 Wireless Innovation Forum

The Wireless Innovation Forum (WINNF) provides an Information Process Architecture (IPA) which attempts to clarify cognitive radio and extend the concept as a disruptive technology in communications. The WINNF is an industry group focused on advanced use of spectrum, including SDR, CR, and DSA, for industrial use. The forum is not focused on any particular market place or regulatory concern but rather takes a broad viewpoint. As such, WINNF is attempting to balance solidification of CR concepts without impeding future progress.

The goals of WINNF in balancing advancement and exploitation of CR for a broad range of applications means that described systems must be general. As such, WINNF presents a structure for CE in Figure 2.5. This architecture is based on defining the information processing elements central to CE operation. Each element in the architecture provides an encapsulation of information processing in a similar manner as the task encapsulation discussed in Section 2.3. Furthermore, the interconnection of these elements through a *systems services layer* provides a generalized architecture for constructing a CE.

Each element in Figure 2.5 represents a subset of the components discussed above in Section 2.3. For example, the Data Storage and Management accomplish the same functionality as that provided by the knowledge base as discussed above. The WINNF views the data management element as central to the cognitive engine, since the WINNF views learning as the function that confers cognition. The Data Communication element provides the same functionality as the radio-interface component. The rest of the components are combined into the application processing. This element ultimately determines the behavior of the CE based on the confluence of environment information and user guidance. This includes the radio modeling functionality of the objective analyzer, the environment information collection of the sensor interface, the action determination of the optimizer, and the goal modeling of the ranker. These processes are separated into those that operate independently of the system user and those that do not. The former are referred to as autonomous processes and functionality in this domain is said to have undergone a *cognitive transformation*; that is, these processes have transitioned from the realm of human control to the domain of the cognitive engine. These functions are those that provide the core of the CE as discussed above.

WINNF leaves further details of the architecture intentionally vague. Rather the WINNF has defined this architecture as "a working paradigm rather than a technology" [52]. The stated goal of the architecture is the facilitation of further development.

2.5.3.2 European Telecommunications Standards Institute

The standards of the European Telecommunications Standards Institute (ETSI) Reconfigurable Radio Systems (RRS) grew chiefly out of the more industrial world of cellular networks and tend to consider the operation of networks of radios. This prospective shapes their approach to the definition and use of CEs. In this capacity, the CE takes the form of a network level entity that automates many of the functions of human designers so that adjustments to the network may be made in real time. Note that these entities are rare if ever allowed direct control over the real deployed networks of cellular operators; instead the application of CEs here more often serves to advise human controllers by analyzing data and offering suggested knob settings. Thus, the CEs referred to by the ETSI RRS represent a specialized instantiation of the concept designed to advise humans with regard to cellular networks.

While the ETSI RRS has published many documents, covering the industrial use of all aspects of reconfigurable radio, the one most relevant to the topic of cognitive engines is ETSI TR 102 682 [53]. This document discusses the entities that control of reconfigurable radio functionality in networks, which we are treating as cognitive engines. The language used by the ETSI may initially seem rather far from the components discussed here, although, in fact, both refer to the same concepts. Additionally, since the ETSI is focused on providing standards rather than implementations of CE, only logical diagrams are provided. However, these diagrams provide a clear example of CE architectures adapted for cellular networks. We will discuss ETSI's application of the CE concept, including the components covered above, in terms of the two primarily controlling entities defined by ETSI: the Dynamic Spectrum Management (DSM) block and the Dynamic Self-Organizing Network Planning and Management (DSONPM) block.

Dynamic Spectrum Management Block: The DSM block adjusts the spectrum use within the network on the order of hours or days. The functionality provided by this block encompasses the typical DSA problem but also includes maintenance of spectrum assignment knowledge and band trading information. The reader must recall that since this entity is designed for use within a cellular network, its users will likely have licenses for particular bands and are therefore less concerned with discovering frequencies on which they may operate and more focused on managing their own resources efficiently. Thus, the while finding and using

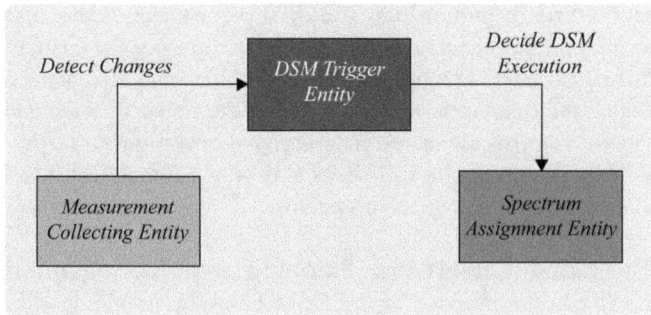

Figure 2.6   DSM block. ©2009 ETSI. Reprinted, with permission, from Reference 53.

unlicensed (or secondary) spectrum is a portion of the DSM block's responsibilities, the lion's share rests in efficient and legal use of licensed spectrum.

The DSM block consists of only four components: a sensor, an optimizer, a knowledge base, and a radio. As discussed above, the goals of mobile network operators focus more on managing stable licensed spectrum than on using unstable secondary spectrum. Furthermore, recall that the DSM operates on the time scale of hours or days and that operators tend to prefer determinism over real-time optimality. This results in optimization based primarily on expert systems, and sometimes mathematical models, rather than on stochastic searches. The knowledge base component of the DSM block exists primarily to index rules and regulations that may be applied by the optimizer. Additionally, the knowledge base is implemented in a manner that allows for easy updates as spectrum deals are made or changed. The radio and sensor components provide their typical function of interfacing to the radio management and sensor elements. However, recall that the DSM block is operating at the network level, meaning that these components broadcast knob settings to or aggregate information from a number of distributed radios.

Figure 2.6 displays the conceptual architecture of the DSM block provided by the ETSI [53]. In this figure, the block labeled "Measurement Collecting Entity" is the sensor component, responsible for interfacing with physical radios throughout the network that provides the relevant sensing data. The block labeled "DSM Trigger Entity" contains the combination of the expert-system optimizer component and the supporting knowledge base component. As the knowledge base holds spectrum status, polices, and agreements that are directly by the expert system, the two logical components are implemented together as this triggers entity. However, the logical separation aids the design and operation of the entity by distinguishing the machinery that applies rules (the optimizer) from the rules themselves (the knowledge base), allowing for updates to the knowledge base that need not affect the internal operation of the optimizer. Finally, the block labeled "Spectrum Assignment Entity" is the radio component, responsible for interfacing with the relevant radios throughout the network to apply the decisions of the DSM block.

The DSM block presents a CE architecture tailored for use in network management. While this particular architecture emphasizes determinism, it provides a good example of the breadth of the CE concept. The primary function of the DSM, offloading network design tasks from humans, places it squarely within the realm of a CE, despite that the traditional aspects of learning or stochastic optimization are minimized. In fact, the DSM block should be recognized as implementing a simplified cognition loop. The phases of awareness and adaptation are largely intact, but the learning phases occurs externally. That is, learning for this CE operates in the human network managers working alongside the DSM block and affects the operation of the DSM at the point that the rules governing its behavior are updated. Such *human in the loop* style learning reduces concerns about surrendering vital operations to be the sole domain of artificial entities. Thus, the DSM block provides a view of CE architecture adapted for rapid adoption into cellular network management systems.

Dynamic Self-organizing Network Planning and Management Block: The DSONPM block extends the purpose of the DSM block by providing suggestions on the reconfiguration of the network. Thus, this block represents a step toward the type of CE typically seen in academic research. The ETSI categorizes the operations of this block as occurring at the application layer, the network layer, and the combination of PHY and MAC layers. At the application layer, the actions of this block attempt to refine the Quality of Service (QoS) class assignment per cell. On the network layer, the distribution of traffic, handover parameters, interference control, and logical element connections are all adjusted. With regard to the PHY and MAC layers, the DSONPM block orchestrates the number of transceivers used in sensing, the particular transceiver technologies employed, the spectral usage, and radio parameters.

Accomplishing this variety of tasks requires a broader range of components than the DSM block. Here sensors deliver information that is then refined by objective analyzers. This is then applied to optimization, which allows more randomness in operation and output than the DSM block since the result is destined for review by human controllers. The nature of this optimization requires the determination of several subjective measures, for example, the QoS, which provides a place for use of a ranker. Once a good solution is found and reviewed by human overseers, it is deployed to the radio component. Additionally, the ETSI discusses the application of artificial learning to bypass the need for running optimization, which would then take place in a knowledge base component. A controller component coordinates the operation of these components, triggering their actions when needed. Finally, a user interface component communicates the suggested actions to the human controller.

Figure 2.7 depicts the architecture of the DSONPM block. The figure provided by the ETSI is very similar to the original vision of a CE, discussed in Section 2.5.4, with differences narrowing focus to the cellular application. The sensors embedded in the network deliver information to the input of the DSONPM, filtered through the operation of an objective analyzer labeled "Context Acquisition." This information travels alongside material from the knowledge base in the form of profiles and policies. (Note that the knowledge base as discussed here is split into the blocks labeled "Profiles Management," "Policies Derivation," and "Learning" to highlight the individual tasks of this component.) This wealth of information is then processed by the optimizer, with aid from the ranker component, in the so-called "Optimization Process." Finally, the knob settings are deployed to the network (and the knowledge base for learning purposes) through the radio interface component, here called "Behavior Configuration." This configuration of a CE is highly typical: the optimizer, supported by the ranker, is the central component, with the radio component handling output, the sensor providing raw input to be processed by the objective analyzer, and the knowledge base recording data [53].

The other two components, the controller and the user interface, are not shown here, but this does not indicate that they are least important. On the contrary, these components are so fundamental to this process that they are assumed to be present. The controller must coordinate the flow of information throughout this system, accessing the relevant input data for the given

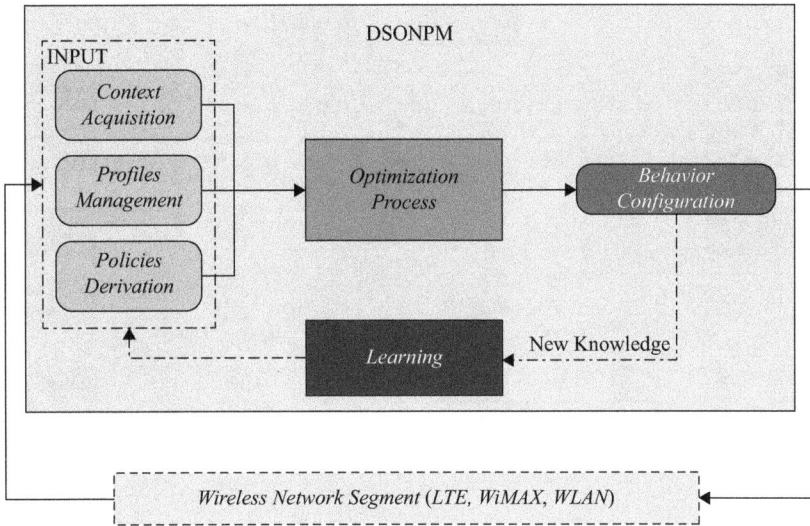

Figure 2.7   DSONPM block. ©2009 ETSI. Reprinted, with permission, from Reference 53.

task and ensuring that the correct instance of each other component is used. A controller in such a CE is responsible for configuring components to meet a certain need in the context of the current application. In the case of the DSONPM block, this would take the form of requesting appropriate profiles or activating the relevant calculations for context acquisition. Though seemly simple, these activities and operations often consume much of the effort of the CE designer. Likewise, user interaction represents a non-trivial consideration that depends greatly on the precise use case for the CE. Even in the standardized material provided by the ETSI, the topic of user interface for the DSONPM block is not considered because its details would depend too greatly on the implementer of the entity. Thus, while the DSONPM is designed for advising human operators in the vast majority of its deployments, no further information is offered by the ETSI on how this should be done.

Where the DSM block exemplifies a CE architecture designed for rapid adoption into current cellular networks, the DSONPM block represents a larger step. This architecture is built on incorporating as many of the purely academic elements of CE into a system suitable for industrial application as possible. As such the DSONPM architecture provided by the ETSI can be thought of as a mapping of a full CE to a realizable application in cellular networks. Moreover, this architecture provides a striking example of how readily CEs may be adapted for industrial use, assuming appropriated concessions are made to the realities of business practices.

2.5.4 Original CE Architecture

The original architecture for a CE provides an example of an efficient implementation of CE components that blazed the trail for subsequent realizations. This concept grew out of the process of constructing initial cognitive radios. During implementation, the benefits of packaging the complex reasoning functionality separately from the complex radio functionality had clear

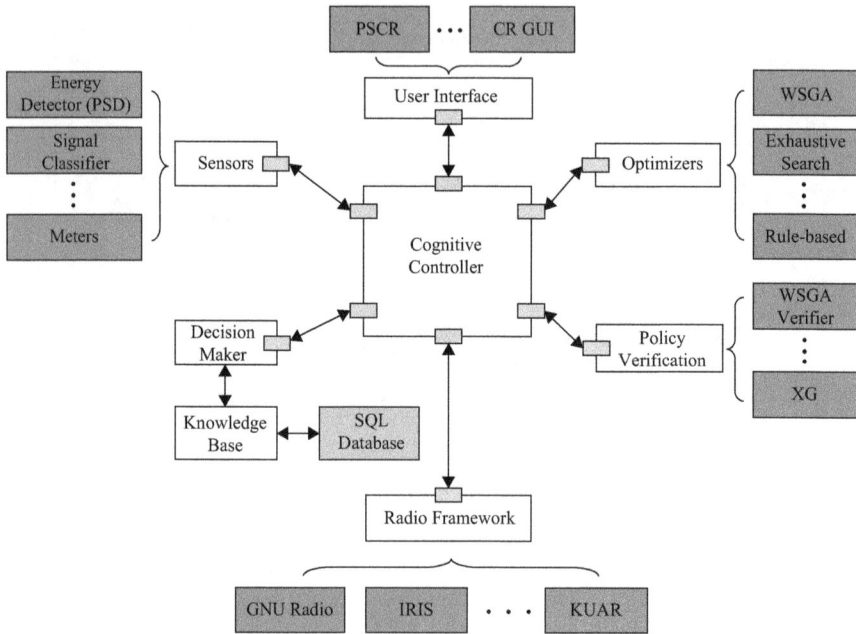

Figure 2.8 Original CE logical architecture. ©2007 Thomas W. Rondeau. Reprinted, with permission, from Reference 54.

benefits. At this stage, the definition of the above components was just beginning to crystallize; those working on early CEs were concerned with functionality first and developed more general definitions based on functional knowledge of CEs. In this way, the entire original concept of CE was an invention largely born out of necessity.

Figure 2.8 displays the logical organization of the original CE. This diagram displays the components, as defined by the tasks of a CE. The components discussed in Section 2.3 are a streamlined set of these components, refined by the experience gained in the years following the realization of the initial concept. Many components (the sensor, user interface, controller, and radio) are largely unchanged from their original conception; the tasks associated with these components are both fundamental and straightforward enough to remain unaltered. Initially, the optimizer component was very focused on meeting a given QoS because this measure was the primary metric with which a user's goals would be represented in the context of telecommunications at the time. However, the application of a CE has since been generalized well beyond QoS, requiring the introduction of the objective analyzer to capture a user's measures of interest and the ranker to capture the user's definition of good (which still may be a QoS, but also includes other possibilities). The original components of knowledge base, SQL database, and policy verification have all been combined into the knowledge base component. While this initially seems like lumping a great deal of functionality into a single block, the tasks of many of these components are intertwined to a degree that separating them into individual tasks is not entirely straightforward. Finally, the functionality of the decision maker has been incorporated into the controller component. Thus, the logical diagram of the original CE realization has provided much of the foundation for subsequent work on CEs.

Cognitive Engine - Software Architecture

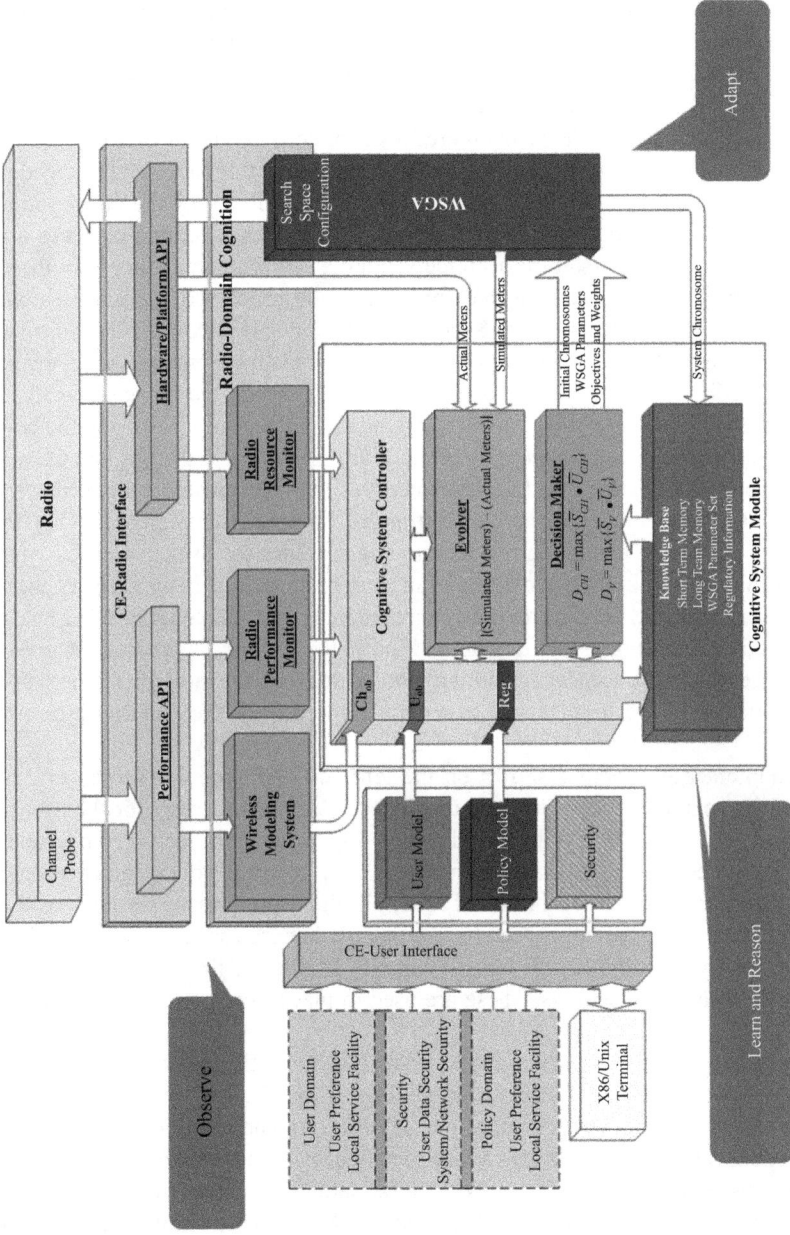

Figure 2.9 Original CE functional architecture. ©2014 Thomas W. Rondeau. Reprinted, with permission, from Reference 54.

While the logical diagram for the original CE provides a straight forward division among the components, the functional diagram, shown in Figure 2.9 is slightly more complex. The functional diagram provides a more true-to-implementation visualization of the original CE than the logical diagram. Contrast this to Cognitive System Enabling Radio Evolution (CSERE), for which the logical architecture and functional architecture are one and the same. Note that the components shown in the logical view are represented in the functional diagram as well, but the functional diagram provides more information about the sub-components and particular choices that suit implementation.

The functional architecture shown in Figure 2.9 can be considered a direct implementation of the cognition loop. In fact, Figure 2.10 displays the nested cognition loop that directly describes the operation of the original CE. The actions of this CE are driven by the desire to maintain a given QoS. Thus, developing an understanding of the current situation provides a natural beginning to operation. In pursuit of this understanding, environmental data from the sensors and radio performance data are collected from radio and processed by various monitoring systems. In this language introduced by Section 2.3, the block labeled the CE-Radio Interface would be termed the radio component and the Wireless Modeling System, Radio Performance Monitor, and Radio Resource Monitor would all operate within the objective analyzer. The scenario synthesized by these components is passed to the Decision Maker by the controller to determine the nature of the action to be taken, if any. Learning in the original CE occurred through the application of case-based decision theory to determine whether the current scenario matched a previously encountered one (see Section 2.4.1.4 for a discussion of CBDT). If action was necessary, adaptation could begin by employing the Wireless System Genetic Algorithm (WSGA) (a GA-based optimizer) to determine an appropriate action. In the case that the current scenario was sufficiently close to a previously encountered one, the WSGA could be loaded with previously used knob settings as starting point. Finally, the knob settings would be saved in the knowledge base (for future scenario matching) and loaded onto the radio (for current use). The loop would repeat in this fashion, the CE ever maintaining a QoS level, for the entire course of operation.

The user interaction modules shown on the left side of Figure 2.9 served to guide the execution of the cognition loop. First, the user interface provides a way for the user to indicate their desired level of QoS. Second, the user may enter their security preferences and relevant security details. Third, the user may enter any relevant policy information. In the scope of CR, policy information holds an especially high importance because of the legal issues surrounding the use of unlicensed (or secondarily licensed spectrum). The policy information expected by the original CE largely consists of a set of limitations, preventing certain combinations of knob settings. These are used to place limits on the space searched by the WSGA.

This structure can be thought of as a specific realization of the logical architecture. Where the logical architecture defines classes of modules that may be implemented, the functional architecture represents the combination of one selection for each component. Further, notice that the functional architecture provides a more detailed decomposition of tasks often defining modules as sub-elements for components or instantiating new blocks altogether to provide some additional functionality. This is fairly typical in CEs, due to the limitations of real implementations when compared to conceptual organizations. The main components of a CE can be thought of as providing a framework for CE realization and the additional functionalities fill in the gaps as necessary.

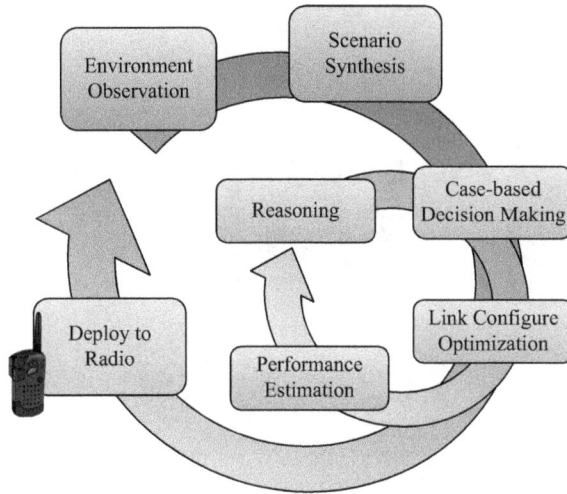

Figure 2.10   Nested cognition loop. ©2012 Radio Science Press, Belgium. Reproduced, with permission, from Reference 4.

2.5.5 CSERE Architecture

One possible organization for a CE is to simply construct the individual components and provide connections between them. This approach is distinct from typical approaches in that functional diagrams tend to reorganize structures for the sake of implementation efficiency, as seen in Section 2.5.4. Ignoring such efficiency concerns and instead implementing separate modules that perfectly match the logical diagram enables simple component swapping for the purpose of changing tasks or refining the approach to a problem. Such a CE is useful primarily for the research and development of CE techniques, including the development of meta-optimization approaches in which the CE applies AI (or other like means) to refine its own operation. To explore some of these topics we will discuss the CSERE project of Virginia Tech (VT).

Figure 2.11 displays the star architecture of CSERE. Here, each component is relegated to its own separate module, implemented in a separate file (or library); that is, there is no blending of component tasks into functional blocks as we've seen in other CE architectures. Rather, all of the ranker functionality, for example, is handled completely separately from the optimization functionality. This makes the architecture of CSERE very easy to understand; once you understand the components used in a CE, you understand the architecture of CSERE. Furthermore, the functional abstraction along component lines allows designers to easily alter CSERE for their own purposes. This organization allows a designer to simply redefine a fitness function, for example, without implementing an entirely new GA to serve as their optimizer. Thus, the simple star architecture of CSERE supports serves to directly support efforts to explore and develop the CE concept.

The cost of such a simplified architecture is the increased complexity of the controller. Isolating functionality according to components has the clear drawback of requiring more data passing between independent modules. As we have discussed throughout this chapter some components work very closely with another component to accomplish their tasks: the objective analyzer tends to require sensor data, the optimizer tends to require interaction with the ranker. Thus, as can be seen in other CE architectures, these components are typically implemented

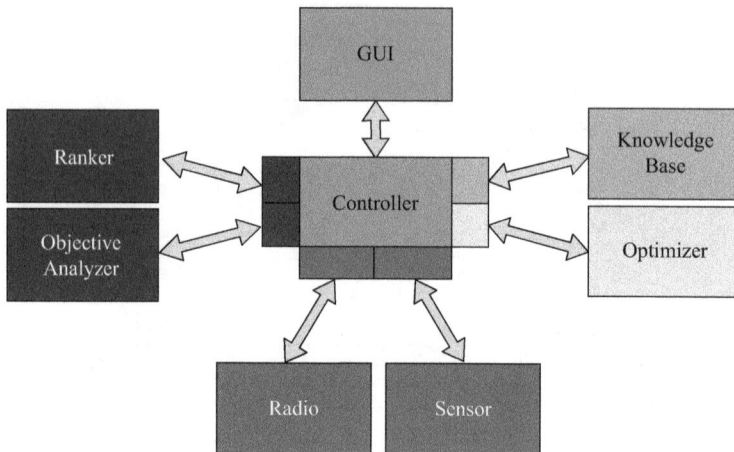

Figure 2.11   CSERE architecture. ©2012 IEEE. Reprinted, with permission, from Reference 34.

within the same functional unit. Separating these elements requires that data be passed between them during the course of their operation. Furthermore, enforcing the abstraction of modules along component organization, so that the implementation of one module need not depend on that of another, requires a common interface for each component to communicate. These tasks fall on the controller, which is now responsible for ensuring that each block has the appropriate data at the appropriate time and in a format that makes sense.

These increased responsibilities make the controller more central to the operation of the CE. The controller in CSERE is required to handle all of the data that are given to or taken from any given component. Extending this data management responsibility to making decisions about the components would allow the controller to refine the operation of the CE as a whole without introducing significantly more complexity. As discussed throughout this chapter, the purpose of a CE is to apply available information to design a communications system. The exact nature of this information depends on the techniques applied by the components, with various techniques having different requirements or benefits (see Section 2.4). Swapping between techniques allows the controller to align the employed techniques with the current situation, rather than forcing the designer to guess which techniques will apply. Furthermore, the encapsulation enforced by the components encompasses key elements of the communication system design process including the radio model and the definition of goodness. The radio model is defined by the parameters made available for reasoning by the objective analyzer and the radio interface provided by the radio component. The definition of goodness determines the CE's search for knob settings and the current application level goals is provided by the ranker component. Altering any of these provides the controller knobs on the operation of the CE itself.

The architecture displayed in Figure 2.11 is simply a starting organization. In practice, a CSERE controller selects components from a set of libraries that catalog various options for each component. An advanced controller, perhaps implemented as a CE itself, may apply several instances of the same components; that is, interfaces to multiple sensors may be used in parallel to balance their abilities and limitations. These components may all by run on a single computer (or embedded platform) or on separate computational platforms that communicate through some auxiliary network. The encapsulation of component tasks and abstraction along those lines allows the controller to alter this basic architecture as needed to serve the goals and situation of the CE.

2.6 Information Flow in Cognitive Engines[4]

All of the tools and architectures for a CE exist to transform available information into a working radio system. That is, a CE encodes information into the determination of knob values. Indeed, this is the purpose of each individual component within the CE as well. Figure 2.12 illustrates this process. Considering this process first at the level of the entire CE (instead of examining a single component), note that the information-processing entity applies two separate sources of information to the final outcome: stimulus information and built-in information. The CE obtains stimulus information from its sensors, and built-in information is embodied in the design of the CE. This represents any factors that limit or control its operation. For example, a CE may work with a single pre-defined model for a radio system where knob values are set from a limited pool of options; the bounds on this option pool constitute information built into the CE by its designer. Both the stimulus information and the built-in information are encoded into the radio system. The tools of information theory can be applied to investigate how these types of information flow[5] through a CE.

The amount of information, in bits, that any random variable contains is the minimum number of bits that are required to specify its value. This metric, denoted as $H(X)$ where X is the random variable involved and known as entropy, is based on the probability of the variable assuming any particular value, with the general result that the more random the variable, the more bits required to represent it. The mathematical representation of this, for discrete random variables, is shown in (2.1). In this equation, x_i is a potential value of X and $p(x_i)$ is the probability with which X takes on that value. Note that this metric can be used to quantify the amount information contained in an action simply by viewing the choice of a particular action as a random variable.

$$H(X) = -\sum_{i=1}^{n} p(x_i) \log (p(x_i)) \tag{2.1}$$

The common way to examine the informational relationship between variables is the mutual information. Mathematically, the mutual information between two signals, A and B, is given as:

$$I(A;B) = H(A) - H(A|B) \tag{2.2}$$

Conceptually, this is the number of bits required on average to describe signal A, minus the number bits that are still required to describe A once B is known. More simply, the mutual

---

[4]The material in this section is largely taken from Nicholas J. Kaminski's dissertation [55].

[5]The methods we present for information flow analysis here do not perfectly agree with those typically associated with the moniker. We feel that the methods we present are more readily applicable to CE design than others; however, this is no way negates the application of more traditional information flow analysis. Interested readers are directed to Reference 56.

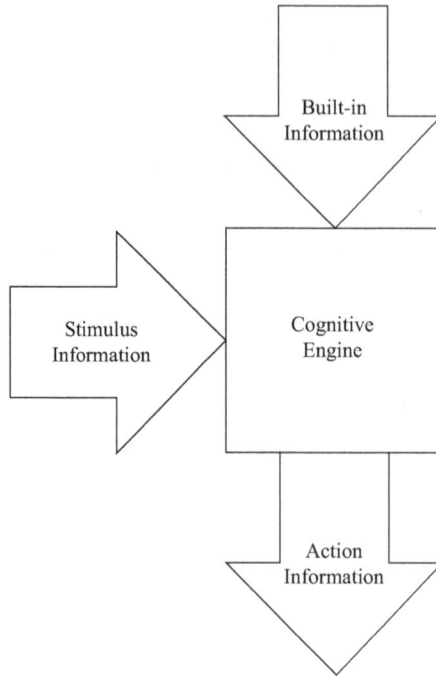

Figure 2.12   Encoding of information into action. ©2014 Nicholas J. Kaminski. Reprinted, with permission, from Reference 55.

information is the number of bits of information that B contains about A. Note that the mutual information is symmetric, therefore the amount of information that A contains about B is the same as the amount of information that B contains about A [57]. This measure can be very powerful, but is not well suited for comparing arbitrary systems, since the magnitude is not normalized. Instead, we will use the uncertainty coefficient, given by (2.3), which normalizes mutual information by the entropy of one of the two signals. Mathematically, this is written as:

$$U(A;B) = \frac{I(A;B)}{H(A)} \tag{2.3}$$

The uncertainty coefficient gives the fraction of bits needed to specify A that may be predicted using knowledge about B. Conceptually, this provides the proportion of information in A that comes from B. This measure is limited to values between 0 and 1, providing a clear measure of the accuracy with which signal A (one particular source of information) can be predicted given knowledge of signal B (the knob settings selected). We will use this measure to examine the degree to which some outcome in the CE depends on some other information. That is, the uncertainty coefficient provides CE designers with a valuable tool for determining the factors that influence the potentially stochastic outcome of a CE and its subprocesses.

2.6.1 Example Use of Uncertainty Coefficient

As an instructive example for the use of the uncertainty coefficient, we will examine three simplified CEs, consisting primarily of an optimizer component, operating in an AWGN environment. Each CE will select an output power level and modulation scheme based on a measurement of the noise level in the environment provided by a sensor component. The design of the sensor component is the same for each CE, simply providing a quantized measurement of the current noise level. However, the optimizer component that forms the core of these CEs differs in each case. For purposes of this example, each CE will select appropriate knob settings (power level and modulation scheme) from a common set (i.e., all CEs have the same options for power level and modulation scheme) for use in an open channel with common noise power across all frequencies following a Gaussian distribution. The uncertainty coefficient will be applied to determine how information is used in the optimization component.

In order to calculate the uncertainty coefficient, each CE must first be run through several iterations of the selection scenario and its decisions must be cataloged along with the noise measures that preceded them. Recall that the uncertainty coefficient is fundamentally an information theoretical tool that gives the potential for one data set to encode another, thus the data must be available for calculation of the coefficient. Once the data are collected, we can calculate the relative frequency of each value for both the noise measurement and knob settings. These relative frequencies allow for the calculation of the individual entropies for each of these data sets, the joint entropy of both together, the resulting mutual information, and finally the uncertainty coefficient. Each of these calculations proceeds as discussed above.

The product of these calculations reveals how information is used by the optimizer to determine knob settings. This examination is not concerned with the exact amount of information used by the optimizer or the specific values produced by the sensor; rather, the uncertainty coefficient is useful for determining the degree to which the result of the optimizer depends on the stimulus information provided by the sensor. In this way, the uncertainty coefficient provides a view of how information flows through the system, that is, it reveals what information decisions are based upon. Thus, the method displayed here allows the quantification of information use within a CE.

Each of the three CEs to be examined here employs a different optimizer technique to determine an action. The first, depicted in Figure 2.13a, is based on a completely random search, randomly selecting an action from the available options. In this CE, the action selected is independent of the noise level. The second CE, Figure 2.13b, applies the approach of an expert system, using the noise level to select an action from a table. This provides an example of a rule-based system. The third CE, Figure 2.13c, applies a GA to select its action. The GA searches through a space of candidate knob settings under the guidance of a fitness function. As we will see, each of these approaches to optimization uses available information in a different way.

The GA-based CE is clearly the most complex of the three exhibits here. Recall from Section 2.4.2 that searching in a GA proceeds by the pseudo-genetic evolution of high fitness candidate settings. In this case, the fitness of any given candidate is determined as a weighted sum of the transmit power required, the estimated bit rate based on the sensed noise level, and the modulation index. This definition for fitness is logically encapsulated in the ranker component of this CE and depends on values provided by the objective analyzer component. Due to the intertwined nature of the optimizer, ranker, and objective analyzer components, it is most proper to say that the uncertainty coefficient here relates the information flow through

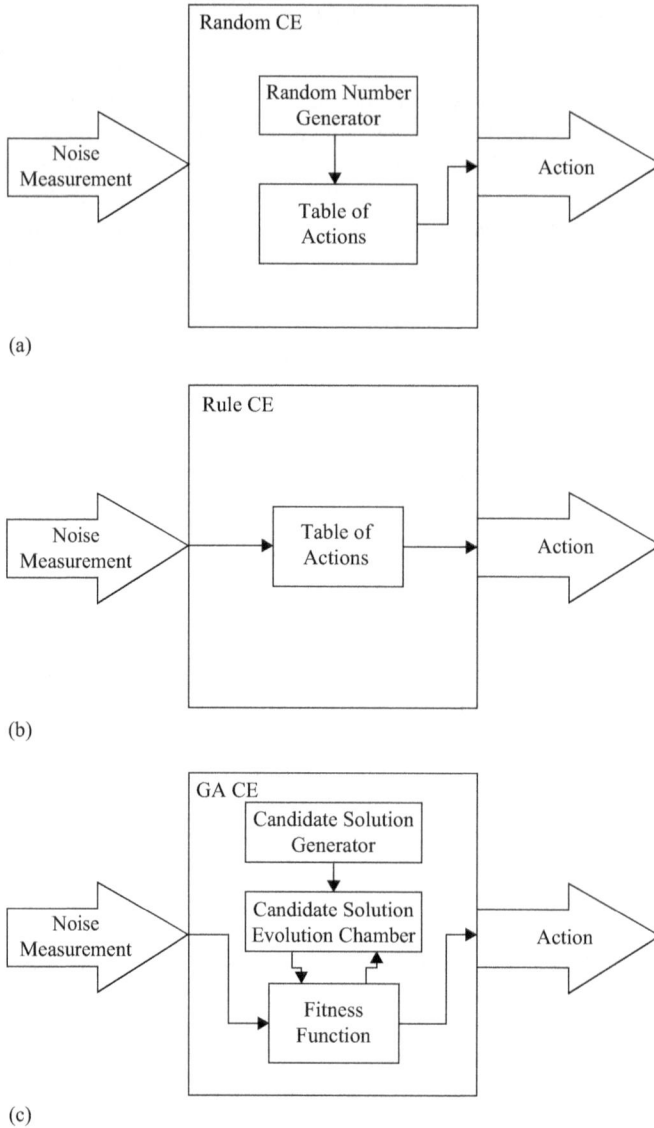

Figure 2.13 Tested cognitive engines. (a) Random CE. (b) Rule-based CE. (c) Genetic algorithm CE.

the CE, excluding the sensor component here. Note that this is still analogous to the other CEs that do not require ranker or objective analyzer components.

The uncertainty coefficient relating the sensed noise value to the selected knob settings for each CE appears in Table 2.1. These numbers represent the informational connection between 100,000 instantaneous noise power measurements and the knob settings selected in response. As such, these numbers represent statistical trends for the particular CEs examined here, rather than measures for the decisions that specifically occurred here. The figures shown here were

Table 2.1 Total action results under 100,000 tests.

CE under test	Uncertainty coefficient
Random	0.008529
GA	0.103133
Expert System	1.000000

calculated by normalizing the mutual information between the knob settings and the noise by the information contained in the knob settings; see (2.3).

Recall that the uncertainty coefficient provides the accuracy with which the sensed noise value can be predicted from knowledge of the selected knob settings alone. That is the metric quantifies the flow of information from the sensor component through the optimizer component (with the objective analyzer and ranker) into the selected knob settings. An uncertainty coefficient value of one indicates that all of the information about the sensed value are encoded in the selected knob settings and a value of zero indicates no sensed information appears in the knob settings. Values in between these extremes relate the proportion of sensed information that is represented in the knob settings. Note, for example, the extremely low uncertainty coefficient for the random CE. This logically flows from the selection of knob settings based on a random number generator, without consideration of the sensed knob value. The expert system CE, on the other hand, determines knob settings based on looking up settings based solely on the noise value. In this case, in particular each sensor output is mapped into a unique knob setting, thus, knob settings represent a simple coding of sensor values. In fact, this coding is such that the sensor value may be perfectly decoded from the knob setting.

The GA CE provides an interesting example of a case in between the extremes of full sensor value representation and independently selected knob values. Recall that the output of any component is based on both built-in and stimulus information. Additionally, recall that the operation of a GA depends on the use of random number generation and random pseudo-genetic operations to search through a space to discover knob settings with a high fitness value. The sensor value (our stimulus information in this case) is used only to determine the fitness of a candidate solution. Thus, the definition of the fitness function (encapsulated in the ranker) determines the use of stimulus information in the determination of knob settings. In this particular case, the fitness function clearly emphasizes the use of built-in information, in the form of internal states of random number generators, over the sensed noise value. This may be desirable when attempting to dampen the effect of a highly dynamic environment or detrimental when attempting to track environmental changes. In either of these cases, the uncertainty coefficient provides a useful method for quantifying this connection.

While the process shown here applies the uncertainty coefficient in a simplified scenario, this tool may be applied to examine information flow in more complex scenarios as well. As shown above, the uncertainty coefficient may be applied to determine the proportion of stimulus information represented in the result of a process. This proportion provides an indication of the factors that cause the variability that appears in a process's result. Calculation of this metric simply requires that the input and output of a process be recorded, allowing the method to be applied to any process with a defined input and output.

The methods for information flow analysis presented here provide the tools necessary to understand the passage of information in a CE. Specifically, the uncertainty coefficient approach offers a window into quantifying the importance of information sources. Recall that a CE is fundamentally applying available information to design a communication system. Understanding which information is applied and to what degree allows the CE designer to determine the factors that will most influence the behavior of her CE. As revealed in the example above, the uncertainty coefficient identifies the differing application of information by differing methods. Therefore, this tool provides the machinery necessary for prospective designers to understand the information use variations between individual techniques or even entire CE architectures.

2.7 Conclusion

This chapter has presented a brief overview of the topics central to CE design. Fundamentally, CEs are focused on the autonomous design of communication systems. This central task can be decomposed into several sub-tasks, each of which may then be encapsulated by an individual CE component. Additionally, these components are chiefly concerned with the manipulation and application of information. Also, several possible techniques may be applied to realize any of the components, each with their own abilities and limitations. The combination of these components presents several challenges, as well as, opportunities to adjust CE operation. Finally, since the communication system design central to CE is based on information processing, tracking the flow of this information provides a method for understanding the factors that drive the behavior of an individual component or entire CE.

Bibliography

[1] T. Rondeau and C. Bostian, *Artificial Intelligence in Wireless Communications*. Artech House, 2009.

[2] J. Ramiro and K. Hamied, Eds., *Self-Organizing Networks*. New York, NY: Wiley, 2012.

[3] (2013, November) Software Communications Architecture Specification – Appendix A: Glossary. Joint Tactical Networking Center. [Online]. Available: http://jpeojtrs.mil/sca/Documents/SCAv4_0/SCA_4.0.1_20121001_App_A_Glossary.pdf

[4] C. W. Bostian and A. R. Young, "Cognitive Radio: A Practical Review for the Radio Science Community," *Radio Science Bulletin*, vol. 342, 2012, pp. 15–25.

[5] B. F. Skinner, "The Operational Analysis of Psychological Terms," *Behavioral and Brain Sciences*, vol. 7, 1984, pp. 547–553. [Online]. Available: http://journals.cambridge.org/article_S0140525X00027187

[6] I. Pavlov, *Conditioned Reflexes*. Courier Dover Publications, Mineola, NY, 2003.

[7] J. Anderson, *Cognitive Psychology and Its Implications*. Worth Publishers, New York, NY, 2010.

[8] S. Russell and P. Norvig, *Artificial Intelligence: A Modern Approach*, 3rd ed. Prentice Hall, Englewood Cliffs, NJ, 2010.

[9] K.-L. Yau, P. Komisarczuk, and P. Teal, "Applications of Reinforcement Learning to Cognitive Radio Networks," in *Communications Workshops (ICC), 2010 IEEE International Conference on*, May 2010, pp. 1–6.

[10] C. Watkins, "Learning from Delayed Rewards," Ph.D. dissertation, King's College, London, England, May 1989.

[11] M. Mohri, A. Rostamizadeh, and A. Talwalkar, *Foundations of Machine Learning*. MIT Press, 2012.

[12] W. McCulloch and W. Pitts, "A Logical Calculus of the Ideas Immanent in Nervous Activity," *The Bulletin of Mathematical Biophysics*, vol. 5, no. 4, 1943, pp. 115–133. doi: 10.1007/BF02478259

[13] G. E. Hinton, "Connectionist learning procedures," *Artificial Intelligence*, vol. 40, no. 1–3, 1989, pp. 185–234. doi: 10.1016/0004-3702(89)90049-0

[14] S. Haykin, *Neural Networks and Learning Machines*. Prentice Hall, 2009.

[15] A. Fehske, J. Gaeddert, and J. Reed, "A New Approach to Signal Classification Using Spectral Correlation and Neural Networks," in *New Frontiers in Dynamic Spectrum Access Networks, 2005. DySPAN 2005. 2005 First IEEE International Symposium on*, November 2005, pp. 144–150.

[16] V. Tumuluru, P. Wang, and D. Niyato, "A Neural Network Based Spectrum Prediction Scheme for Cognitive Radio," in *Communications (ICC), 2010 IEEE International Conference on*, May 2010, pp. 1–5.

[17] S. Haykin, "Cognitive radio: brain-empowered wireless communications," *Selected Areas in Communications, IEEE Journal on*, vol. 23, no. 2, 2005, pp. 201–220.

[18] C. Clancy, J. Hecker, E. Stuntebeck, and T. O'Shea, "Applications of Machine Learning to Cognitive Radio Networks," *Wireless Communications, IEEE*, vol. 14, no. 4, 2007, pp. 47–52.

[19] R. Duda, P. Hart, and D. Stork, *Pattern Classification*. John Wiley & Sons, 2001.

[20] A. Markov, "The Theory of Algorithms," *Trudy Mat. Inst. Steklov*, vol. 42, 1954, pp. 3–375.

[21] Y. Zhang, "Dynamic Spectrum Access in Cognitive Radio Wireless Networks," in *Communications, 2008. ICC '08. IEEE International Conference on*, May 2008, pp. 4927–4932.

[22] A. Raschella, J. Perez-Romero, O. Sallent, and A. Umbert, "On the Use of POMDP for Spectrum Selection in Cognitive Radio Networks," in *Cognitive Radio Oriented Wireless Networks (CROWNCOM), 2013 8th International Conference on*, July 2013, pp. 19–24.

[23] J. Gittins, "Bandit Processes and Dynamic Allocation Indices," *Journal of the Royal Statistical Society. Series B (Methodological)*, vol. 41, no. 2, 1979, pp. 148–177.

[24] Q. Zhao, B. Krishnamachari, and K. Liu, "On Myopic Sensing for Multi-channel Opportunistic Access: Structure, Optimality, and Performance," *Wireless Communications, IEEE Transactions on*, vol. 7, no. 12, 2008, pp. 5431–5440.

[25] Y. Gai, B. Krishnamachari, and M. Liu, "On the Combinatorial Multi-Armed Bandit Problem with Markovian Rewards," in *Global Telecommunications Conference (GLOBECOM 2011), 2011 IEEE*, December 2011, pp. 1–6.

[26] W. Jouini, D. Ernst, C. Moy, and J. Palicot, "Multi-armed Bandit Based Policies for Cognitive Radio's Decision Making Issues," in *Signals, Circuits and Systems (SCS), 2009 3rd International Conference on*, November 2009, pp. 1–6.

[27] I. Gilboa and D. Schmeidler, "Case-Based Decision Theory," *The Quarterly Journal of Economics*, vol. 110, no. 3, 1995, pp. 605–639. [Online]. Available: http://qje.oxfordjournals.org/content/110/3/605.abstract

[28] V. Pareto, "The New Theories of Economics," *Journal of Political Economy*, vol. 5, no. 4, 1897, pp. 485–502. [Online]. Available: http://www.jstor.org/stable/1821012

[29] E. Zitzler and L. Thiele, "Multiobjective Evolutionary Algorithms: A Comparative Case Study and the Strength Pareto Approach," *Evolutionary Computation, IEEE Transactions on*, vol. 3, no. 4, 1999, pp. 257–271.

[30] B. Phruksaphanrat and A. Ohsato, "Efficient Linear Combination Method for Multi-objective Problems with Convex Polyhedral Preference Functions," in *Industrial Technology, 2002. IEEE ICIT '02. 2002 IEEE International Conference on*, vol. 1, 2002, pp. 149–154.

[31] M. Bazaraa, J. Jarvis, and H. Sherali, *Linear Programming and Network Flows*, 4th ed. Wiley, 2010.

[32] M. Bazaraa, H. Sherali, and C. Shetty, *Nonlinear Programming: Theory and Algorithms*, 3rd ed. Wiley, 2006.

[33] Z. Quan, S. Cui, and A. Sayed, "Optimal Linear Cooperation for Spectrum Sensing in Cognitive Radio Networks," *Selected 20 Topics in Signal Processing, IEEE Journal of*, vol. 2, no. 1, 2008, pp. 28–40.

[34] A. Young, N. Kaminski, A. Fayez, and C. Bostian, "CSERE (Cognitive System Enabling Radio Evolution): A Modular and User-friendly Cognitive Engine," in *Dynamic Spectrum Access Networks (DYSPAN), 2012 IEEE International Symposium on*, October 2012, pp. 59–67.

[35] T. R. Newman, B. A. Barker, A. M. Wyglinski, A. Agah, J. B. Evans, and G. J. Minden, "Cognitive Engine Implementation for Wireless Multicarrier Transceivers," *Wireless Communications and Mobile Computing*, Nov 2007.

[36] D. E. Goldberg, *Genetic Algorithms*. Pearson Education, Reading, MA, 2006.

[37] P. Kumar and P. Varaiya, *Stochastic Systems: Estimation, Identification, and Adaptive Control*. Prentice Hall, Englewood Cliffs, NJ, 1986.

[38] J. Hilbe and A. Robinson, *Methods of Statistical Model Estimation*. Chapman and Hall/CRC, 2013.

[39] H. Kim and K. Shin, "Efficient Discovery of Spectrum Opportunities with MAC-Layer Sensing in Cognitive Radio Networks," *Mobile Computing, IEEE Transactions on*, vol. 7, no. 5, 2008, pp. 533–545.

[40] Y. Chen, G. Yu, Z. Zhang, H.-H. Chen, and P. Qiu, "On Cognitive Radio Networks with Opportunistic Power Control Strategies in Fading Channels," *Wireless Communications, IEEE Transactions on*, vol. 7, no. 7, 2008, pp. 2752–2761.

[41] T. Yucek and H. Arslan, "A Survey of Spectrum Sensing Algorithms for Cognitive Radio Applications," *Communications Surveys Tutorials, IEEE*, vol. 11, no. 1, 2009, pp. 116–130.

[42] D. Cabric, S. Mishra, and R. Brodersen, "Implementation Issues in Spectrum Sensing for Cognitive Radios," in *Signals, Systems and Computers, 2004. Conference Record of the Thirty-Eighth Asilomar Conference on*, vol. 1, November 2004, pp. 772–776.

[43] Y. Zeng and Y.-C. Liang, "Covariance Based Signal Detections for Cognitive Radio," in *New Frontiers in Dynamic Spectrum Access Networks, 2007. DySPAN 2007. 2nd IEEE International Symposium on*, April 2007, pp. 202–207.

[44] H. Kim and K. G. Shin, "In-band Spectrum Sensing in Cognitive Radio Networks: Energy Detection or Feature Detection?" in *Proceedings of the 14th ACM International Conference on Mobile Computing and Networking*, ser. MobiCom '08. New York, NY: ACM, 2008, pp. 14–25. [Online]. Available: http://doi.acm.org/10.1145/1409944.1409948

[45] K. Bonne Rasmussen and S. Capkun, "Implications of Radio Fingerprinting on the Security of Sensor Networks," in *Security and Privacy in Communications Networks and the Workshops, 2007. SecureComm 2007. Third International Conference on*, September 2007, pp. 331–340.

[46] P. Qihang, Z. Kun, W. Jun, and L. Shaoqian, "A Distributed Spectrum Sensing Scheme Based on Credibility and Evidence Theory in Cognitive Radio Context," in *Personal, Indoor and Mobile Radio Communications, 2006 IEEE 17th International Symposium on*, September 2006, pp. 1–5.

[47] J. Unnikrishnan and V. Veeravalli, "Cooperative Sensing for Primary Detection in Cognitive Radio," *Selected Topics in Signal Processing, IEEE Journal of*, vol. 2, no. 1, 2008, pp. 18–27.

[48] E. Peh, Y.-C. Liang, Y. L. Guan, and Y. Zeng, "Optimization of Cooperative Sensing in Cognitive Radio Networks: A Sensing-Throughput Tradeoff View," *Vehicular Technology, IEEE Transactions on*, vol. 58, no. 9, 2009, pp. 5294–5299.

[49] J. Ma, G. Zhao, and Y. Li, "Soft Combination and Detection for Cooperative Spectrum Sensing in Cognitive Radio Networks," *Wireless Communications, IEEE Transactions on*, vol. 7, no. 11, 2008, pp. 4502–4507.

[50] S. Mishra, A. Sahai, and R. Brodersen, "Cooperative sensing among cognitive radios," in *Communications, 2006. ICC '06. IEEE International Conference on*, vol. 4, June 2006, pp. 1658–1663.

[51] Y. Zhao, L. Morales, J. Gaeddert, K. Bae, J.-S. Um, and J. Reed, "Applying Radio Environment Maps to Cognitive Wireless Regional Area Networks," in *New Frontiers in Dynamic Spectrum Access Networks, 2007. DySPAN 2007. 2nd IEEE International Symposium on*, April 2007, pp. 115–118.

[52] *IPA Information Process Architecture Volume I Document WINNF-09-P-0020*, Wireless Innovation Forum Std., Rev. 1.0.0, 2010.

[53] *Reconfigurable Radio Systems (RRS); Functional Architecture (FA) for the Management and Control of Reconfigurable Radio Systems TR 102 682*, ETSI Std., Rev. 1.1.1, 2009.

[54] T. W. Rondeau, "Application of Artificial Intelligence to Wireless Communications," Ph.D. dissertation, Electrical and Computer Engineering, Virginia Tech, Blacksburg 2007.

[55] N. J. Kaminski, "Social Intelligence for Cognitive Radios," Ph.D. dissertation, Virginia Polytechnic Institute and State University, Blacksburg, VA, 2014.

[56] P. Malacaria and J. Heusser, "Information Theory and Security: Quantitative Information Flow," in *Formal Methods for Quantitative Aspects of Programming Languages*, ser. Lecture Notes in Computer Science, A. Aldini, M. Bernardo, A. Di Pierro, and H. Wiklicky, Eds. Springer Berlin Heidelberg, 2010, vol. 6154, pp. 87–134.

[57] T. M. Cover and J. A. Thomas, *Elements of Information Theory*, 2nd ed. John Wiley & Sons, 2006.

CHAPTER 3

RF Platforms for Cognitive Radio

3.1 Introduction

In our mental concept of a cognitive radio, a cognitive engine (or cognitive controller) sends commands ("turn knobs") to and receives information about current channel conditions ("reads meters") from a radio frequency (RF) platform. Simultaneously, the platform transmits data generated by the radio's user and receives data intended for that user. These activities require that the RF platform be able to perform a variety of sensing and communications functions while simultaneously exchanging information with and responding to commands from both the cognitive engine and the user.

The idea that the RF functions of a cognitive radio might not be provided by a software-defined radio may, at first glance, seem heretical to the reader. Cognitive radio and software defined radio are closely linked in most people's minds, since Mitola's pioneering concept of a cognitive radio assumed a software-based RF platform which would have almost unlimited reconfigurability. Ideally, the receiver and transmitter would consist of nothing more than an analog-to-digital converter (ADC) and a digital-to-analog converter (DAC) connected on their analog sides to the antenna through a circulator (a three-port device that isolates received power from transmitted power) and on their digital sides to a digital signal processor (DSP) system. See Figure 3.1. Such an architecture is unworkable because the receiving ADC must have an impossibly large dynamic range in order to handle both the strongest signals that might be present in the theoretically infinite input bandwidth and the weakest signals that the radio is designed to receive. See Reference 1 for a discussion of these issues.

But even in cases where the dynamic range problem could be handled by limiting the receive bandwidth with band-pass filters (BPFs) between the antenna and the ADC, a fully software-based cognitive radio would be difficult to achieve in practice, since it would require the cognitive engine to generate radio software in real time and load it into a computational platform that has the speed and capacity to run it, all while meeting the user's timing requirements. For these reasons, RF platforms suitable for use in cognitive radios include radio frequency integrated circuits (RFICs), packaged radio front ends, and software defined radios. These differ in the interfaces they present to the cognitive engine and the user, the amount of software development that is required, and in the nature and physical location of the hardware where that software will run, as well as in their RF and computational performance and in their reconfigurability. We will begin this chapter by discussing some general considerations, then RF specifications and performance issues, and then the principal kinds of RF platforms available to a cognitive radio designer.

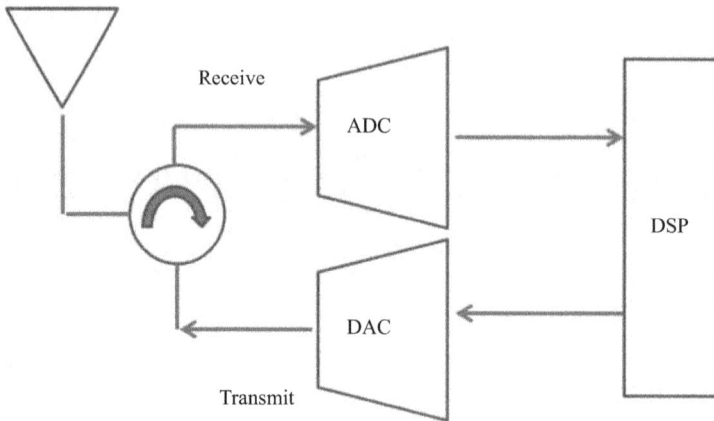

Figure 3.1   Architecture of an ideal software defined radio platform for cognitive radio.

Our goal in writing this section is to give the reader a useful understanding of RF platforms for cognitive radio, emphasizing their characteristics and specifications and how these factors relate to radio performance, particularly to the ability to recognize an open channel in dynamic spectrum access (DSA) applications. The important problems that real circuits and components cause for this process by raising the noise level and introducing spurious signals are often overlooked in the literature. Where possible, we will describe how parameter choices under the reader's control affect specifications and performance, but the underlying design techniques for analog RF front ends, RFICs, and software defined radios are beyond the scope of this book. Consult References 2, 3.

3.2 Preliminary Considerations in Choosing an RF Platform

Besides the obvious factors of cost, size, power supply requirements and application-compatible architecture (for example, will the radio operate on one channel at a time, or is it intended for simultaneous operation on multiple channels?), factors to consider in choosing an RF platform include the following:

Architecture: RF platform architecture is determined by the intended application – single-channel or multi-channel, narrowband or wideband, etc.

Standard RF Specifications: Operating frequency range, sensitivity, dynamic range, interference rejection, receive bandwidth, transmit output power, transmit spectrum, phase noise. Note that these must satisfy both the operational requirements of the cognitive radio being designed and the governing regulatory requirements and desired standards compliance.

Medium Access Control (MAC) and Performance Considerations: Tuning rates (how fast can the transmitter and receiver switch between channels or otherwise change frequency?), capabilities for full-duplex and half-duplex operation, time required for switching between transmit and receive, time required for reconfiguration, modulation types and data rates supported, MAC layer timing requirements.

Input, Output, and Control Interfaces (Application Programming Interface (API)):
What interfaces does the RF platform use for radio communications, control, and sensing functions? How will these be made available to the cognitive engine and the user? What are their voltage, impedance, and frequency requirements? Signals from the receiver and to the transmitter may be analog RF or intermediate frequency (IF) waveforms, analog baseband waveforms, complex IF or complex baseband waveforms, in-phase (I) or quadrature (Q) digital samples, or fully demodulated bit streams.

Computational Requirements: Is the RF platform self-contained, or does it require code running on an external processor to provide radio functions?

3.3 RF Architectures

The transmitter and receiver architectures used in RFICs or supported by software defined radio (SDR) platforms and packages are important factors in choosing an RF platform. Obviously, an architecture is required that is compatible with the applications for which the radio will be designed.

Design issues common to almost all currently used transmitter and receiver architectures involve local oscillators, filters, and, to a lesser extent amplifiers. A receiver or transmitter with wide frequency coverage requires a local oscillator with a similar tuning capabilities and filters that can operate over this frequency range, passing wanted signals and eliminating unwanted ones. Most practical local oscillators require at least one inductor, and constraints on its physical size and losses limit both upper and particularly lower operating frequencies. Local oscillators can interact with signals in the environment to generate unwanted intermodulation products. Local oscillators generate harmonics and spurs, and these will propagate through wideband filters and amplifiers, pushing the designer toward tunable narrow-band circuits, which add to cost and limit radio flexibility. Many of these problems will have been dealt with by the designers of RFIC platforms, but it is important that SDR users and designers be aware of them and not make naïve assumptions about the performance and capability of a paper SDR design. See Reference 3 for a detailed treatment of all these issues.

3.3.1 Receivers

The superheterodyne was the dominant receiver architecture from its invention in 1918 to the early years of the 21st century. Intended to receive a single channel at a time, it typically incorporated a filter with a relatively wide pass band to eliminate image problems and signals outside of the band of interest, followed by a mixer which down-converted the incoming signals to a fixed IF for further amplification and filtering. (Up-conversion was used for some applications.) A narrow-band IF filter provided the required channel selectivity. While conversion to a second IF was employed in some designs, single-conversion was usually sufficient. The output of the last IF stage drove a demodulator stage. The simple modulations used for analog voice transmission did not usually require that the incoming signal's phase be preserved or tracked.

The superheterodyne lives on in receiver architectures that perform analog amplification and possibly filtering at one or more intermediate frequencies. The waveform in the IF passband may be digitized. Brannon terms this a *single-carrier receiver*, "a traditional radio receiver deriving selectivity in the analog filters of the IF stages." See Figure 3.2. This architecture

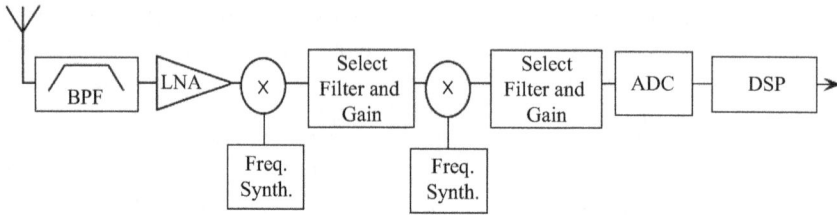

Figure 3.2   Single-carrier receiver architecture. LNA – low noise amplifier © Analog
Devices. Reproduced, with permission, from Reference 4.

often lacks the flexibility required for cognitive radio applications since it is difficult to change
the channel bandwidth without switching in additional filters or filter components. Brannon [4]
distinguishes the single-carrier receiver from a *multi-carrier receiver* that "processes all signals
within the band with a single RF/IF analog strip and derives selectivity within the digital filters
that follow the analog to digital converter." Modern multi-carrier receivers, or at least those
intended for low-cost high-volume markets, typically employ a direct-conversion or zero-IF
(ZIF) architecture in which the chosen incoming RF passband is coherently down-converted to
baseband by mixing with in-phase (I) and quadrature (Q) local-oscillator circuits and the I and
Q baseband signals are digitized. A digital baseband processor then performs channel coding,
modulation mapping, and digital filtering [3, 5]. This is a useful architecture for cognitive
radio applications in that most or all of the bandwidth-determining filtering is done under
software control. See Figure 3.3. Figure 3.4 illustrates a ZIF receiver used in conjunction with
a heterodyne transmitter as discussed below.

3.3.2 Transmitter

Transmitter architecture essentially reverses the block diagram of a multi-carrier or ZIF
receiver: Figure 3.4 includes a heterodyne transmitter. The architecture without an intermediate
frequency is called a *direct launch* (or *direct-launch frequency converter*) transmitter. Its data
input consists of bits corresponding to the I and Q states of the desired waveform; these pass
through shaping filters and modulate the amplitude and phase of an RF carrier generated by
the transmit local oscillator. The carrier is filtered to meet any regulatory spectral requirements
and amplified for transmission. The amplifiers may include both a low-power variable gain
amplifier (VGA) and a higher-power power amplifier (PA). The PA may be on a chip with
the other components, or it may be physically separate. Practical designs often add additional
control loops for waveform purity. Figure 3.5 illustrates a direct-launch transmitter and a ZIF
receiver on the same chip.

Responding in part to the needs of LTE systems, modern transmitter design is evolving
toward an architecture called Direct RF DAC, where a modulated RF waveform is synthesized
directly from a digital baseband signal, moving almost the entire transmitter signal-processing
function to the digital domain. Figure 3.6 presents a conceptual diagram illustrating direct
synthesis of two 20 MHz LTE component carriers with 365 MHz carrier separation. For more
information on transmitter design, see Reference 3 for heterodyne transmitters and References 6, 7, and Reference 8 for Direct RF DAC technology.

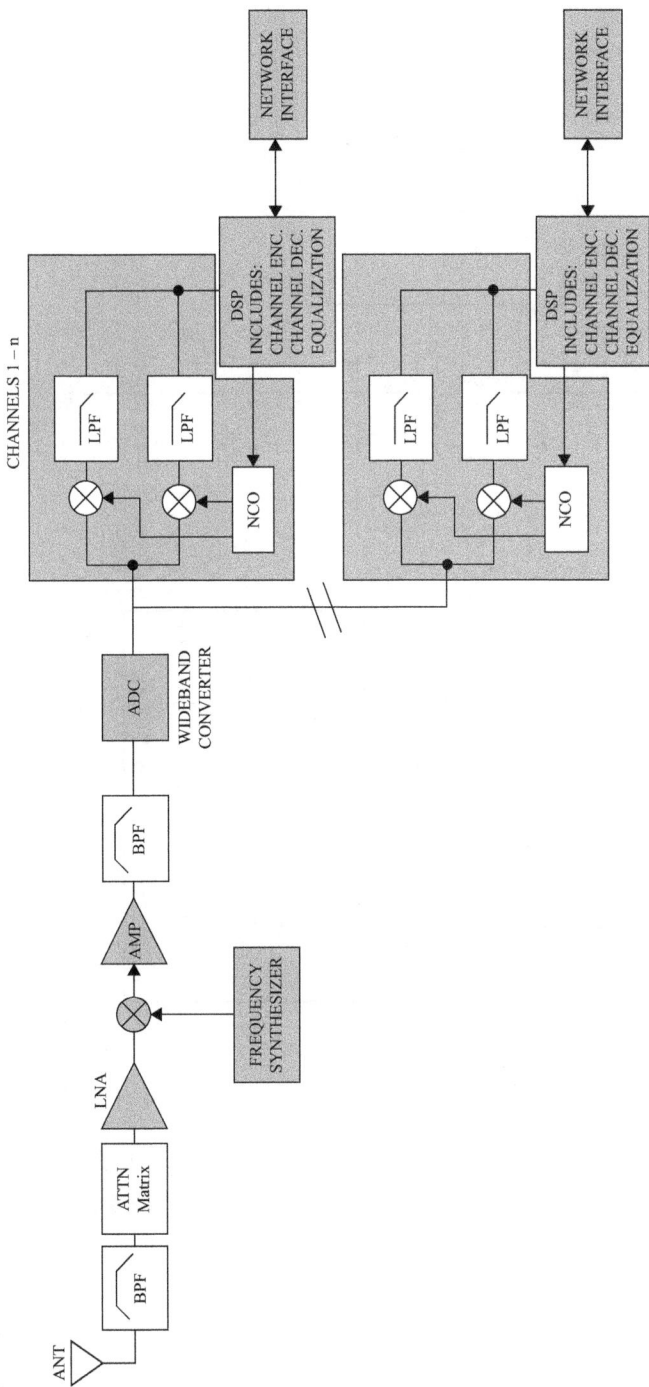

Figure 3.3 Typical multi-carrier receiver architecture. © Analog Devices. Reproduced, with permission, from Reference 4.

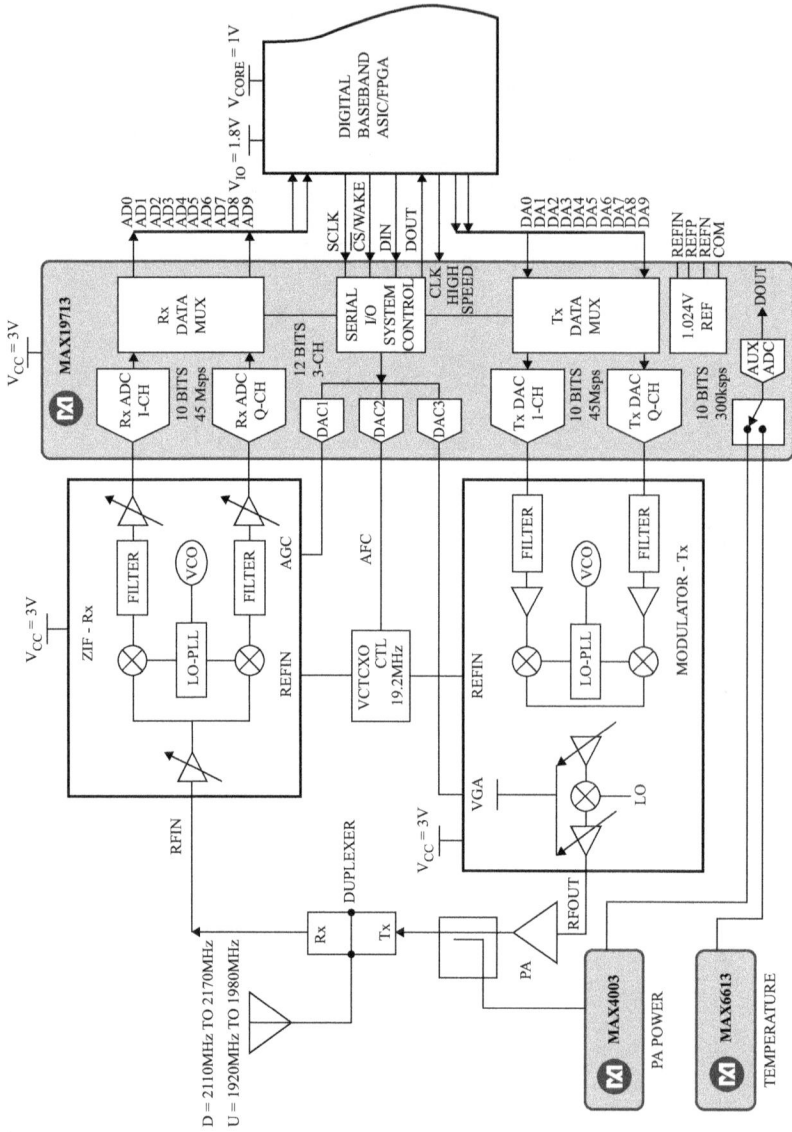

Figure 3.4 Block diagram of a chip with a typical ZIF radio and a heterodyne transmitter. Reproduced from Reference [5], © 2013 Maxim Integrated Products, Inc. Used with permission; Maxim integrated does not endorse, and is not responsible for, the content of this publication.

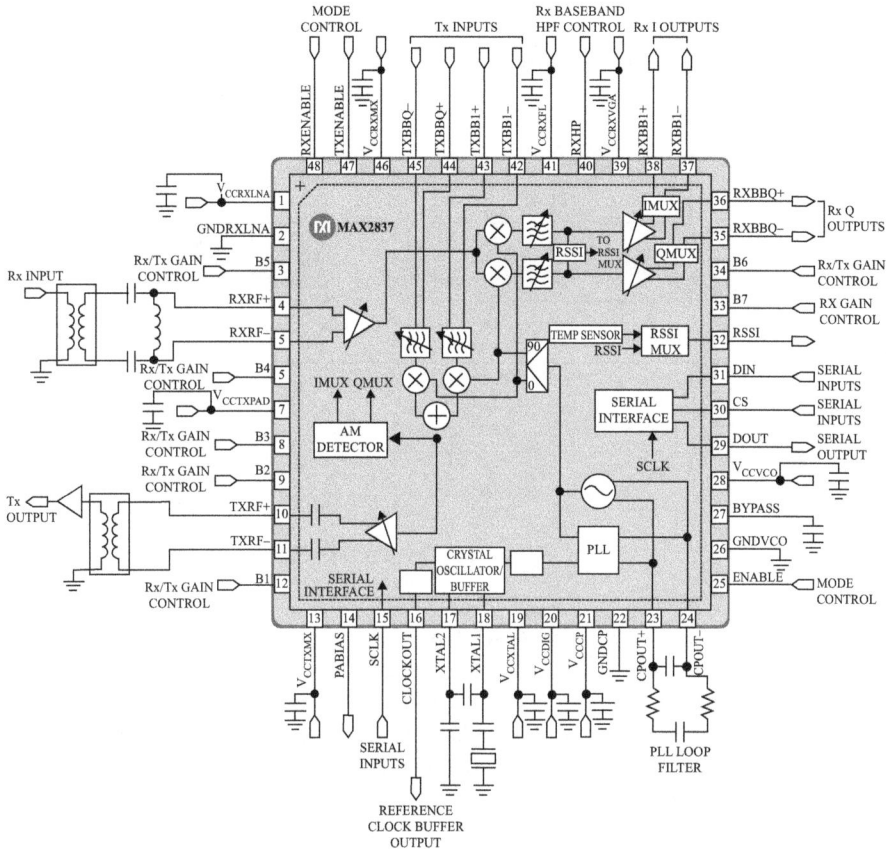

Figure 3.5 Block diagram of an integrated radio front end using the Maxim 2837, a direct-launch transmitter and ZIF receiver chip. Reproduced from https://www.maximintegrated.com/en/products/comms/wireless-rf/MAX2837.html [Accessed 30 Aug 2016], © Maxim Integrated Products, Inc. Used with permission; Maxim integrated does not endorse, and is not responsible for, the content of this publication.

3.4 Receiver RF Specifications

3.4.1 Introduction

In the simplest and most idealized concept of a cognitive radio, the RF platform is capable of transmitting and receiving at any frequency over a wide range of spectrum, using arbitrary receive and transmit bandwidths, doing these things without generating spurious signals that interfere with itself or with other users, and accommodating input signals over a suitably large dynamic range. This idealized behavior is, of course, impossible. The limits over which a real radio can perform its functions are described by a set of specifications. With a reconfigurable radio, the RF specifications themselves will almost certainly depend on the configuration, and this dependence may well limit the potential reconfigurability of radio for a given application.

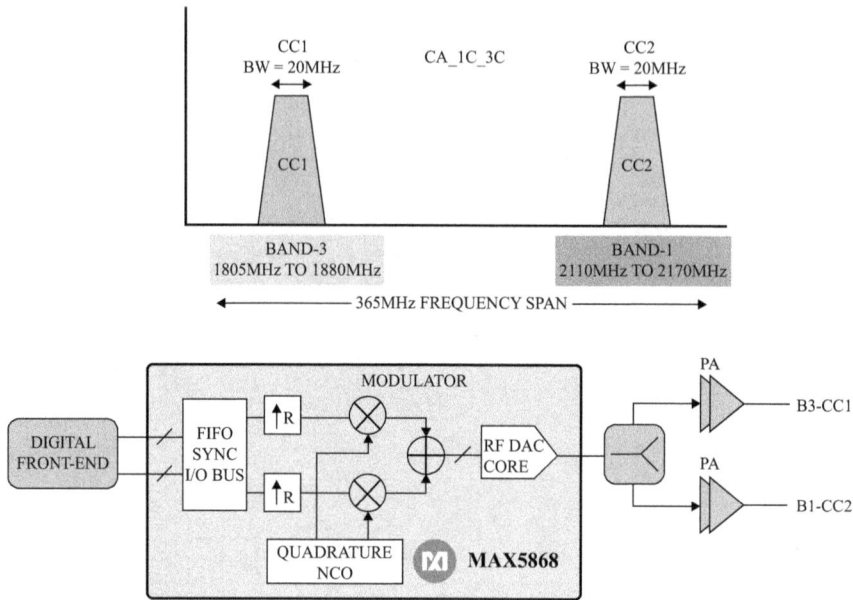

Figure 3.6   Use of a Maxim MAX5688 RF DAC to synthesize simultaneously two 20 MHz component carriers with 365 MHz carrier separation. Reproduced from Reference 8. Available online at https://www.maximintegrated.com/en/app-notes/index.mvp/id/6063 [Accessed 30 Aug 2016], © 2015 Maxim Integrated Products, Inc. Used with permission; Maxim integrated does not endorse, and is not responsible for, the content of this publication.

In this section, we will describe the most important receiver RF specifications for cognitive radio – those governing strong and weak signal behavior – and explain the underlying factors determining them without going deeply into the underlying mathematics and physics or the governing design details. For detailed information about these, we recommend the reader to consult References 2, 3. In the material which follows, we will indicate quantities in dB by upper case letters, and the corresponding quantities as numerical ratios in lower case letters.

3.4.2 Noise, Noise Performance, and Weak Signal Behavior

The fundamental limiting factor in radio operation is noise. The antenna receives ambient noise from the environment, and this noise is amplified and added to by the radio itself. Interference and unwanted internally generated signals degrade performance in a manner similar to noise, and, depending on their spectral and statistical characteristics and the situation, they may be lumped with the ambient noise or treated separately.

For most purposes of analysis, noise is treated as if it was Gaussian noise generated by resistor at physical temperature T degrees Kelvin (K). The available noise power in a rectangular bandwidth B Hz from such a source is

$$P_n = kTB \tag{3.1}$$

where k is Boltzmann's constant (1.38×10^{-23} J/K). An antenna, or any other two-terminal circuit element, can be assigned a noise temperature T_A so that the available noise power delivered by the antenna to a matched load through lossless filter with noise bandwidth B_n is

$$n_A = kT_AB_n \qquad (3.2)$$

For a detailed treatment of noise bandwidth and information about how to compute it for a variety of filters, see Reference 2 of Chapter 5. Note that the antenna noise temperature is not the physical temperature of the antenna, but rather a way quantifying the amount of RF noise that the antenna picks up from its environment and delivers to a radio. The concept of antenna noise temperature was developed at a time of rapid development of radar and microwave point-to-point technology, and in these applications the antenna noise temperature was typically close to 290 K. For this reason, an antenna noise temperature of 290 K was chosen as a standard value for comparison purposes, even though practical antennas may have much higher (if operating in the HF frequency range) or lower values (for large satellite dishes pointing toward the cold background of outer space). A useful number for rule of thumb calculations in that an antenna with a 290-K noise temperature delivers -174 dBm of noise power to a matched load in a 1-Hz bandwidth. This may be calculated from

$$N_A(\text{dBm}) = 10\log_{10}(n_A \text{ in mw}) = 30 + 10\log_{10}(k) + 10\log_{10}(T_A) + 10\log_{10}(B) \quad (3.3)$$

The theory and terminology of receiver noise were developed in the analog era, and the receiver's output was considered to be at the input to the demodulator. In almost all modern receivers, an analog front end (filters, amplifiers, and possibly mixers) drives an ADC instead of a demodulator. The ADC adds noise, and more noise may be created or the existing noise level may be reduced by the digital signal processing that follows the ADC. We will initially ignore these digital effects and focus our attention on the analog front end.

Any real receiver generates noise internally. In the case of an analog front end, this noise appears at the receiver's output in combination with amplified and down converted noise that entered the front end from the antenna. For purposes of analysis, any two-port *analog* device that generates internal noise can be replaced by an equivalent noiseless device with a fictitious noise source at its input that produces the same noise power at the noiseless device output as would appear at the output of the original device if it were connected to a noiseless signal source. (This is the output noise power that would appear if all of the output noises were internally generated, that is, if no input noise was present.) The noise temperature of this fictitious source is called the equivalent noise temperature of the device, T_{Eq}. Mathematically, if the device is connected to an antenna with noise temperature T_A, the total noise power at the output of the device is given by

$$P_{N,output} = gkT_AB + gkT_{Eq}B \qquad (3.4)$$

where g is the power gain of the device (expressed as a ratio, not in dB) and B is the noise bandwidth. The first term represents amplified noise from the antenna and the second term represents internally generated noise, calculated as if it was generated by a fictitious source (resistor) at the input whose noise temperature is T_{Eq}.

Since the power gain of the device is g, the effective input noise power is

$$P_{N,input} = k(T_A + T_{Eq})B = kT_sB \qquad (3.5)$$

where T_s is called the system noise temperature of the device. Front end specifications usually list a related quantity called the noise figure (or noise factor) rather than noise temperature. This is the factor by which the front end's internally generated noise degrades the carrier-to-noise ratio available at the antenna terminals. Its value depends on the antenna noise temperature itself; if this is very large, added receiver noise may have little effect, and, if it is small, receiver noise will dominate. The standard noise figure (the one that appears on data sheets) is specified for an antenna noise temperature of 290 K. As a ratio (not dB) for an analog receiver this is given by

$$nf_{SR} = \frac{(c/n) \text{ at receiver input for an antenna with } T_A = 290 \text{ K}}{(c/n) \text{ at the receiver output for an antenna with } T_A = 290 \text{ K}} = 1 + \frac{T_R}{290} \qquad (3.6)$$

where T_R is front end's effective input noise temperature.

In dB

$$NF_{SR} = 10 \log_{10}(nf_{SR}) \qquad (3.7)$$

An ideal receiver would have a standard noise figure of 0 dB. Good practical values are in the 1.0–2.0 dB range. Consumer products often have standard noise figures of 3.0 dB or higher. Noise figure typically increases with operating frequency.

If the antenna noise temperature is 290 K, then the carrier-to-noise ratio at the front end output is easily calculated from the true value at the antenna terminals. Thus,

$$(C/N)_{out} = (C/N)_{in} - NF_{SR} \qquad (3.8)$$

where all quantities are in dB.

Note that this equation is often misused. It is valid only when the antenna noise temperature is 290 K, and this is often assumed but rarely true in practice. For more information on antenna noise temperature, see Reference 2. The actual noise figure nf_{Act} describes the C/N degradation for an arbitrary antenna noise temperature T_A. It is given by

$$nf_{ActR} = \frac{(c/n) \text{ at receiver input}}{(c/n) \text{ at receiver output}} = 1 + (nf_{SR} - 1)\left(\frac{290}{T_A}\right) \qquad (3.9)$$

where all quantities are in ratio rather than dB. In dB

$$NF_{ActR} = 10 \log_{10}(nf_{ActR}) \qquad (3.10)$$

$$(C/N)_{out} = (C/N)_{in} - NF_{ActR} \qquad (3.11)$$

The actual noise figure and the receiver noise temperature are related by

$$T_R = 290(nf_{ActR} - 1) \qquad (3.12)$$

The noise floor or minimum detectible signal (MDS) of the receiver in W is given by

$$mds = k(T_R + T_A)B_N = kT_S B_N \qquad (3.13)$$

where B_N is the receiver noise bandwidth in Hz. This is commonly specified in dBm and, assuming (possibly incorrectly) that the antenna noise temperature is actually 290 K, it is given by Reference 2

$$MDS = NF_{SR} - 174 + 10 \log_{10}(B_N) \text{ dBm} \qquad (3.14)$$

The term *minimum detectable signal* is a holdover from the early days of radio when the most robust form of modulation was off–on keyed Morse code (called "continuous wave modulation (CW)"). A skilled operator could "read" (understand) CW signals at a 0-dB carrier-to-noise ratio, and thus the MDS was a useful indication of receiver's sensitivity. But digital demodulators usually require a carrier-to-noise ratio significantly greater than 0 dB in order to produce useful output, and "minimum detectable signal" is not a meaningful term without further qualification. Manufacturers typically specify *sensitivity*, which is the input carrier level in dBm required for a specified symbol error rate (SER) for a given modulation type and data rate.

The problem of quantifying the weak signal performance of a receiver becomes more complicated when, as is almost always the case in modern designs, an ADC follows an analog front end. Unless one of them is clearly dominant, the noise contributions of the ADC and analog front end must be combined.

For a detailed analysis of noise in ADC systems, see References 9 and 2, some of whose results we summarize here. The ADC must accurately sample the values of its input signal and represent these samples in binary form. It will make timing errors (the times at which samples are taken exhibit random variations) and quantizing errors (the analog sample values must be represented as fixed-length binary numbers). These errors appear in the ADC output as noise that would manifest itself as incorrect sample values of the input signal even if no noise (processed antenna noise and noise added by the analog stages ahead of the ADC) was present at the ADC input. In a real receiver, noise is present at the ADC input, and how this noise combines with the ADC's internally generated noise is not easily analyzed.

A common tool for studying the performance of an ADC is to compute a fast Fourier transform (FFT) from the I and Q bit streams from the output and plot the resulting spectrum out to half the sampling frequency (i.e., $F_s/2$). The resulting picture is analogous to a spectrum analyzer plot taken at the output of an analog RF front end, and the noise floor, amplitude of the wanted signal, amplitudes of intermodulation products, etc. can readily be identified and various signal-to-noise ratios calculated. Of these, the most useful for our purposes is SINAD, the dB ratio of signal to the sum of noise and distortion. Normally this is specified for a sinusoidal input signal whose peak is at the maximum input value, and is approximately equal to the ADC's dynamic range (the dB difference between the maximum signal level and the noise floor on the FFT plot) [9]. The SINAD derivation assumes that the input signal is sampled at the minimum required rate. Data sheets specify SINAD under these conditions and the given value must be adjusted for the signal level and sampling rate that will be used.

Manufacturers' recommended design procedures compute the signal-to-noise ratio (SNR) at the detector by treating the ADC and the analog front end (AFE) as cascaded linear networks whose SNR values, expressed as ratios, add reciprocally, as in (3.15).

$$snr_{Overall} = \frac{1}{\left(\dfrac{1}{snr_{AFE}} + \dfrac{1}{sinad_{ADC}}\right)} \qquad (3.15)$$

In dB,

$$SNR_{Overall} = -10\log_{10}\left(10^{-\left(\frac{SNR_{AFE}}{10}\right)} + 10^{-\left(\frac{SINAD_{ADC}}{10}\right)}\right) \qquad (3.16)$$

The derivation of these equations assumes that the noise bandwidth of the AFE is equal to the Nyquist bandwidth (twice the sampling frequency) of the ADC and that SNR_{AFE} is computed for the maximum (full-range) input voltage to the ADC. Calculations made using these equations give the same result as those following the approach in Reference 4, based on noise voltage calculations. Knowing $SNR_{Overall}$ and the corresponding input signal power (based on the signal that provides the full-range input voltage to the ADC), we can calculate an overall noise floor and the corresponding minimum detectable signal for a receiver. We will present an example of this calculation below. Whether or not this quantity is useful for predicting receiver performance in detecting an open channel depends on the kind of processing that will be done with the ADC output data.

Manufacturer's recommended design procedures typically specify ADC noise characteristics by working backward from the demodulator input, starting with the SNR that a particular demodulator requires to deliver a stated SER. Using numbers from an example presented in Reference 5, a 64-QAM signal requires an 18 dB SNR for demodulation with a 1×10^{-5} SER. The ADC is then treated as a linear device, so that its input and output SNR values, as ratios, are related to the ADC's SINAD by the equations above. For an allowed SNR degradation of 0.6 dB, the SNR delivered by the AFE, $SNR_{in} = 18.6$ dB, $SNR_{out} = 18$ dB, and $SINAD_{ADC} = 26.89$ dB. This is preliminary value which must be adjusted to compensate for a number of practical problems.

If multiple carriers are present, they may combine in phase and overload the ADC, so compensation for the composite waveform's peak-to-average power ratio (PAPR) is necessary to avoid overdriving the ADC on peaks. A typical number is 12 dB. In addition, the ADC may exhibit gain and voltage offset errors, and, if an automatic gain control (AGC) stage is ahead of the ADC, the AGC may make inaccurate adjustments in gain. Typically, 2 dB is added to compensate for both of these effects. We must also allow for interfering signals stronger than the wanted signal. These will be digitally filtered out downstream from the ADC, but they must not overdrive it. The example in Reference 5 allows 12 dB headroom for interferers. Finally, we must subtract any processing gain (improvement in SNR from spreading input noise over a larger bandwidth) which results from oversampling. We will assume an oversampling factor of 2 and subtract 3 dB to obtain the required SINAD.

$$SINAD_{Compensated} = \overbrace{26.829}^{\text{Uncompensated SINAD}} + \overbrace{12}^{\text{PAPR}} + \overbrace{2}^{\text{Gain Offset}}$$

$$+ \overbrace{2}^{\text{AGC Error}} + \overbrace{12}^{\text{Interferer Headroom}} - \overbrace{3}^{\text{Processing Gain}} = 51.89 \text{ dB} \qquad (3.17)$$

The dynamic range of a receiver is the dB difference between the strongest and weakest signals that the receiver can accommodate, subject of course to the definitions of "strongest" and "weakest." The SINAD is one measure of the dynamic range of an ADC, where the strongest signal is the full-range input voltage and the noise level includes all of the noise sources plus distortion from harmonics (see below) [9]. The dynamic range can be used to calculate the effective number of bits (ENOB) which is the number of bits that an ideal data converter would need to achieve that dynamic range.

$$ENOB = \frac{SINAD - 1.763}{6.02} \qquad (3.18)$$

For this example, ENOB $= 8.32$ bits, meaning that a 9-bit ADC is required.

The application note on which this example is based [5] ultimately specified a dual 10-bit ADC with a SINAD of 51.86 dB. If it is used with an analog front end with 16 dB gain, a maximum output of 2 V p-p (0.707 V rms) and a noise figure of 10.5 dB, we can estimate the receiver noise floor as follows.

First, we calculate the AFE input signal for the stated output, the equivalent AFE input noise, and the AFE equivalent input SNR.

$$s_{AFEout} = \frac{0.707^2}{50} = 0.04\,w \tag{3.19}$$

$$s_{AFEin} = \frac{0.04}{10^{86/10}} = 1.005 \times 10^{-10}\,w \tag{3.20}$$

$$\begin{aligned} n_{AFEin} &= k \times T_A \times B \times nf \\ &= 1.38 \times 10^{-23} \times 2.5 \times 10^6 \times 10^{10.5/10} \\ &= 1.16 \times 10^{-13}\,w \end{aligned} \tag{3.21}$$

$$snr_{AFEin} = \frac{1.005 \times 10^{-10}}{1.16 \times 10^{-13}} = 865.21 \tag{3.22}$$

$$SNR_{AFEin} = 10 \log_{10}(865.21) = 29.372\,\text{dB} \tag{3.23}$$

The overall effective input SNR is 29.35 dB which, for all practical purposes, is equal to that of the AFE alone, so the noise floor is also that of the AFE, 1.16×10^{-13}w or -99.3 dBm.

The reader must keep in mind that the noise floor calculated thus far ignores internally generated signals from the analog front end, as we will discuss in the next section. The RF platform's receiver noise floor is only a first, and perhaps not always very useful, measure of a cognitive radio's ability to identify empty channels.

The geographic range of a radio system is determined by many factors in addition to the receiver sensitivity. These include the transmitter output power, the gains of the transmitting and receiving antennas, and the propagation characteristics of the radio path in between the two antennas. The dependence of the received signal on these quantities is specified in a link budget, the details of which are beyond the scope of this book. See Reference 10 for a simple explanation and Reference 11 for a comprehensive treatment.

3.4.3 Strong Signal Behavior[1]

The response of the radio platform to strong signals is critical to cognitive radio behavior, and it may well be the single most important limiting factor in cognitive radio performance. This is particularly true if the input stages of the RF platform must admit a wide spectral bandwidth to facilitate frequency agility and spectrum sensing. This creates problems both in the analog (RF amplifiers, mixers, etc.) and ADC domains in that internally generated waveforms will

---

[1] Some material in this section is reprinted from Bostian and Sweeney [12]. Reproduced with permission.

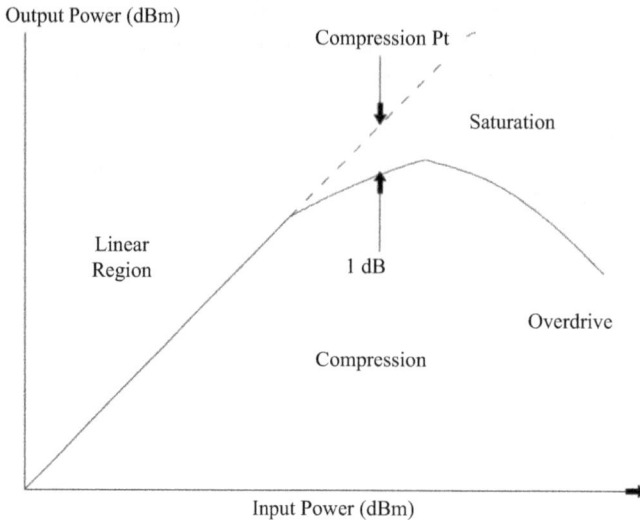

Figure 3.7 Output power versus input power for a two-port analog device. ©1999 John Wiley & Sons, Inc. All rights reserved. Reprinted, with permission, from Reference 12.

appear to the receiver as input signals, interfering with wanted signals and making empty channels look occupied. These important practical considerations are often overlooked.

We will begin by considering the radio platform's AFE. In a linear two-port device super-position holds, and the output waveform is linearly proportional to the input waveform. No frequencies appear in the output waveform that were not present in the input waveform. But any real two-port device will, when driven hard enough (i.e., when given an input waveform of sufficiently large amplitude), become nonlinear. While nonlinear operation may be desirable for some applications like high-efficiency power amplifiers, in receivers it creates unwanted signals called intermodulation products, which interfere with the wanted signals. To visualize the concept of nonlinearity, consider a two-port device with a sinusoidal input at frequency f_1. A typical plot of input power versus the output power delivered at frequency f_1 would appear as in Figure 3.7.

At low power levels, the relationship between output and input powers is linear. As the input power increases past a certain point, the slope decreases and the input–output relationship becomes nonlinear. This onset of nonlinearity is called *gain compression* (or just *compression*). A common measure of large signal handling ability is the *input 1 dB compression point*, the input power level at which the output power has fallen 1 dB below the extrapolation of the linear relationship. Increasing the input power beyond the 1 dB compression point causes the curve to fall farther and farther below a linear relationship. This part of the curve is called the *compression region*. The output power reaches a peak and then decreases. The peak is called *saturation* and the region beyond the peak is called the *overdrive region*.

One of the reasons that output power falls below the linear value after the device enters compression is that output power is being developed at frequencies other than the input fre-quency. So long as the input is a single-frequency sinusoid at frequency f_1, these frequencies are harmonics of f_1 (i.e., $2f_1$, $3f_1$, ...) and, in the traditional superheat architecture, easy to

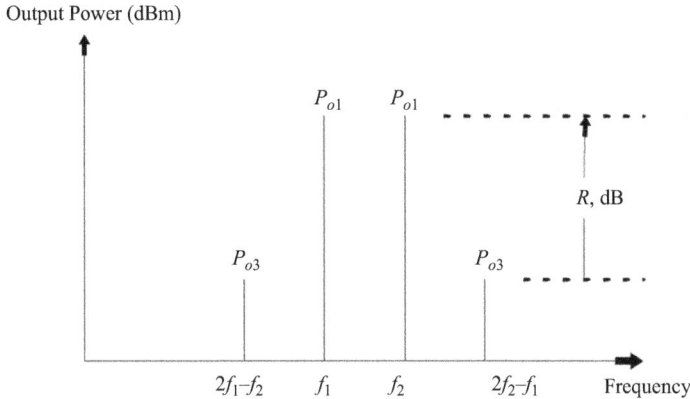

Figure 3.8   Conceptual spectrum analyzer display of the output of a nonlinear two-port device whose input is driven by two equal-amplitude sinusoids at frequencies f_1 and f_2. Third-order product appear at frequencies $2f_1-f_2$ and $2f_2-f_1$. R is called the rejection. ©1999 John Wiley & Sons, Inc. All rights reserved. Reprinted, with permission, from Reference 12.

filter out. If the input is a more complex waveform containing two or more frequencies, then the results intermodulation products can interfere with and distort the wanted output signal. This is particularly true for third-order products in most architectures, and for second-order products in ZIF receivers and all receivers whose analog front end passband is an octave or more.

In receiver specifications, intermodulation products are usually described by a two-tone test in which a device is given an input signal with two equal-amplitude sinusoids at closely spaced frequencies f_1 and f_2. These represent the practical situation of a wanted signal and an equal-amplitude adjacent-channel interferer. Nonlinearities cause outputs to appear at all the sums and positive differences of integer combinations of f_1 and f_2. The most important are the *third-order products* at frequencies $2f_1-f_2$ and $2f_2-f_1$. These third-order frequencies are close f_1 and f_2 and may interfere with the wanted signal. See Figure 3.8.

To describe the process mathematically, we can plot the output power at one of the third-order frequencies ($2f_1-f_2$ or $2f_2-f_1$) versus the input power at one the input frequencies (f_1 or f_2). If the powers are plotted in logarithmic units (dBm or DB), the curve will be a straight line with a slope of 3, at least for reasonable values of input power [13]. On the same graph, we can also plot output power at one of the input frequencies. (If we were plotting output power for a second-order product, at low levels it would have a slope of 2.) Below the compression region, this will be the straight line. If we extrapolate the straight line far enough, it will cross the straight-line extension of the first-order curve at the so-called *third-order intercept point*. The corresponding input power is called the third-order input intercept point (IIP3) and the corresponding output power is called the third-order output intercept point (OIP3). IIP3 and OIP3 are measures of the strong signal handling capabilities of two-port device.

It is worth emphasizing that the third-order intercept point is a graphical construction. The device cannot operate at that point. The intercept point represents an extrapolation of linear operation into the nonlinear region.

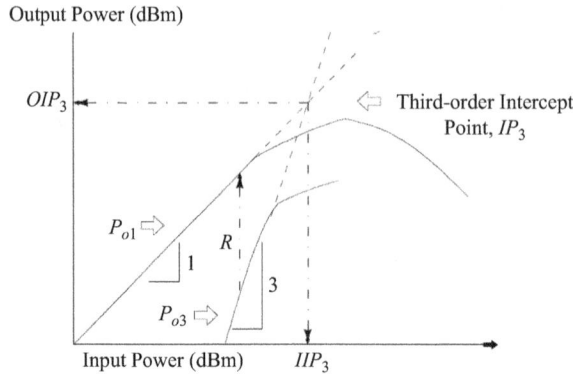

Figure 3.9 The third-order intercept and related concepts. P_{o1} is the output power at one of the wanted frequencies and P_{o3} is the output power at one of the unwanted third-order product frequencies. ©1999 John Wiley & Sons, Inc. All rights reserved. Reprinted, with permission, from Reference 12.

On a plot of output power in decibel units versus input power in decibel units, the linear portion of the curve representing the first order output power has a slope of unity and the linear portion of the curve representing the third-order output power has a slope of 3. See Figure 3.9. Since these two lines have known slopes and a know intersection point, it is easy to represent them geometrically. The values if IIP3 and OIP3 are easy to measure, based on this geometrical relationship.

To find the values of the intercept points, it is necessary to drive the two-port device to a point where the third-order products are measurable and look at the output on a spectrum analyzer. Figure 3.8 sketches the resulting display. If P_{o1} and P_{o2} are the output powers in dBm at one of the wanted frequencies and at one of the third-order frequencies, respectively, then the rejection R is given by

$$R = P_{o1} - P_{o3} \tag{3.24}$$

P_{o1} is related to the input power P_i at one of the wanted frequencies, f_1 or f_2 by the gain, G, of the device.

$$P_{o1} = P_i + G \tag{3.25}$$

When the device is operating under conditions corresponding to a rejection R, the input and output intercept points can be calculated from

$$IIP_P = \frac{R}{2} + P_i \tag{3.26}$$

$$OIP_3 = \frac{R}{2} + P_{o1} \tag{3.27}$$

$$OIP_3 = IIP_3 + G \tag{3.28}$$

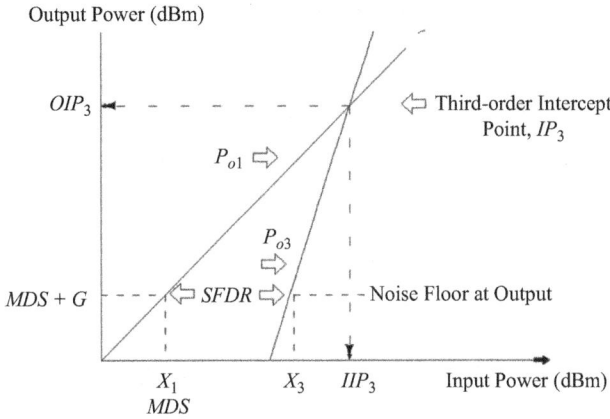

Figure 3.10 The concept of spurious-free dynamic range as the difference between the
x coordinates of the points where the third-order and the first-order output
powers cross the noise floor at the receiver output. ©1999 John Wiley & Sons,
Inc. All rights reserved. Reprinted, with permission, from Reference 12.

Continuing with our treatment of analog front end components, we will now discuss
dynamic range. This is the dB difference between the strongest signal and the weakest signal
that the receiver can handle. There are multiple measures of dynamic range, depending on how
the strongest signal and the weakest signal are defined.

Spurious-free dynamic range (SFDR) makes the weakest signal the receiver noise
floor kT_SB_N.

The strongest signal used in the SFDR definition is the input power that, in a two-tone
intermodulation test, makes the power in either of the third-order products at the receiver
output equal to the noise floor. This concept is illustrated in Figure 3.10.

Using the equation for a straight line

$$y - y_1 = m(x - x_1) \tag{3.29}$$

where (x,y) is a general point on the line and (x_1, y_1) is a particular point on the line and m
is the slope, we can write the equations for the lines describing the first-order and third-order
output powers P_{o1} and P_{o3}. (Remember that all powers here are in dBm.)

$$P_{o1} - OIP_3 = P_i - IIP_3 \tag{3.30}$$

$$P_{o3} - OIP_3 = 3(P_i - IIP_3) \tag{3.31}$$

At the point $P_i = x_1$, the input power corresponds to the MDS and the first-order output
power is $MDS + G$. G is the overall gain of the receiver. By definition, x_3 is the value of

the input power which makes the third-order output power $MDS + G$. We can express these mathematically and solve for the SFDR.

$$G + MDS = x_1 - IIP_3 + OIP_3 \qquad (3.32)$$

$$x_1 = MDS + G + IIP_3 - OIP_3 \qquad (3.33)$$

$$G + MDS = 3(x_3 - IIP_3) + OIP_3 \qquad (3.34)$$

$$x_3 = \frac{MDS + G}{3} + IIP_3 - \frac{OIP_3}{3} \qquad (3.35)$$

$$SFDR = x_3 - x_1 = \frac{2}{3}(IIP_3 - MDS) \qquad (3.36)$$

Spurious signals (spurs) are also generated in the receiver ADC. These may be distortion products, analogous to those described above for analog front end components, or their causes may be inherent in the ADC architecture, examples of which from Reference 14 include coupling from system clocks and mismatch between sub-converters in an interleaved ADC architecture. If the ADC input is a single frequency sinusoidal carrier at f_1, second and third harmonics (at frequencies $2f_1$ and $3f_1$) typically dominate and define the device's SFDR, which, in an FFT computed from the ADC output, is the dB difference between the carrier level and the level of the strongest spur (excluding the spur at DC, if present). See Figure 3.11. The level of the third-order products in an ADC output varies significantly with input frequency and with the characteristics of the analog front end. See Figure 3.12 that shows the third-order product level in BC (dB below the wanted carrier output) for a typical commercial ADC.

Correct calculation of the overall noise figure and third-order intercept point for a series of cascaded stages is an important part of radio platform design that is often overlooked in papers about cognitive radio design. Consider a series of cascaded networks as shown in Figure 3.13, where the gain G (dB) and the standard noise figure NF_{SR} for each stage are known.

The overall noise figure as a ratio is easily calculated from the radio (not dB) values of the individual stage's noise figure and gain.

$$nf_{SR\,Overall} = nf_1 + \frac{nf_2 - 1}{g_1} + \frac{nf_3 - 1}{g_1 g_2} + \cdots + \frac{nf_n - 1}{g_1 g_2 \cdots g_{n-1}} \qquad (3.37)$$

Here,

$$nf_1 = 10^{\frac{NF_1}{10}} \qquad (3.38)$$

$$g_1 = 10^{\frac{G_1}{10}} \qquad (3.39)$$

The overall noise figure is given by

$$NF_{SR\,Overall} = 10 \log_{10}(nf_{SR\,Overall})$$

$$= 10 \log_{10}\left(nf_1 + \frac{nf_2 - 1}{g_1} + \frac{nf_3 - 1}{g_1 g_2} + \cdots + \frac{nf_n - 1}{g_1 g_2 \cdots g_{n-1}}\right) \qquad (3.40)$$

FFT PLOT (32,768-POINT DATA RECORD)

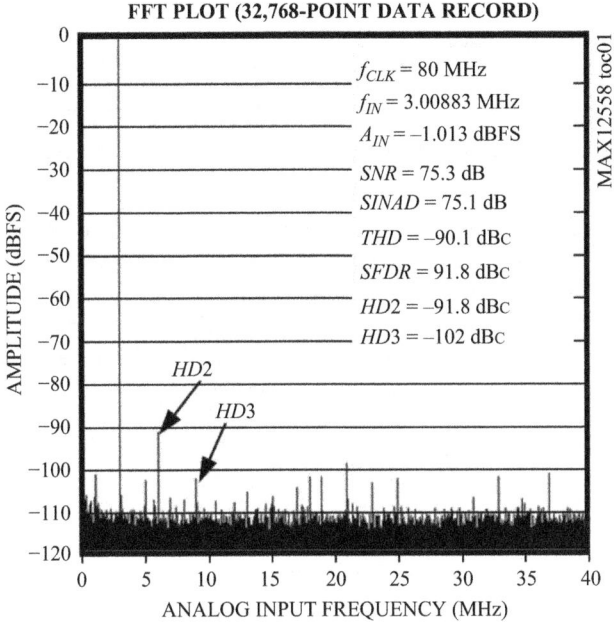

f_{CLK} = 80 MHz
f_{IN} = 3.00883 MHz
A_{IN} = −1.013 dBFS

SNR = 75.3 dB
$SINAD$ = 75.1 dB
THD = −90.1 dBc
$SFDR$ = 91.8 dBc
$HD2$ = −91.8 dBc
$HD3$ = −102 dBc

MAX12558 toc01

Figure 3.11   FFT computed from a sample ADC output. ©2004 Maxim Integrated Products. Reprinted, with permission, from Maxim 12558 Data Sheet. Maxim Integrated does not endorse, and is not responsible for the content of this publication.

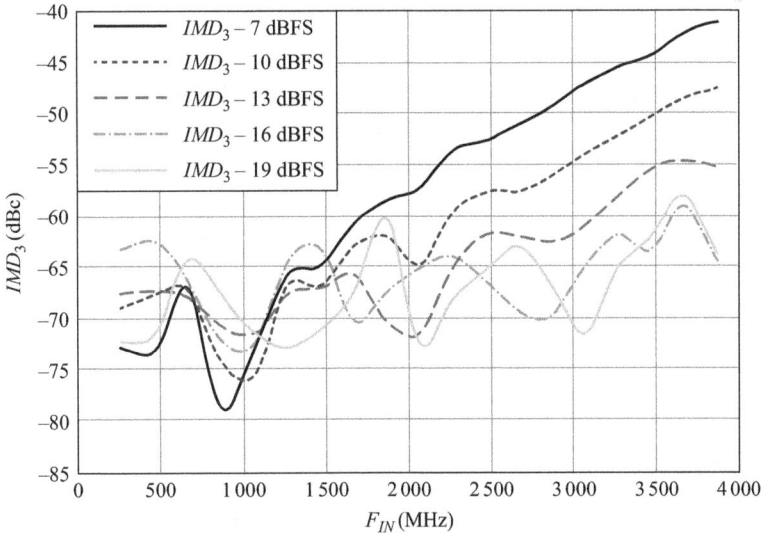

Figure 3.12   Third-order intermodulation product level for a typical ADC. From Reference 14, Courtesy Texas Instruments.

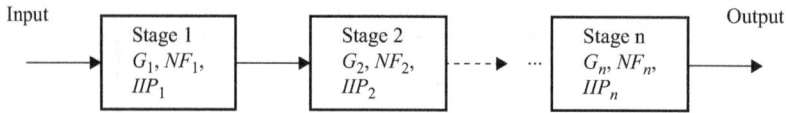

Figure 3.13   Cascaded stages with specified gain, noise figure, and IIP.

The third-order intercept point calculation is more complicated. See Reference 15. We must assume that the interfaces between all stages are 50 ohms and that all third-order products add in phase. We project the output intercept points of the individual stages through to the output of the last stage and add the projected values reciprocally. We then divide the results by the overall gain to obtain the third-order input intercept point. Not all writers about cognitive radio design seem to be aware of the need for this calculation!

$$OIP_i = IIP_i + G_i \tag{3.41}$$

$$oip_i = 10^{\frac{OIP_i}{10}} \tag{3.42}$$

$$oip_{Overall} = \frac{1}{\dfrac{1}{oip_1}} \text{ MW for one stage} \tag{3.43}$$

$$oip_{Overall} = \frac{1}{\dfrac{1}{oip_2} + \dfrac{1}{g_2 oip_1}} \text{ MW for two stages} \tag{3.44}$$

$$oip_{Overall} = \frac{1}{\dfrac{1}{oip_3} + \dfrac{1}{g_3 oip_2} + \dfrac{1}{g_3 g_2 oip_1}} \text{ MW for three stages} \tag{3.45}$$

$$oip_{Overall} = \frac{1}{\dfrac{1}{oip_n} + \sum\limits_{k=1}^{n-1} \dfrac{1}{oip_k \prod\limits_{j=k+1}^{n} g_j}} \text{ MW for } n \text{ stages} \tag{3.46}$$

The last formula can be rewritten in a simpler form to calculate $OIP_{Overall}$.

$$OIP_{Overall} = 10 \log_{10}(oip_{Overall}) \tag{3.47}$$

$$= -10 \log_{10}\left(\sum_{k=1}^{n} \frac{1}{oip_k g_{k+1} g_{k+2} \cdots g_n}\right) \text{dBm} \tag{3.48}$$

Blocking: Blocking, also called *de-sensitizing* or *desensing*, occurs when an unwanted strong input signal drives the analog portion of a receiver front end into the gain compression region of Figure 3.7. The unwanted signal is called a *blocking signal* or a *blocker*, and its frequency may be so far removed from the wanted signal that the only indication of the blocker's presence is a loss of receiver sensitivity. Blocking is often said to occur when the front end gain is reduced by 1 dB from its linear value. The input power level at which this occurs cannot be calculated from general principles, and there is no standard specification

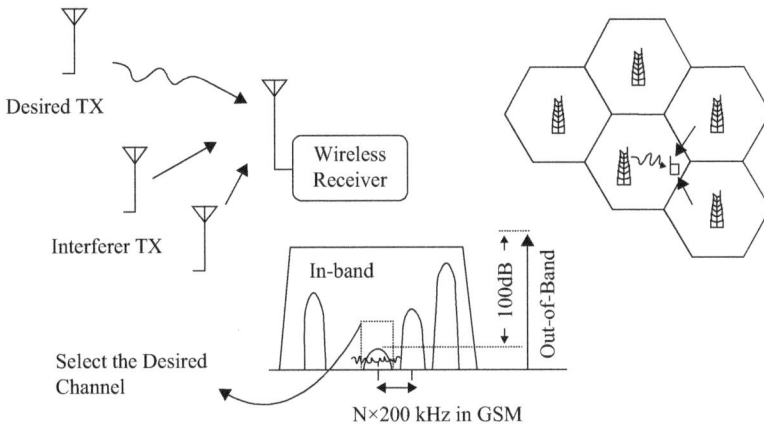

Figure 3.14   Blockers in a wireless environment. ©2011 IEEE. Reprinted, with permission, from Reference 1.

that describes a receiver's susceptibility to blocking. Instead, standards describe the levels and frequency offsets of blockers that receiver must accommodate. See References 1 and 16 for a discussion of blocking issues.

Blocking is a particular problem in cellular systems where receivers must demodulate wanted signals at 99 dBm while rejecting in-band blockers at −23 dBm 3 MHz away and out-of-band blockers a few hundred MHz away at power levels of 0 dBm. See Figure 3.14 from Reference 1. Transmitter phase noise and local oscillator feed through can also cause blocking.

The primary solution to out-of-band blocking in cellular receivers has been the insertion of surface acoustic wave (SAW) filters ahead of the receiver ADC. These are both expensive and not tunable – a particular problem for cognitive radio designers who require great frequency agility. Architectures and techniques that improve blocking performance without requiring SAW filters are discussed in References 1, 17 and 18.

Designers of cognitive radios for cellular applications should pay close attention to blocking issues. These are rarely mentioned in the cognitive radio literature.

3.5 Transmitter RF Specifications

Commercial transmitters are typically designed for particular standards, operating in relatively limited portions of the spectrum and restricted to a small number of modulation formats. Their specifications reflect these characteristics, often being defined for and measured under conditions governed by the standards.

Any real transmitter radiates a combination of wanted and unwanted signals. The wanted signal is an imperfect version of an ideal waveform. The ideal waveform is usually a sinusoid with particular instantaneous values of amplitude and phase, and specified transitions from each amplitude and phase state to the next one. We represent these states as two-dimensional vectors in rectangular coordinate system whose axes are the I and Q components of the RF waveform. Both the ideal and the actual waveforms can be plotted in this way, and their vector difference is called the error vector (EV). See Figures 3.15 and 3.16.

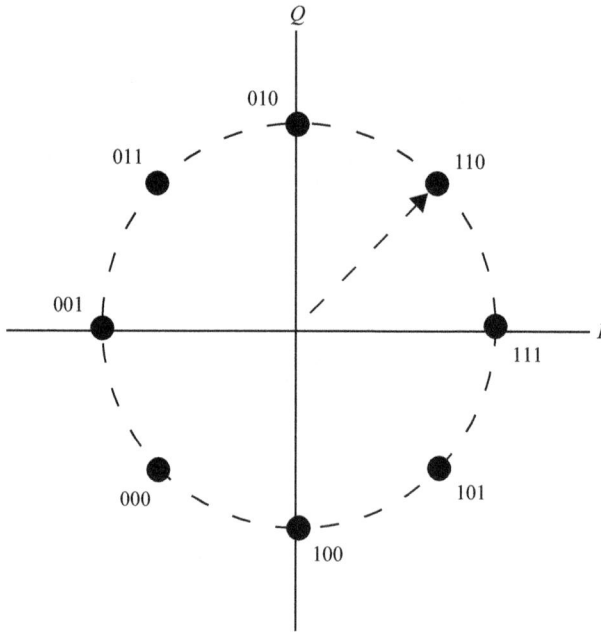

Figure 3.15   Polar representation of an eight-PSK waveform.

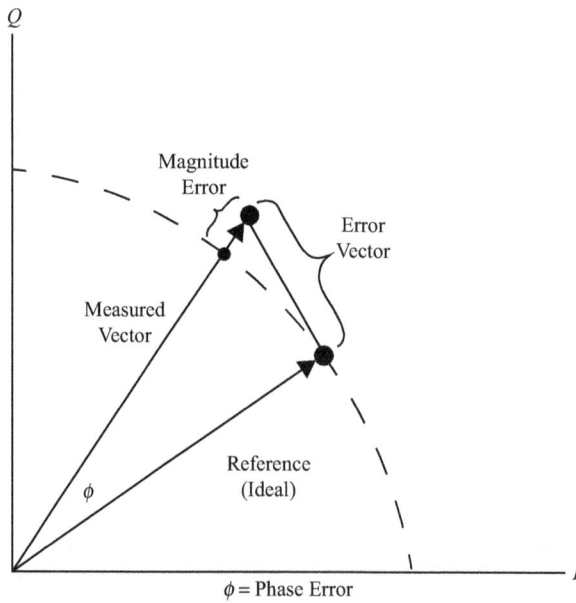

Figure 3.16   Transmitter waveform error and resulting error vector.

The absolute error vector magnitude (EVM) is the normalized rms value of an ensemble of individual error vector magnitudes $e(k)$ measured over some large number of randomized symbol state transitions [3].

$$EVM = \frac{\sqrt{\sum_n |\bar{e}(k)^2|}}{n} \qquad (3.49)$$

More commonly found on data sheets is a normalized value where the average power in a large number of error vectors is divided by the average power in the corresponding ideal waveform vectors. This may be expressed in dB or as a percentage [19] and the power values may be rms or peak.

$$EVM(dB) = 10\log_{10}\left(\frac{P_{error}}{P_{ideal}}\right) \qquad (3.50)$$

$$EVM(\%) = 100\sqrt{\frac{P_{error}}{P_{ideal}}} \qquad (3.51)$$

Typical values depend on output power level and are on the order of a few percent. EVM may be visualized as either a measure of the noise and distortion that the transmitter adds to an ideal transmitted waveform or as simply a measure of the deviation of the transmitted waveform from the ideal. Its main value in cognitive radio design is an indicator of the best performance that a transmitter can deliver to the user when the cognitive engine configures the transmitter for a particular waveform.

Potentially more important for cognitive radio applications is noise radiated by the transmitter at frequencies other than the intended frequency. This can cause interference to other radios, and it will limit a cognitive radio's ability to recognize an empty channel when its own transmitter is operating. Such unwanted radiation falls below limits specified in standards and/or dictated by government regulators. It may be specified in several ways.

Since most radio systems operate in standardized frequency channels, it is particularly important that transmitters not cause unacceptable interference to links operating in channels immediately adjacent to the one being used (adjacent channels) or to the next pair of channels farther out from the channel in use (alternate channels). A transmitter's potential to cause such interference is measured by the adjacent channel power ratio (AC PR or AC PR1) and alternate channel leakage or alternate channel power ratio (AC PR2), given by

$$ACPR1 = 10\log_{10}\left(\frac{P_{adjacent}}{P_{intended}}\right) \qquad (3.52)$$

$$ACPR2 = 10\log_{10}\left(\frac{P_{alternate}}{P_{intended}}\right) \qquad (3.53)$$

Both are expressed in decibels relative to the carrier (or the transmitted signal level at the center frequency of the wanted channel), dBc.

The noise power radiated by the transmitter at frequencies farther from the carrier than nearby channels is obviously important. This can be specified in several ways; one is output noise density in dBm/Hz, and another is output power measured with respect to the carrier (i.e., in dBc) in some stated bandwidth. Other sources of radiated noise include local oscillator (LO) leakage and image frequencies; the latter happen when the LO and IF waveforms combine in the mixer to create an unwanted output at either $f_{RF} + 2f_{LO}$ or $f_{RF} - 2f_{LO}$ (one of these is the wanted frequency and the other is the image frequency). Radiated noise levels depend strongly on transmitter output power (this must be remembered when interpreting the specifications), and they can exceed the -174 dBm/Hz thermal noise entering the cognitive radio's companion receiver from a 290-K antenna by many dB.

3.6 MAC and Performance Considerations

In most cognitive radio applications the transmitter and receiver(s) will share the same antenna. If the radio transmits and receives at the same time (full-duplex operation), the designer must take care to keep the transmitted signal from leaking into the receiver at power levels which will cause harmful interference, blocking (a situation where a strong signal overdrives the analog front end and/or the ADC, causing at least a rise in the apparent noise level and a loss of weak signal sensitivity), or even damage. Devices that separate the transmit and receive signal paths are called duplexers and diplexers; the exact difference is a matter of some dispute. Typically, *duplexer* means that the transmitter and receiver are operating at nearly the same frequency, and *diplexer* means that the transmitter and receiver frequencies are in different bands. Full-duplex operation is commonly achieved in the frequency domain and is called frequency division duplexing (FDD) because the transmit and receive frequencies are sufficiently different to allow filters to separate them. This method is built into many commercial RF platforms, but it will not work for single-channel receiver cognitive radios that transmit and receive on closely adjacent frequencies. Thus, it cannot be used in DSA schemes in which the cognitive radio must periodically stop transmitting and listen to ensure that the channel is open, unless a separate sensing receiver is provided.

In time division duplexing (TDD), the transmitter and receiver use the same frequency but do not operate simultaneously. This may be obvious to the user as in a traditional public safety push-to-talk radio system, or it may be invisible to the user making a video call over WiFi. The difference is in how fast the transmit/receive and receive/transmit changeovers occur. Values from milliseconds to microseconds are achievable, but the underlying timing and synchronization requirements can be difficult to meet, particularly with software defined radios running under a non-real-time operating system.

In order to search for open channels, the receiver portion of a cognitive radio used in DSA applications may be required to sweep a portion of the spectrum at regular intervals, playing the role of a spectrum analyzer. The process is simpler when users occupy a set of discrete channels which are regularly spaced in frequency and more complicated when continuous frequency sweeping is required. For the first case, RFIC-based RF platforms can be configured like swept-tuned spectrum analyzers, measuring the energy in a resolution bandwidth filter ahead of the detector, as shown in Figure 3.17 from Reference 20. Sufficient time must be allowed for the front end to change frequency and for the associated transients to die out. Frequency synthesizers' *hop time* provides a first estimate of the time required to set the receiver up for a measurement if it is already operational. If the receiver is turned off during transmission as in

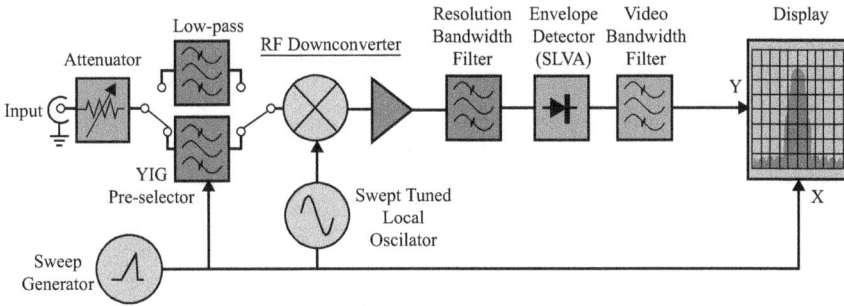

Figure 3.17   Block diagram of a swept-tuned spectrum analyzer. ©2014 Tektronix. Reprinted, with permission, from Reference 20. All Rights Reserved.

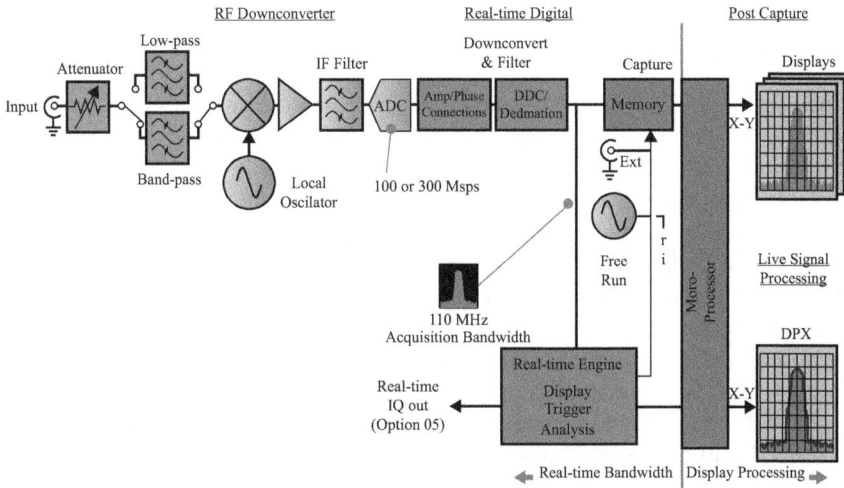

Figure 3.18   Block diagram of an SDR-based real-time spectrum analyzer. ©2014 Tektronix. Reprinted, with permission, from Reference 20. All Rights Reserved.

half-duplex operation, the receiver wake-up time must pass before frequency sweeping begins, and this can be significantly longer than the hop time.

An SDR-based RF platform can be configured like a real-time spectrum analyzer which samples the RF waveform for successive fixed-time intervals called frames and computes the FFT of each frame. See Figures 3.17–3.19 from Reference 20. Effectively, the receiver simultaneously examines all frequencies in its RF or IF passband within resolution limits imposed by the ADC specifications. For a general discussion of specialized MAC development for cognitive radio, see Reference 21.

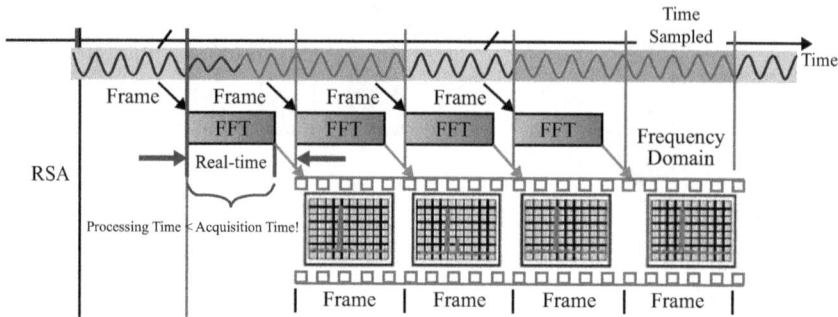

Figure 3.19   Operation of real-time spectrum analyzer. © 2014 Tektronix. Reprinted, with permission, from Reference 20. All Rights Reserved.

3.7 Radio Frequency Integrated Circuits

3.7.1 Introduction[2]

As distinguished from SDR systems, where radio operation is determined by software blocks, an RFIC is controlled by reading from and writing into memory registers within the device itself. Instead of generating a software emulation of a radio communication chain that implements data sources, mixers, filters, and modulators, the cognitive engine writes control bits into registers to set the desired communication parameters. Radio reconfiguration is a simple matter of changing a few register values. In contrast, for radio reconfiguration, SDR platforms must build and load new software blocks or radio chains. As alternative to SDR, we are interested in the frequency agile electronically controllable full RF solutions in low-cost small form factor packages. Currently available products are capable of transmitting and receiving a variety of FISK-based modulations in multiple bands, typically between 200 and 900 MHz, or in the 2.4 GHz band. Data rates can vary between 1 and 600 kbps. While these are low for commercial applications, they are adequate for CR experimentation. Frequency, transmit power, data rate, and modulation type are all user configurable, and the modules provide information about received signal strength indication (RSSI), link quality, and cognitive radio (CR) status. Additionally, users can customize the data packet structure. Examples of these flexible and reconfigurable RFIC-based RF platforms include the RFM22B and RFM69CW from Hope RF Electronics [23], which will be discussed next. Others include the Maxim7032 [24] and the Motorola RFIC4 and RFIC5, which offer higher performance but which may not be commercially available [25].

3.7.2 Example: RFM69CW

The Hope RF RFM69CW is a low-cost highly configurable transceiver capable of transmitting and receiving FSK, GFSK, MSK, GMSK, and OOK waveforms between 290 and

---

[2]Material in Sections 3.7.1 and 3.7.2 appeared earlier in Reference 22. ©2013 IEEE, Reprinted, with permission, from Reference 22.

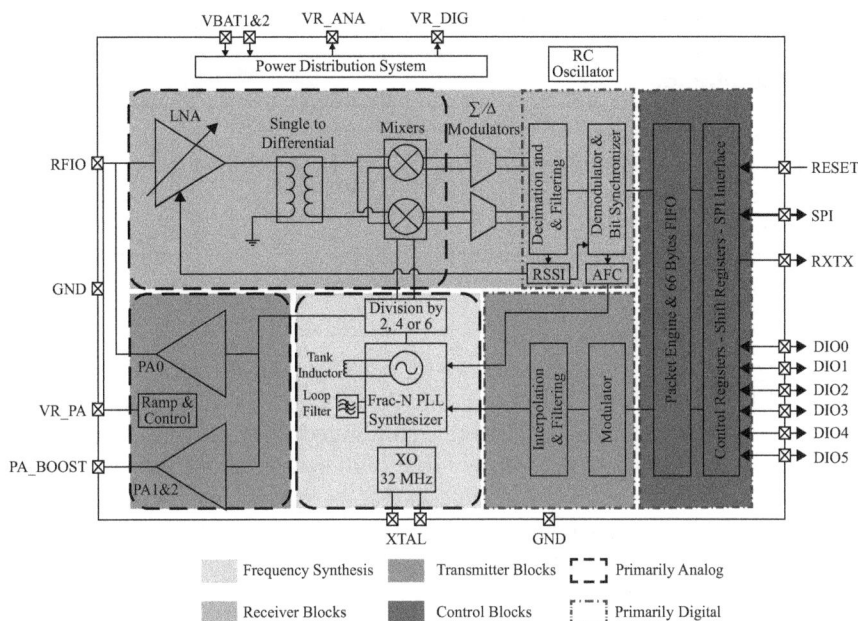

Figure 3.20   A block diagram of RFM69HCW. Courtesy HopeRF. Reproduced, with permission, from RFM69HCW Data Sheet http://www.hoperf.com/upload [Accessed 30 Aug 2016].

Table 3.1   RFM69CW receiver specifications

Specification	
FSK sensitivity at 4.8 kbps (dBm)	−114
OOK sensitivity at 4.8 kbps (dBm)	−112
IIP_2 at lowest LNA gain (dBm)	+75
IIP_2 at highest LNA gain (dBm)	+35
IIP_3 at lowest LNA gain (dBm)	+13
IIP_3 at highest LNA gain (dBm)	−18
Channel filter bandwidth (kHz)	2.6–500

1020 MHz with a maximum transmitter output power of +20 dBm. Receiver sensitivity is advertised down to −120 dBm at 1.2 kbps. The module is marketed as a fully contained radio solution, providing all the necessary mixing, filtering, tuning, and A/D components to transmit and receive packeted bit data. The RFM69HCW is sold in three separate versions designed for operation in the 433, 868, and 915 MHz band. These differ in their output impedance matching and filter networks. Figure 3.20 presents a simplified block diagram of the RFIC, Table 3.1 and 3.2 give its RF specifications. The RFM69HCW is an updated version of the RFM22B transceiver, which is discussed in References 22 and 26.

The module's operation is configured (and reconfigured) by values stored in its internal registers. Figure 3.21 shows some of the memory registers that can be set to configure the radio operation (and subsequently read to determine what the current configuration is).

Table 3.2   RFM69HCW transmitter specifications

Specification	
RF output power (dBm)	−18 to +13
Phase noise at 50 kHz offset (dBc/Hz)	−99 to −95, depending on band
Adjacent channel power at 25 kHz offset (dBm)	−37

The primary input/output (I/O) mechanism for the RF module is a first in first out (FIFO) buffer. Serial data bytes are written to the transmit FIFO buffer in succession for transmission. When the buffer is full, the accumulated bytes are transmitted. Received data are likewise stored serially in the receive FIFO buffer. Continued reads will transfer all the data out of the buffer to the user. From the user perspective, it appears that the transmit and receive FIFO buffers are one and the same, as they are both access by writing to and reading from the same register address; however there are in fact two FIFO buffers, one for transmit and one for receive and internal RFIC controls ensure proper access to the appropriate FIFO.

A serial peripheral interface (SPI) bus is the primary method of interaction with the RFM22B, and four SPI lines are used to send and receive data to and from the module. General-purpose input/output (GPIO) is used for secondary signaling, controlling a transmit/receive (T/R) switch and providing a path for reading hardware interrupts.

3.7.3 Computational Support for RFICs

A cognitive radio's computational platform usually supports both the cognitive engine (CE) and either the DSP functions required by an SDR or the control functions required by an RF application-specific integrated circuit (ASIC). Moving away from SDR and employing an RFIC reduces the computational load. No DSP-intensive calculations are required; the computational platform unlike the microcontroller-based needs only support the high-level cognition functionality and RFIC configuration and control. High-end laptops are no longer required.

Microcontroller-based platforms like the Arduino [27] and chipKIT [28] are the lowest end option for RFIC-based cognitive radios. These boards generally have the same form factor or footprint, but vary widely in types of microcontroller chips, and target applications ranging from robotics to home automation to remote sensing. They are popular due to their low cost (the Arduino itself is less than $30 USD), and the ease with which users can build working projects very quickly.

Single board computers are the next level above microcontroller chips. These are often designed as development platforms for set-top boxes (cable boxes) and mobile systems. They feature small form factors, low power consumption, and a wealth of I/O options, as well as the processing power to support intensive graphics operations. Unlike the microcontroller-based platforms discussed above, these platforms are full computers. Their operating systems provide file management, task scheduling, and drivers for I/O and peripherals. Current choices include Raspberry Pi [29], BeagleBone Black [30], Gumstix Overor [31], and DuoVero [32]. The Raspberry Pi is notable because of its extreme low cost ($35 USD).

Addr	R/W	Function	D7	D6	D5	D4	D3	D2	D1	D0	POR
0E	R/W	I/O Port Configuration	Reserved	extitst[2]	extitst[1]	extitst[0]	itsdo	dio2	dio1	dio0	00h
0F	R/W	ADC Configuration	Adcstart/adcdone	adcsel[2]	adcsel[1]	adcsel[0]	adcref[1]	adcref[0]	adcgain[1]	adcgain[0]	00h
10	R/W	ADC Sensor Amplifier Offset	Reserved	Reserved	Reserved	Reserved	adcoffs[3]	adcoffs[2]	adcoffs[1]	adcoffs[0]	00h
11	R	ADC Value	adc[7]	adc[6]	adc[5]	adc[4]	adc[3]	adc[2]	adc[1]	adc[0]	—
12	R/W	Temperature Sensor Control	tsrange[1]	tsrange[0]	entsoffs	entstrim	tstrim[3]	tstrim[2]	tstrim[1]	tstrim[0]	20h
13	R/W	Temperature Value Offset	tvoffs[7]	tvoffs[6]	tvoffs[5]	tvoffs[4]	tvoffs[3]	tvoffs[2]	tvoffs[1]	tvoffs[0]	00h
14	R/W	Wake-Up Timer Period 1	Reserved	Reserved	Reserved	wtr[4]	wtr[3]	wtr[2]	wtr[1]	wtr[0]	03h
15	R/W	Wake-Up Timer Period 2	wtm[15]	wtm[14]	wtm[13]	wtm[12]	wtm[11]	wtm[10]	wtm[9]	wtm[8]	00h
16	R/W	Wake-Up Timer Period 3	wtm[7]	wtm[6]	wtm[5]	wtm[4]	wtm[3]	wtm[2]	wtm[1]	wtm[0]	01h
17	R	Wake-Up Timer Value 1	wtv[15]	wtv[14]	wtv[13]	wtv[12]	wtv[11]	Wtv[10]	Wtv[9]	Wtv[8]	—
18	R	Wake-Up Timer Value 2	wtv[7]	wtv[6]	wtv[5]	wtv[4]	wtv[3]	wtv[2]	wtv[1]	wtv[0]	—
19	R/W	Low-Duty Cycle Moe Duration	ldc[7]	ldc[6]	ldc[5]	ldc[4]	ldc[3]	ldc[2]	ldc[1]	ldc[0]	00h
1A	R/W	Low Battery Detector Threshold	Reserved	Reserved	Reserved	lbdt[4]	lbdt[3]	lbdt[2]	lbdt[1]	lbdt[0]	14h
1B	R	Battery Voltage Level	0	0	0	vbat[4]	vbat[3]	*vbat[2]	vbat[1]	vbat[0]	—
1C	R/W	IF Filter Bandwidth	dwn3_bypass	ndec[2]	ndec[1]	ndec[0]	filset[3]	filset[2]	filset[1]	filset[0]	01h
1D	R/W	AFC Loop Gearshift Override	afcbd	enafc	afcgearth[2]	afcgearth[1]	afcgearth[0]	1p5 bypass	matap	ph0size	40h
1E	R/W	AFC Timing Control	swait_timer[1]	swait_timer[0]	shwait[2]	shwait[1]	shwait[0]	anwait[2]	anwait[1]	anwait[0]	0Ah
1F	R/W	Clock Recovery Gearshift Override	Reserved	Reserved	crfast[2]	crfast[1]	crfast[0]	crslow[2]	crslow[1]	crslow[0]	03h
20	R/W	Clock Recovery Oversampling Ratio	rxosr[7]	rxosr[6]	rxosr[5]	rxosr[4]	rxosr[3]	rxosr[2]	rxosr[1]	rxosr[0]	64h
21	R/W	Clock Recovery Offset 2	rxosr[10]	rxosr[9]	rxosr[8]	stallctrl	ncoff[19]	ncoff[18]	ncoff[17]	ncoff[16]	01h
22	R/W	Clock Recovery Offset 1	ncoff[15]	ncoff[14]	ncoff[13]	ncoff[12]	ncoff[11]	ncoff[10]	ncoff[9]	ncoff[8]	47h
23	R/W	Clock Recovery Offset 0	ncoff[7]	ncoff[6]	ncoff[5]	ncoff[4]	ncoff[3]	ncoff[2]	ncoff[1]	ncoff[0]	AEh

Figure 3.21 Sample of memory registers set and read on the RFM22B. Courtesy HopeRF. Reproduced with permission.

3.8 Platforms for Software Defined Radio

3.8.1 Introduction

All of the signal-processing operations that take place between the antenna terminals and the output of a receiver and between the input to a transmitter and its antenna terminals can be represented as mathematical functions. For a radio system using digital modulation, appropriate combinations of these functions map analog signals from the antenna to binary signals at the receiver output and binary signals at the transmitter input to analog signal sent to the transmitting antenna. These functions can all be performed by digital electronics under software control, so in principle a receiver can be made from a computer with an ADC at its input, and transmitter can be made from a computer with a DAC at its output. And, indeed, the first software receivers and transmitters were simply digital emulations of analog radios, with many of the same internal blocks. More recent architecture employ digital signal-processing algorithms that have no simple analog counterparts.

While it is possible to build radios entirely from digital components, cost and performance requirements normally dictate a hybrid approach, with analog circuits closest to the antenna to do amplification, filtering, and frequency conversion at carrier frequencies, and digital circuits closest to the information sink or source to do filtering, modulation/demodulations, and frequency conversion at intermediate frequencies and baseband. Conceptually, then, an RF platform consists of the necessary analog components to do RF signal processing, ADC and DAC, and the necessary electronics to do the required digital signal processing. Physically, these components may all be part of a single piece of hardware (controlled by a cognitive engine in the case of a cognitive radio) or it may be divided between a board that performs the RF functions and some of the waveform-independent digital functions plus analog-to-digital and digital-to-analog conversion, and a computer platform that performs the required digital signal processing. In a cognitive radio, this may well be the same computer that runs the cognitive engine code.

3.8.2 Packaged RF Front Ends and All-in-one Platforms

A large number of prototype radios have been built using the Universal Software Radio Peripheral (USRP) or the Wireless Open-Access Research Platform (WARP) families of products. Our informal November 2014 survey of the *IEEE Explore* database produced 83 entries for USRP-based CR and 21 entries for WARP-based prototypes and no entries involving both platforms. USRP offer the user a great deal of PHY and MAC layer flexibility along with an understandable and necessary trade off in the RF performance. WARP board RF interfaces are based on RFIC designed and manufactured for compliance with the IEEE 802.11 series of standards, and these typically offer optimized RF specifications with perhaps less PHY and MAC layer flexibility. Many competitors are entering the market. See, for example, the BladeRF SDR platform at the low-cost end [33] and products growing out of the DARPA RF-FPGA program [34]. See also the Triad Semiconductor RF-FPGA [35] and the Arria V FPGA RF [36] product lines.

3.8.2.1 The USRP Family of Products and GNU Radio

Possibly the most common SDR platform used by cognitive radio researchers combines the GNU Radio Software package with the family of Universal Software Radio Peripheral (USRP)

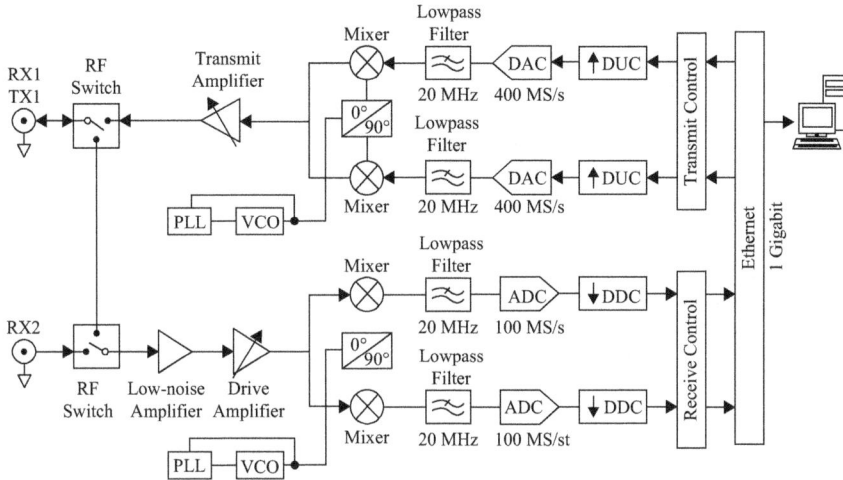

Figure 3.22   Typical Ettus USRP-based architecture. Reproduced with permission from https://www.ettus.com/kb/detail/usrp-bandwidth [Accessed 28 Oct 2014].

products made by Ettus Research. The lower cost USRPs are typically mother boards incorporating a field programmable gate array (FPGA), ADC and DAC, and the necessary interface and control circuitry for radio operation plus connectors or sockets for analog RF daughter boards. In the terminology we are using here, a daughter board is an RF platform, and a mother board is a computational platform with some or all of the hardware and software required for SDR functions. Typically the lower cost mother boards work in tandem with an external general purpose processor (often a high-end laptop) running GNU Radio. Figure 3.22 (from Ettus website) illustrates the typical radio architecture of a USRP-daughter board combination. Blocks to the left of the ADC and DAC are on the daughter board. In this typical architecture, the daughter board translates signals from complex RF to baseband and the mother board processes the result. Figure 3.23 (also from the Ettus website) presents a more comprehensive view of the architecture from the perspective of the mother board. More powerful and more expensive mother boards (e.g., the USRP X310) do not require an external computational platform and offer the cognitive radio designer a complete SDR suitable for integration with a CE.

The Ettus product line has expanded to include a variety of sophisticated all-in-one (RF and computational platforms on a single board) products as well as more advanced mother board/daughter board combinations. These devices allow users to develop SDR quickly and easily, for applications ranging from simple analog transceivers to GSM base stations to advanced commercial and military communications systems.

Ettus products are closely integrated with, but by no means restricted to, GNU Radio software. Quoting the GNU Radio website [37]:

GNU Radio is a free software development toolkit that provides the signal-processing runtime and processing blocks to implement software radios using readily available, low-cost external RF hardware and commodity processors. It is widely used in hobbyist, academic, and commercial environments to support wireless communications research as well as to implement real-world radio systems.

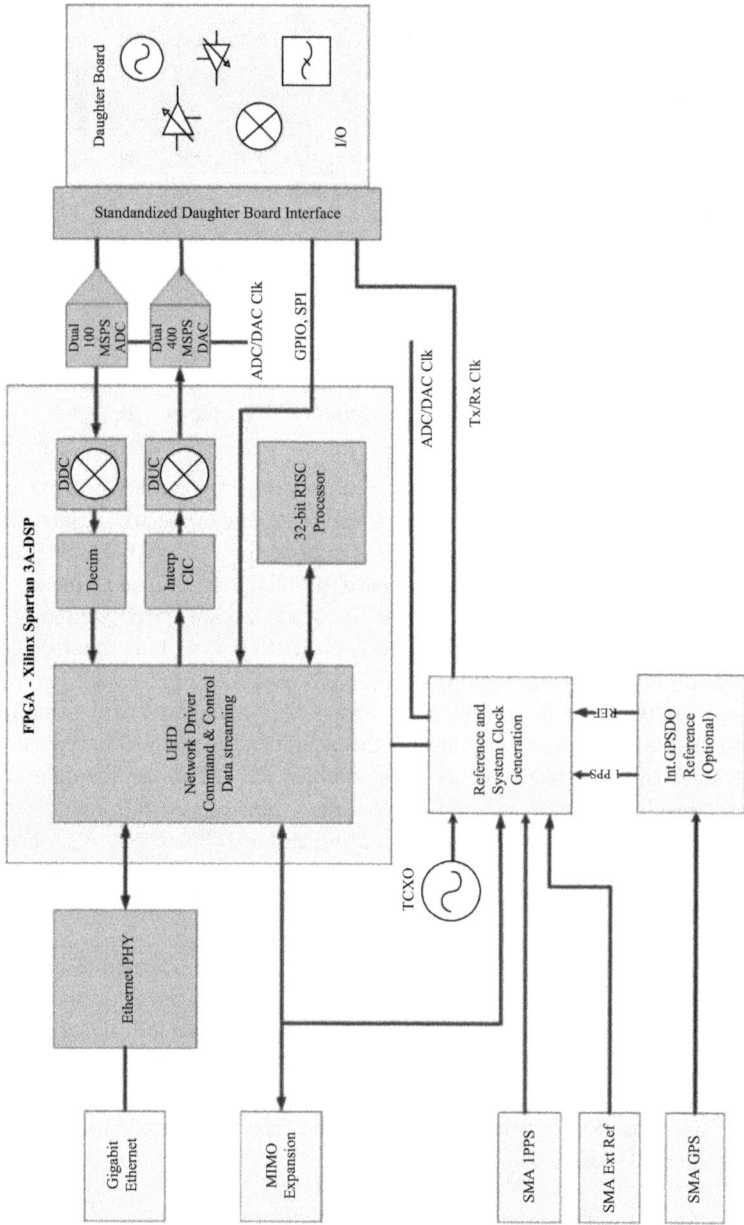

Figure 3.23 Ettus architecture showing N200/N210 mother board. Reproduced with permission from https://www.ettus.com/content/files/07495_Ettus_N200-210_DS_Flyer_HR.pdf [Accessed 28 Oct 2014].

GNU Radio applications are primarily written using the Python programming language, while the supplied, performance-critical signal-processing path is implemented in C++ using processor floating point extensions where available. Thus, the developer is able to implement real-time, high-throughput radio systems in a simple-to-use, rapid-application-development environment.

3.8.2.2 RF Daughter Boards

Ettus RF daughter boards range in sophistication from simple RF pass-through devices which implement basic transmit–receive switching functions and condition signal voltages for the mother board's ADCs and DACs to sophisticated transceiver front ends implementing a variety of analog and digital signal-processing functions. Their input and output interfaces and waveforms depend on the particular model, and the mother boards have the necessary hardware and are easily configured for these. In this section, we will describe the WBX and SBX series. The reader should consult the Ettus web site for full details on available products and their specifications.

The WBX is a popular and well-documented example of the Ettus RF daughter board family. A full-duplex transceiver, it covers 50 MHz to 2.20 GHz and can be further customized by the user to optimize its performance for a particular application. The WBX offers two quadrature front ends, one for transmitting and one for receiving, and it is capable of full-duplex operation on different transmit and receive frequencies. Its advertised typical specifications are: 40 MHz of bandwidth, output power of up to 100 mW (20 dBm), and a noise figure of 5 dB. Individual device receiver performance varies with frequency and RF gain; see Figure 3.24 for an example [38]. More detailed specifications are available in Reference 39 and appear here in Table 3.3.

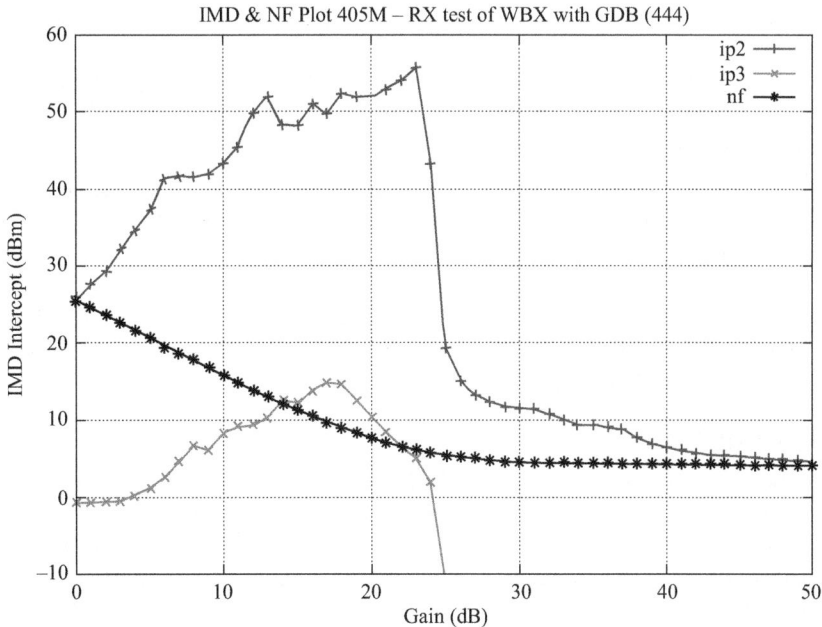

Figure 3.24   Measured performance of example WBX at 405 MHz. Courtesy Ettus.

Table 3.3   Representative RF specifications for selected USRP products

Device	Ettus WBX	Ettus WBX + N200/N210 USRP	Ettus SBX
Specification			
Frequency range (MHz)	50–2200		400–4400
Max transmit power (dBm)	15–20	15	16–20
Power range control (dB)	25		32
Maximum RX bandwidth (MHz)	40		
NF (dB)	5–7	5	5–7 f < GHz 7–10 3 < f < 4 GHz 10–13 4 < f < 4.4 GHz
IIP2 (dBm)	40–55		
IIP3 (dBm)	5–7	0	0
LO suppression (dBc)		50	
Phase noise (dBc/Hz)		−80 at 1.8 GHz, 10 kHz sep.	

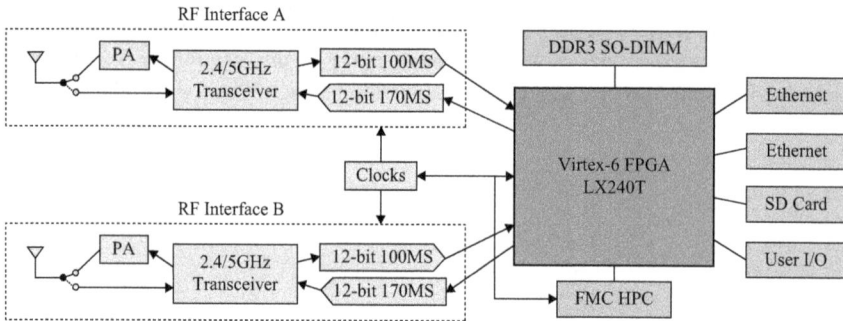

Figure 3.25   WARP v3 hardware architecture. From https://warpproject.org/trac/wiki/
HardwareUsersGuides/WARPv3 [Accessed 26 Feb 2016]. Reproduced with
permission.

A similar product, the WBX-120, offers similar RF specifications but with a 120-MHz instantaneous bandwidth. The table also gives specifications for the WBX in combination with the N200/N210 USRP as well as for the SBX [39].

3.8.2.3 WARP Board

Wireless Open-Access Research Platform (WARP) products are similar to USRPs in that they offer a mother board that accepts two RF interfaces and provide an FPGA and a variety of control and computational functions. WARP products originated at Rice University and are now manufactured and sold by Mango Communications. Since they operate in the popular WiFi bands and are built around RFICs that meet the IEEE 802.11 specifications, WARP products may offer features of particular interest to researchers working on the networking aspects of cognitive radio. Figure 3.25 from Reference 40 illustrates the architecture of WARP v3, the latest generation of WARP hardware.

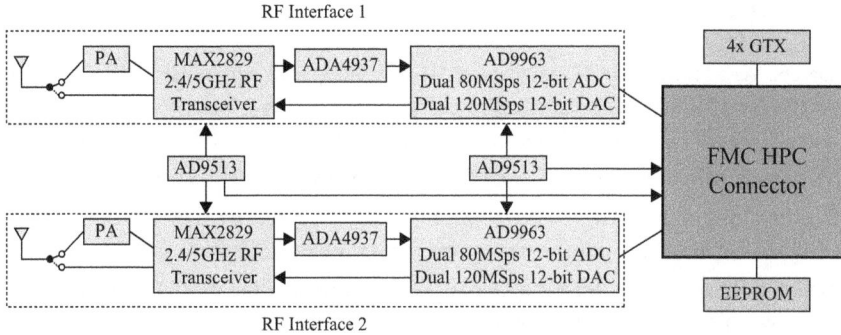

Figure 3.26   Block diagram of Mango FMC-RF-2X245. From http://warpproject/trac/ wiki/HardwareUserGuides/FMC-RF-2X245 [Accessed 26 Feb 2015]. Reproduced with permission.

Table 3.4   Selected RF specifications for MAX2829 RFIC used in Mango FMC-RF2X245 RF daughter board.

Specification	
Frequency range (MHz)	4900–5350
Maximum transmit power (dBm)	−2
NF (dB)	4.5
IIP3 (dBm)	−24
RX spurs at LNA input (dBm)	−50
Transmitter EVM % for 54 Mbps 802.11g OFDM waveform	1.5

Figure 3.26 shows a representative WARP RF daughter board, the Mango FMC-RF-2X245 [41].

Data sheets for the Maxim MAX2829 transceiver and the Anadigics AWL6951 PA RFICs used by the FMC-RF-2x245 are available from their manufacturers. They provide extremely detailed information, since the devices must comply with IEEE 802.11 requirements. It is not possible to list all of the numerous RF specifications here. Table 3.4 provides a small representative sample of the MAX2829 specifications for IEEE 802.11a low-band operation with the receiver in the high-gain setting only. The chip offers several modes of operation.

3.9 Conclusion

In this chapter, we have presented an overview of the functions and specifications for cognitive radio RF platforms. Typically, these combine analog and digital components and signal-processing systems and offer the user a wide range of options trading off cost, performance, the development effort required, and required computational support. RFICs are attractive for simple applications typically requiring channelized communications with FSK-based waveforms, while SDRs offer power and flexible performance in return for commensurate investment in software development.

Bibliography

[1] H. Darabi, A. Mirzaei, and M. Mikhemar, "Highly Integrated and Tunable RF Front Ends for Reconfigurable Multiband Transceivers: A Tutorial," *IEEE Trans. Circuits Syst. I*, vol. 58, no. 9, 2011, pp. 2038–2050. [Online]. Available: http://dx.doi.org/10.1109/tcsi.2011.2162460

[2] K. McClaning, *Wireless Receiver Design for Digital Communications*. SciTech Publishing, 2012.

[3] G. Hueber and R. Staszewski, *Multi-Mode/Multi-Band RF Transceivers for Wireless Communications: Advanced Techniques, Architectures, and Trends*. Wiley-IEEE Press, 2011.

[4] B. Brannon. Basics of Designing a Digital Radio Receiver (Radio 101). [Online]. Available: http://www.analog.com/static/imported-files/tech_articles/480501640radio101.pdf

[5] D. Anzaldo. (2013, May) Application Note 5519 "Navigate the AFE and Data-Converter Maze in Mobile Wireless Terminals". Maxim Integrated. Retrieved August 19, 2014. [Online]. Available: http://www.maximintegrated.com/en/app-notes/index.mvp/id/5519

[6] S. Overhoff. (2015) Application Note 5446 "Direct-Sampling DACs in Theory and Application". Maxim Integrated. [Online]. Available: https://www.maximintegrated.com/en/app-notes/index.mvp

[7] A. Kuckreja. (2015) Application Note 5317, "Implementing a Direct RF Transmitter for Wireless Communications". Maxim Integrated. [Online]. Available: https://www.maximintegrated.com/en/app-notes/index.mvp

[8] D. Anzaldo. (2015) Application Note 6063. "LTE_Advanced Release-12 Shapes New eNodeB Transmitter Architecture: Part 2, Analog Integration Challenge". Maxim Integrated. [Online]. Available: https://www.maximintegrated.com/en/app-notes/index.mvp/id/6063

[9] J. Reed, *Software Radio*. Prentice-Hall PTR, 2002.

[10] T. Pratt, C. Bostian and J. Allnutt, *Satellite Communications (Second Edition)*. John Wiley & Sons, January 2003.

[11] L. Barclay, Ed., *Propagation of Radiowaves*. Institution of Engineering and Technology (IET), January 2012. [Online]. Available: http://dx.doi.org/10.1049/pbew056e

[12] C. W. Bostian and D. G. Sweeney, "UHF Receivers," *Wiley Encyclopedia of Electrical and Electronic Engineering*, vol. 22. New York, NY: John Wiley and Sons, Inc., 1999, pp. 613–623.

[13] D. Henkes and S. Kwok, "Intermodulation: Concepts and Calculations," *Applied Microwave & Wireless*, vol. 9, no. 4, 1997.

[14] P. Plisch, "Maximizing SFDR Performance in the GSPS ADC: Spur Sources and Methods of Mitigation," *Texas Instruments*, 2013.

[15] S. E. Wilson, "Evaluate the Distortion of Modular Cascades," *Microwaves*, vol. 20, no. 3, 1981, p. 67.

[16] R. Tanner and J. Woodard, Eds., *WCDMA – Requirements and Practical Design*. Wiley-Blackwell, Jan 2004. [Online]. Available: http://dx.doi.org/10.1002/0470861797

[17] H. Darabi, D. Murphy, M. Mikhemar, and A. Mirzaei, "Blocker Tolerant Software Defined Receivers," in *ESSCIRC 2014 – 40th European Solid State Circuits Conference (ESSCIRC)*. Institute of Electrical & Electronics Engineers (IEEE), September 2014. [Online]. Available: http://dx.doi.org/10.1109/esscirc.2014.6942018

[18] S. Heinen and R. Wunderlich, "High Dynamic Range RF Frontends from Multiband Multistandard to Cognitive Radio," in *2011 Semiconductor Conference Dresden*. Institute of Electrical & Electronics Engineers (IEEE), September 2011. [Online]. Available: http://dx.doi.org/10.1109/scd.2011.6068724

[19] L. Frenzel, "Understanding Error Vector Magnitude," *Electronic Design*, October 2013.

[20] Real-Time Spectrum Analyzer Fundamentals. Tektronix. Retrieved November 4, 2014. [Online]. Available: http://info.tek.com/www-fundamentals-of-real-time-spectrum-analysis.html

[21] J. Ansari, X. Zhang, A. Achtzehn, M. Petrova, and P. Mahonen, "A Flexible MAC Development Framework for Cognitive Radio Systems," in *2011 IEEE Wireless Communications and Networking Conference*. Institute of Electrical & Electronics Engineers (IEEE), March 2011. [Online]. Available: http://dx.doi.org/10.1109/wcnc.2011.5779123

[22] A. Young and C. Bostian, "Simple and Low-cost Platforms for Cognitive Radio Experiments [Application Notes]," *Microwave Magazine, IEEE*, vol. 14, 2013, pp. 146–157.

[23] (2012) RFM22B FSK Transceiver – FSK Modules – HOPE Microelectronics. Hope Microelectronics Co. [Online]. Available: http://www.hoperf.com/rf_fsk/fsk/RFM22B.htm

[24] (2012) MAX7032 Low-cost, Crystal-based, Programmable, ASK/FSK Transceiver with Fractional-n PLL – Overview. Maxim Integrated. [Online]. Available: http://www.maximintegrated.com/datasheet/index.mvp/id/4755

[25] S. Hasan and S. Ellingson, "Multiband Public Safety Radio Using a Multiband RFIC with an RF Multiplexer-based Antenna Interface," in *Software Defined Radio (SDR) '08, Washington DC*, 2008. [Online]. Available: http://www.ece.vt.edu/swe/mypubs/Hasan_VT_SDR08_Final.pdf

[26] C. Bostian and A. Young, "Cognitive Radio: A Practical Review for the Radio Science Community," *URSI Radio Science Bulletin*, no. 342, pp. 16–26, September 2012.

[27] (2012) Arduino – HomePage. Arduino. [Online]. Available: http://www.arduino.cc

[28] (2012) Digilent Inc. – Digital Design Engineer's Source. Digilent Inc. [Online]. Available: http://www.digilentinc.com/Products/Catalog.cfm?NavPath=2,892&Cat=18

[29] (2012) Raspberry Pi – An ARM GNU/Linux box for $25. Take a Byte! [Online]. Available: http://www.raspberrypi.org

[30] (2014) BeagleBoard.org – Black. BeagleBoard.org. [Online]. Available: http://beagleboard.org/black

[31] (2012) Gumstix Overor COMS Open Source Products. Gumstix, Inc. [Online]. Available: https://gumstix.com/store/index.php?cPath=33

[32] (2014) Gumstix DuoVero COMS. Gumstix, Inc. [Online]. Available: https://store.gumstix.com/index.php/category/43/

[33] Nuand. bladeRF – the USB 3.0 Software Defined Radio. Retrieved January 7, 2016. [Online]. Available: http://www.nuand.com

[34] J. Keller. (2012, August) DARPA RF-FPGA Program Awards Six Contracts to Develop Programmable RF Front-ends. Retrieved January 7, 2016. [Online]. Available: http://www.militaryaerospace.com/articles/2012/08/darpa-rf-fpga.html

[35] R. Wender. (2015) RF-FPGA Early Access Program Announced by Triad Semiconductor. Retrieved January 7, 2016. [Online]. Available: http://www.triadsemi.com/2012/10/24/rf-fpga/early-access-program-announced-by-triad-semiconductor/

[36] Altera. (2013) Arria V FPGA RF Development Kit. Retrieved January 7, 2016. [Online]. Available: http://www.altera.com/products/boards_and_kits/dev-kits/altera/kit-arria-v-rf.html

[37] The GNU Radio Website. [Online]. Available: http://gnuradio.org/redmine/projects/gnuradio

[38] (2014, Mar) WBX Performance Data. Ettus Research. Retrieved October 23, 2014. [Online]. Available: http://files.ettus.com/performance_data/wbx/wbx_imd_and_nf_vs_gain.pdf

[39] S. Johnston. (2011, October) Software Defined Radio Hardware Survey. NEWSDR. [Online]. Available: http://people.bu.edu/mrahaim/NEWSDR/Presentations/NEWSDR_Johnston.pdf

[40] WARP v3 User Guide. WARP Project. Retrieved October 28, 2014. [Online]. Available: http://warpproject.org/trac/wiki/HardwareUsersGuides/FMC']RF']2X245

[41] FMC-RF-2X245. WARP Project. [Online]. Available: http://warpproject.org/trac/wiki/Hardware UsersGuides/FMC-RF-2X245

CHAPTER 4

Cognitive Radio Computation and Computational Platforms

4.1 The Role of Computing and Cognitive Radio Architecture

When software is included in radio development it becomes important to incorporate computational considerations into the design process. Radio performance and quality become affected by considerations like computational devices, programming languages, and memory interfaces. This chapter covers some of the computer architecture principles needed to understand the differences between the main computational devices available and be able to appreciate their strengths and weaknesses. In addition to computing hardware, it is important to understand the basic computational models that can be used to design and structure software for cognitive- and software-defined radio. The computational models discussion is longer than the computer architecture portion of the chapter since it is not typically found in Software Defined Radio (SDR)/Cognitive Radio (CR) literature. This chapter also presents some basic theoretical background material in computing to provide a greater understanding of the computing hardware available and the different paradigms of programming applications.

4.2 Control Flow and Data Flow Computer Architectures

Computing deals with processing data according to certain algorithms and generating associated outputs. Computing architectures fall into two main classes: control flow and data flow computing. In control flow architectures, processing is driven by control sequences, *if-else, looping, etc.*, used in computing devices like microprocessors (General Purpose Processors (GPPs)). In data flow computing, processing is driven by availability of input data, as is the case with Field-programmable Gate Arrays (FPGAs).

4.2.1 Control Flow Computing

Microprocessors, or GPPs, and Digital Signal Processors (DSPs) are examples of control- flow-based devices. To illustrate the concept, Figure 4.1 shows the architecture of a computer with a Princeton, or Von Neumann, memory architecture. It incorporates a centralized controller which can read program instructions by passing the addresses of where programs and data lie in memory. To further illustrate this point, imagine we want to run a program that adds

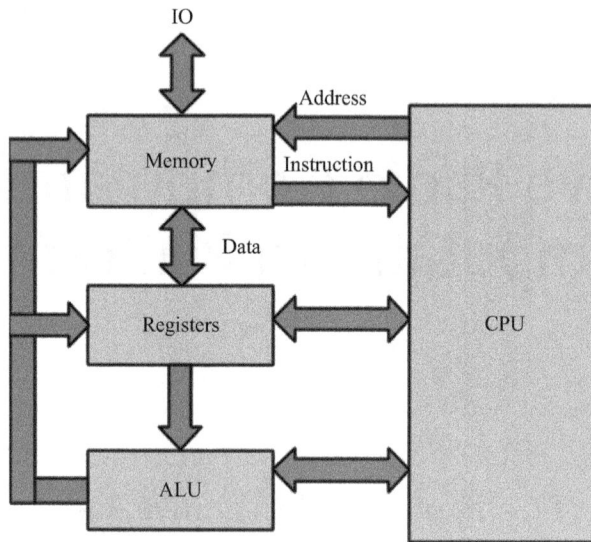

Figure 4.1   Von Neumann architecture, from https://www.utwente.nl/ub/en/services/MAIN/
copyright/cr/thesis/ [Accessed 2008] [1].

two numbers, $A + B = C$, where $A = 1$ and $B = 2$. The data and instructions to execute the
program would lie in memory as shown in Figure in 4.2. By accessing memory, the controller is
able to obtain the individual instructions in addition to accessing the associated data needed to
perform computation. The computer is able to access data through input devices, e.g., keyboard,
microphone etc, and is able to output results of its processing through outputs, e.g., monitor,
printer etc. The computer needs to load information for processing from external memory to
memory locations on-board its CPU known as registers. The Arithmetic Logic Unit (ALU) is
used to compute arithmetic operations on data and is able to save the results back to either
regular memory or the registers.

4.2.1.1 Control and Data Paths

The ALU is also used to compute the memory address where the subsequent program instructions lie. If a program needs to add two numbers and write the output to memory, it needs to
do the following:

1. Fetch the addition instruction from memory.
2. Fetch the first operand.
3. Fetch the second operand.
4. Calculate the addition result from both operands.
5. Save the addition result back to memory.

As we can see, the control and data information share the same common path, meaning in
this view the processor is able to access either data or instructions, but not both at the same
time. This use of a single path from the controller to memory for instructions and data creates a
bottleneck. To mitigate the *Von Neumann bottleneck*, the *Harvard* architecture provides separate

Memory Address	Segment	Memory Content
	Data	
1		A = 1
2		B = 2
3		C
	Program	
4		Read address 1
5		Read address 2
6		Add
7		Store result in address 3

Figure 4.2   Program and data contents in memory.

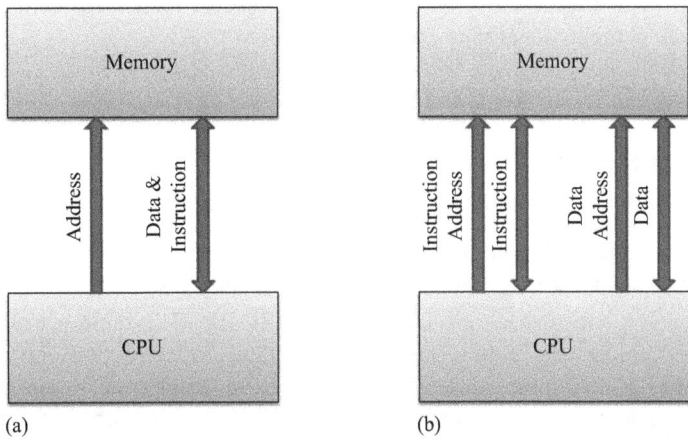

Figure 4.3   The Von Neumann memory architecture (a) and Harvard memory architecture (b).

paths for data and instructions. This way it becomes possible to access both at the same time. See Figure 4.3.

4.2.1.2 Memory Hierarchy

Memory accesses take time, and therefore reducing memory accesses and/ or using them more effectively can greatly impact overall computational performance. Memory hierarchy is used to mitigate memory access bottlenecks and to improve processor execution times. It aims to do this by placing the fastest (more expensive) memory components closest to the processor and the slowest (least expensive memory components) farther from the processor as shown in Figure 4.4. Data that will be accessed by the processor frequently should be in fast close-by memory, e.g., *cache memory*, and data that will be accessed infrequently can be placed farther away from the processor.

By doing so, it becomes possible to make use of the *spatial* and *temporal* locality of data. Spatial locality means that if, for example, a program is computing on a few elements of

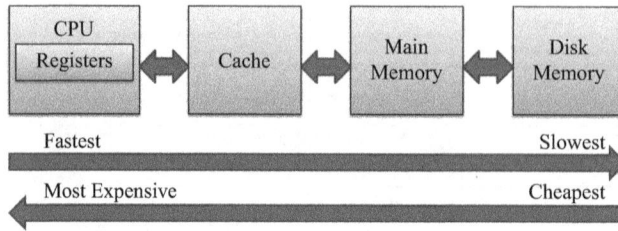

Figure 4.4   Memory hierarchy.

a matrix, then it will probably need to access subsequent elements from the same matrix; therefore, placing these elements in fast memory can improve overall computation time. Temporal locality means if the processor just accessed data elements, then it might need to access them again, e.g., in matrix multiplication. By making use of locality in terms of data and instruction overall computational time can be improved.

For a more detailed discussion of processor architecture design and principles, interested readers should refer to References 2 and 3.

4.2.1.3 Multiprocessors

Multiprocessor architectures can be broken down to four categories [2]:

- Single Instruction stream Single Data stream (SISD): A uniprocessor, where a single instruction is executed at a time and single data point is being processed at a time.
- Single Instruction stream Multiple Data streams (SIMD): The same instruction can be executed on multiple data points. In this architecture, an instruction only needs to be retrieved once and it can be executed on multiple data. In case of a loop in a program where a mathematical operation needs to be executed to a vector with no data dependency, e.g., multiplying each element by a constant, this architecture allows an instruction to be retrieved once and executed on multiple data points. This way a processor is able to forgo retrieving the multiplication operation multiple times. In this architecture, multiple processors are present with their own data memory and they share the same instruction memory.
- Multiple Instruction streams Multiple Data streams (MIMD): Different instructions are able to execute on different data at the same time. This is realized by having different processors that have their own data and instruction memories. In these processors, it is necessary to synchronize execution and data synchronization when necessary since each processor can operate independently. Also these processors are able to support true concurrency for multithreaded applications since different threads can execute on different processors.
- Multiple Instruction streams Single Data stream (MISD): In this architecture, multiple instructions would be realized on a single data stream, e.g., processing of network packets.

4.2.2 Data Flow Computing

While presenting control flow computing, we discussed the memory access overhead of the Von Neumann architecture: memory contains both data and control information and the processor is in constant need of accessing it. While there are techniques to alleviate the overhead of memory accesses, it is still a fundamental aspect of the architecture. Data flow is an alternative

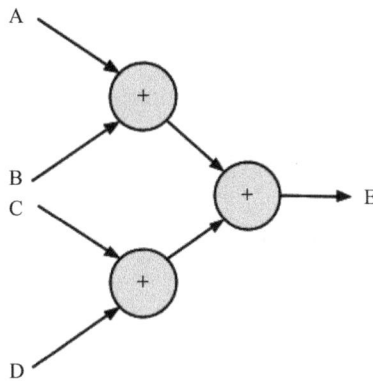

Figure 4.5   Data flow example for $A + B + C + D = E$.

architecture, where instead of being saved in memory the program content is laid out as a path of execution with different operators being executed along the path [4, 5]. Figure 4.5 shows an example of how $A + B + C + D = E$ can be expressed in a data flow context, the path from the inputs to the output contains the effective computational operators needed to execute the program. The operations do not need to be retrieved from memory as they are present in the execution path, and data do not need to be saved in a central memory location as they are copied over from one processing stage to another. Therefore, data flow computers are driven by the availability of data instead of by a central processor. Thus, data flow computing processes data by providing an effective path which defines the necessary computation. Computation is triggered by the availability of input data and data traverses the effective path of computation instead of depending on the reading of instructions necessary to execute programs.

4.3 Overview of Computational Devices (GPP, DSP, FPGA)

In this section, we describe common computational devices and some of their similarities and differences.

4.3.1 Digital Signal Processors

DSPs are control-based computational devices and are typically used in streaming applications requiring signal processing, e.g., audio, video, and image processing. Therefore, they are intended for specific purposes requiring fast Multiply Accumulate (MAC) operations, and they are available in both fixed point and floating-point processors.[1] Typically, they are programmed using the language C or assembly in order to fine tune application performance.

---

[1]In floating point processing, numbers are represented using a mantissa and exponent and a floating-point unit facilitates such computation without user involvement. In fixed-point processing, numbers are represented as integers with fixed bits, e.g., 8/16/32/64 bit, and users would need to scale numbers accordingly, e.g., $0.5 = 255$. Users would need to account for precision in applications; e.g., if range of valid floating point values is [0, 1] versus [0, 100] the scaling needs to be different to mitigate potential overflow and underflow occurrences.

The ability to fine tune applications is important to ensure the proper use of the underlying DSP architecture. Programmers can, for example:

- Ensure that all the MAC units in a processor containing multiple units are being effectively used.
- Have their programs make use of available *shifters* in the processor to perform scaling operations.
- Make sure that accumulation operations are using available accumulator circuits in the processor.

DSPs also have specialized addressing schemes catering to DSP applications, such as in Reference 6:

- Modulo addressing, which allows circular buffers by resetting the buffer pointer once it reaches the last element.
- Reversed bit addressing, used in FFT operations to map between the indices of the input elements and the output results.

4.3.2 General Purpose Processors

A variety of programming languages can be used to target GPPs, including low-level assembly languages, Object-oriented Programming (OOP) languages, and functional languages. GPPs are able to address a wide range of applications, and, while they can be programmed on the same detailed low-level approach as DSPs, they don't necessarily have to be. From an application developer perspective, GPPs have a larger support base, in terms of Operating System, development tools, programming languages, and open-source software which programmers can continue building on, e.g. GNU Radio and OSSIE in the SDR case. While embedded GPPs, e.g., ARM and PowerPC, processors don't have the same level of mature tool support as their laptop/desktop counterparts, x86 and x64 the 32-bit and 64-bit Intel and AMD processors, support has been growing especially in terms of open-source software availability and use. The OpenEmbedded (OE) [7] and Yocto [8] projects allow users to build kernels and file systems for supported embedded targets, meaning that users can build custom operating systems and add to them open-source projects like GNU Radio and OSSIE. The availability of an OS allows users to make use of existing hardware drivers and reuse compatible software. Also by leveraging available software, e.g., GNU Radio, developers are able to use existing functionality as a starting point and only add new functionality as deemed necessary thus accelerating application development. However, application performance can be encumbered by the use of OS's, since extra applications would need to run on the processor and thus compete for computing resources.

4.3.3 Field-programmable Gate Arrays

FPGAs are data flow computational devices. They provide a combination of digital logic components, such as: Look-up Tables (LUTs), registers, boolean logic gates, and memory units. Programmers are able to configure the basic components available on FPGAs to perform the necessary computations in their applications. Some FPGAs even contain MACs and accumulators for use in DSP operations, as in the Xilinx [9] Virtex and Altera Stratix [10] FPGAs. Therefore, FPGAs provide basic mathematical functionality, but if developers need to apply complex multiplication, trigonometric operations, FIR filters, etc., they must either implement those functionality from scratch using the logic available in the FPGA or purchase Intellectual Property (IP) blocks, which are third-party implementations of logic available directly from FPGA vendors [11, 12], third-party companies, or open-source projects, like OpenCores [13]. In addition, Verilog and VHDL language-based code can be generated via Mathwork's Simulink using the HDL coder package [14].

Some FPGAs provide GPP cores, since they are more suited for control-based tasks, to complement their functionality, either as physical (hardcore) processors, such as with the Xilinx Zynq-7000 [15] series which contains ARM-based processors and the Altera Cyclone V SoC series [16]. In addition, it is possible to utilize *softcore* processors, which are implementations of processors that can be downloaded to FPGAs that do not have a hardcore processor. Examples are the Xilinx Microblaze [17] and the Altera NIOS II [18] processors. FPGAs can be programmed using Hardware Description Languages (HDLs), e.g., VHDL and Verilog or can be programmed using software suites such as Mathwork's Simulink [19] and National Instrument's LabVIEW [20].

In GPPs and DSPs, it is possible to create new objects on the fly. These are basically, objects and buffers of varying sizes to accommodate application needs. However, in FPGAs circuit designs are usually more static, meaning that new resources are not typically allocated during system runtime. That is because the process of translating HDL into FPGA circuitry is done offline and FPGAs do not contain a hardware or software layer allowing for the dynamic creation of resources during runtime. FPGA partial reconfiguration is a technique allowing designers to swap resources in and out from the FPGA, which can be used to accommodate application needs during system executions thus allowing some dynamic change of FPGA behavior during runtime, but it does not allow for full-dynamic behavior change in the same sense as is available with DSPs and GPPs. Examples of using this technique can be found in References 21–23.

4.3.4 Alternative Computational Devices

4.3.4.1 Application-specific Integrated Circuits

Application-specific Integrated Circuits (ASICs) are specialized chips or circuits custom designed for application-specific tasks [24]. ASICs are intended to provide better performance and utilize less power than their GPP, DSP, and FPGA counterparts considering that they provide dedicated hardware circuits that are fine tuned to perform certain tasks. However, the Non-recurring Engineering (NRE) cost from designing and fabricating such chips can be prohibitive to justify creating an ASIC for some application, especially given that they are typically not very flexible in terms of expanding or modifying their functionality since it might require modifying the underlying chip design and fabricating new varieties of the chips. While in some high volume applications, the NRE cost might be justified. In low volume scenarios, an FPGA might be provide the hardware speed up factor and flexibility to replace the need for an ASIC. However, when observing other factors such as power consumption and cost, an FPGA might not be the right computation device and at this point the application designer should perform a more thorough analysis of what computation devices would work best for the given application.

4.3.5 Computational Heterogeneity

In some situations, a single computational device might be sufficient. However, considering the processing and power needs of SDR applications, ranging from base stations to mobile

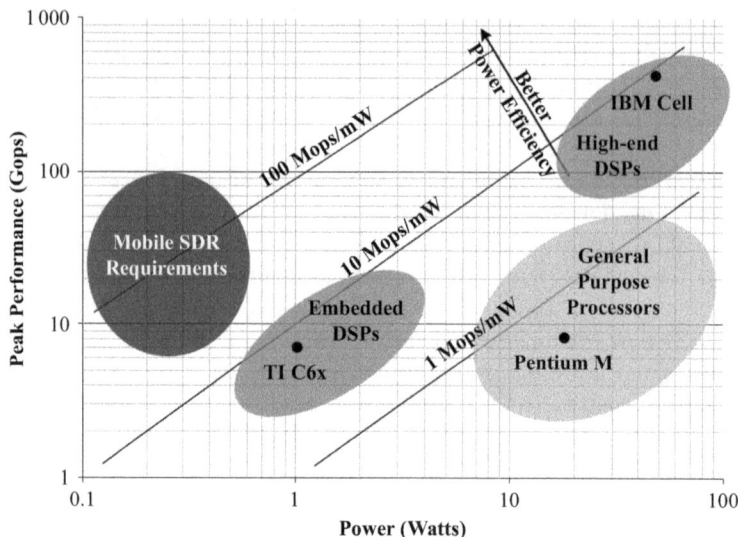

Figure 4.6   Peak performance versus power. ©2006 IEEE. Reprinted, with permission, from Reference 25.

handsets, it might be necessary to combine various computational devices in a single platform. In one situation a GPP and FPGA combination might be appropriate, and in another situation, a GPP and DSP combination might be a better choice. Looking at Figure 4.6, we can see the peak performance of various computing devices versus their power consumption in addition to the performance needed for mobile SDR in Reference 25.[2]

In approaching programming in a heterogeneous environment, it is important to develop simulations of the desired system. This provides a frame of reference to compare the actual system behavior with its desired one. This is especially important in fixed-point arithmetic scenarios, in FPGAs and some DSPs, where the simulations can enable system development by observing and analyzing quantization effects. It is also possible to use software, such as Simulink [19] from Mathworks, to generate fixed-point designs from floating-point ones.

Latency associated with memory transfers can become a system bottleneck. While offloading data processing from one processor to another can speed up computation, the overhead associated with the memory transfer must not cause delays exceeding the processing time saved.

The next chapter offers further discussion on processor choices and how heterogeneous computing environments can be programmed.

4.4 Models of Computation

Models of Computations (MoCs) provide formal semantics for computing; by applying these semantics it is possible to make determinations like deadlock detection, bounded memory

---

[2]The figure is from the paper's conference slides.

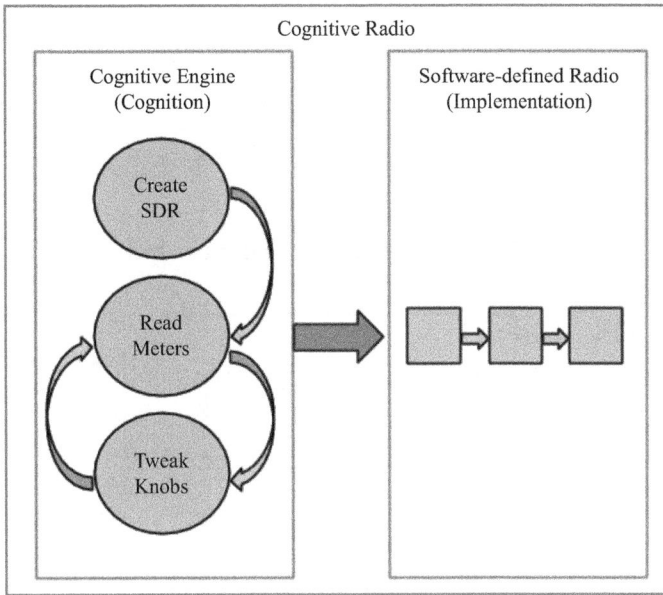

Figure 4.7   Implementation domains of CE and SDR.

execution, and finite periodic schedule execution. Therefore, exploring MoCs for SDR and Cognitive Engine (CE) applications is in essence a process for exploring the semantics of constructing such radios using computers.

Earlier in the chapter, we presented an overview of computational devices. Here, we focus on the software design aspects of radio design. As we discuss more thoroughly in the next section, CEs and SDRs provide different views of the same radio. The next section introduces this difference by presenting reactive and real-time systems, discussing the relevant computational models for each system type.

4.4.1 Reactive and Real-time Systems

CEs monitor, modify, and adapt radio configuration depending on their environment and user needs. They are reactive in nature and represent how individual radio components are stitched together in CRs. The CE represents the layer of intelligence which controls the overall radio definition encompassed in the SDR system as shown in Figure 4.7. However, SDRs embody the real-time component of CRs since they represent the flow of data in the system.

Reactive representations view how a system is interacting with its environment [26]. A car and its driver can represent a reactive system, when the driver approaches a red traffic light, the car must be stopped. The notion of stopping the car in this context is reactive, as can be seen in Figure 4.8, because it defines how the overall car/driver system must react to its environment but does not define the coordinated mechanical motions necessary phys-ically to stop the car. However, the same system can be represented as a real-time system; then it represents the physical act of stopping the car and involves the driver coordinating the necessary motions of braking and stopping the car in a timely fashion, as shown in Fig-ure 4.9. In a CR, the CE represents the reactive component of the radio, where it represents

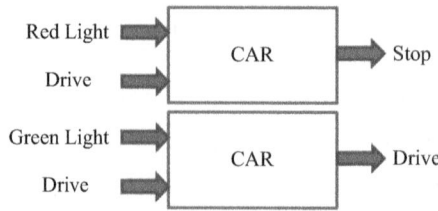

Figure 4.8   Car example reactive domain interplay.

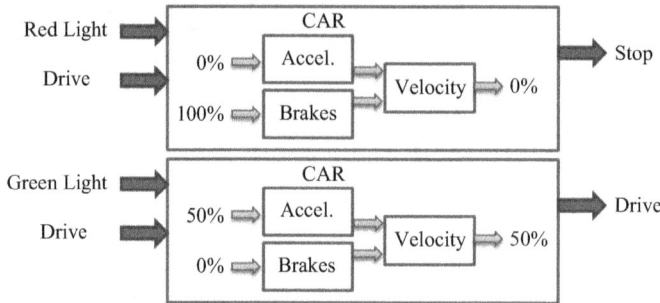

Figure 4.9   Car example real-time domain interplay.

the overall structure of the radio without the detailed implementation specifics. An SDR is then the real-time description of the system, representing the implementation specifics and the time-dependent information. Please note that the term *real-time* here refers to the radio's ability to satisfy the signal-processing sampling rates from a computational perspective, versus the traditional definition where tasks must execute before the elapse of pre-defined *deadlines* [27]. Figure 4.10 is an expanded illustration of a CR, where a CE is the reactive space and an SDR real-time space of CR operation. Therefore, it is necessary to explore and understand both MoC domains, which will include two main model classes, Process Algebra and Data flow MoCs. The authors explore the mapping between radio implementations from the SDR and CE realm in greater detail in Reference 28.

4.4.2 Data Flow Models of Computation

We will begin by explaining some of the basic data flow terminology and then present some of the available data flow models. The following introduction and the Synchronous Data flow model should be sufficient for most readers to understand the real-time SDR perspective, but interested readers are encouraged to read through the other models.

A graph, *G*, is defined as "a triple consisting of a vertex set, *V(G)*, an edge set, *E(G)*, and a relation that associates with each edge two vertices called endpoints" [29]. Data flow graphs are directed graphs, meaning they have unidirectional paths. Nodes, or actors, represent functions and data paths are represented by arcs [30]. Data flow class of MoCs is *data driven*, meaning that computation can proceed as soon as input data are ready, in contrast to being *control driven*, where computation is performed as soon as instructions are available [31]. Data flow programs are composed of *nodes* or *actors*, representing computational functions, connected by *arcs*, representing the First In First Out (FIFO) buffers connecting the nodes [30], and *tokens*

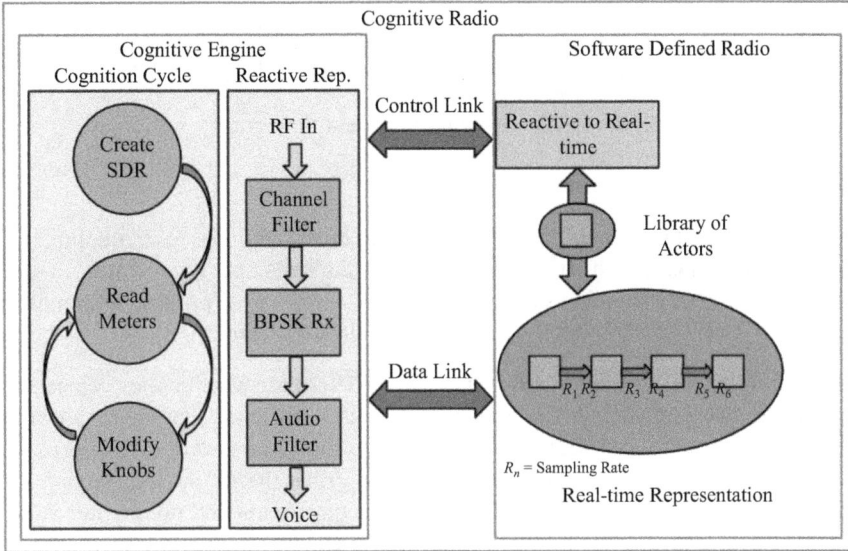

Figure 4.10   Proposed domain definition for CE and SDR.

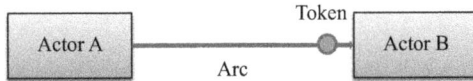

Figure 4.11   Dataflow graph.

represent the data transferred over the arcs. Nodes *fire*, or execute, once they have input data available for processing. Figure 4.11 illustrates this terminology. The data flow MoC essentially realizes the execution of data flow computing paradigm which can be executed using either control-flow-based processors, such as GPPs and DSPs, or data-flow-based devices, such as FPGAs.

4.4.2.1 Kahn Process Networks

In the Kahn Process Networks (KPN) model nodes, or *processes*, communicate using FIFO buffers, or channels, with an infinite theoretical capacity. Processes write to queues asynchronously (non-blocking writes), while reading from channels is performed synchronously (blocking reads). The KPN model is determinate because the writes are asynchronous and the reads are synchronous as discussed in Reference 32. There are no global variables in KPN; therefore, data exchanges between processes must be communicated over connected processes.

4.4.2.2 Synchronous Data Flow

Synchronous Data Flow (SDF) is a special case of KPN, where nodes consume and produce a fixed number of tokens [30]. In Figure 4.12, integers to the right of a node represent the node's *production* rate, the number of tokens generated after the node executes, while integers to the left are node *consumption* rates, the number of tokens consumed once nodes execute.

Figure 4.12   Synchronous data flow graph.

Consumption and production rates do not need to be equal; if a node's consumption is higher than its production rate, then it simply consumes more tokens than it generates. Rates in the SDF model are static, which allows bounded memory execution of graphs, determinism, and deadlock detection [30] as will be discussed in the following section.

Topology Matrix: SDF graphs can be analyzed to determine whether they can be executed in bounded memory and whether a *deadlock* would occur, meaning that the graph would not be able to run because of insufficient input tokens to the actor. A *topology matrix* must be constructed and analyzed to allow such analysis. It describes the arc interconnection between nodes in an SDF graph and describes their relative consumption and production rates; rows correspond to arcs and columns correspond to nodes. The (i, j) topology matrix entry corresponds to node j's relative rate on arc i. A production rate is a positive entry in the matrix and a consumption rate is a negative entry. Looking at the graph in Figure 4.12, the corresponding topology matrix is shown in the following:

$$\Gamma = \begin{bmatrix} 1 & -1 & 0 \\ 0 & 2 & -2 \end{bmatrix} \tag{4.1}$$

Arc 1 corresponds to the first row in the matrix $(0, 0)$ entry, $+1$, is the positive output rate from Actor 1 to Actor 2. The $(0, 1)$ entry, -1, is the negative input rate of Actor 2. The $(0, 2)$ entry is 0 since Arc 1 is not connected to Actor 3. The second row in the topology matrix corresponds to Arc 2, the $(1, 0)$ entry is 0 since Arc 2 is not connected to Actor 1, the $(1, 1)$ entry is $+2$ since it represents the output rate of Actor 2, and the $(1, 2)$ entry is -2 since it is the input rate of Actor 2. The topology matrix is important because it describes the change in buffers between firings.

Schedule Construction: A scheduler is needed to manage the execution of actors and allocate their necessary communication buffers. In SDF, the following is assumed about their execution as described in Reference 33:

- SDF graphs do not terminate, meaning that they should never deadlock.
- If SDF graphs are not connected then they are separate graphs, meaning that there is no execution dependency between those graphs.

A scheduler that manages the execution of SDF graphs must be a Periodic Admissible Parallel Scheduler (PAPS). *Periodic*, because SDF graph execution are non-terminating and cyclic; *admissible*, because actors cannot execute until their input data is available; *parallel* because actors can execute in parallel and should have resources to accommodate such execution. If SDF actors execution will be performed sequentially, then its execution schedule is Periodic Admissible Sequential Schedule (PASS).

The rank of the topology matrix indicates whether the data rates of the graph are *inconsistent*, meaning that the production/consumption actor combination will lead to a deadlock or unbounded connecting buffers. A necessary condition for a schedule to exist is described in Reference 33 where the following equation must hold:

$$rank(\Gamma) = s - 1 \tag{4.2}$$

where s is the number of blocks in the graph

The *firing* or execution sequence of nodes is described by the *firing vector* $v(n)$; each row in the vector indicates whether a node is firing at time n. SDF does not provide a global notion of time, therefore *time n* does refer to relative time from the beginning of execution and not an absolute time. In (4.3), $v(0)$ corresponds to the nodes firing at time $n = 0$. Therefore, node 1 is firing at time 0 since the first row is 1, while nodes 2 and 3 are not firing since the second and third rows are 0. Also at time 1, node 2 is the only firing node and at time 2, node 3 is the only one firing node.

$$v(0) = \begin{bmatrix} 1 \\ 0 \\ 0 \end{bmatrix}, \quad v(1) = \begin{bmatrix} 0 \\ 1 \\ 0 \end{bmatrix}, \quad v(2) = \begin{bmatrix} 0 \\ 0 \\ 1 \end{bmatrix} \tag{4.3}$$

The vector $b(n)$ describes the FIFO buffer sizes at time n. In (4.4), the $b(n+1)$ describes the new buffer requirements between the executions of the graph. Γ is the topology matrix, and $v(n)$ is the firing vector at time n.

$$b(n + 1) = b(n) + \Gamma \times v(n) \tag{4.4}$$

In SDF, graphs must execute with bounded FIFO buffers. Therefore, the buffer sizes must be finite for a graph execution period. Instead of looking at individual firings at each n, we need to look at total node firing during a period, p, instead. The repetition vector, q, is a vector detailing how many times each block should fire during the course of a flowgraph period. Equation (4.5) defines the repetition vectors by looking at the firing behavior for all blocks in a graph during the course of a single cyclic execution period p. Therefore, the q vector is the summation of all the firing vectors in a graph over a period n.

$$q = \sum_{n=0}^{p-1} v(n) \tag{4.5}$$

We can rewrite (4.4) as (4.6) since the schedule is periodic as described in Reference 33.

$$b(np) = b(0) + n\Gamma \times q \tag{4.6}$$

To ensure that SDF graphs execute with bounded buffer sizes, the buffer sizes need to remain the same between executions [33]. This implies that we want $b(np) = b(0)$ implying the condition in Equation 4.7, where **0** is a vector of all 0.

$$\Gamma \times q = \mathbf{0} \tag{4.7}$$

By finding the non-trivial vector q in the nullspace of the topology matrix, Γ, we are able to find the number of firings necessary for each actor in order for the graph to execute with

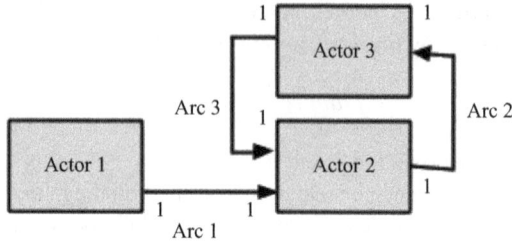

Figure 4.13   SDF graph with deadlock.

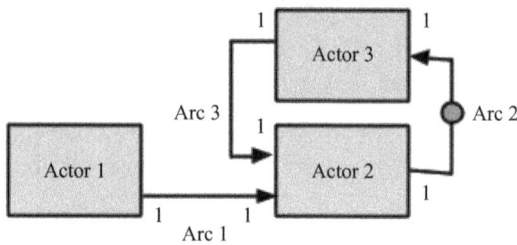

Figure 4.14   SDF graph with no deadlock.

bounded buffer sizes. A sufficient condition for the existence of a PASS is given in Equation 4.8, where $\mathbf{1}^T$ is a row vector of all 1's used to sum the entries of the q vector.

$$p = \mathbf{1}^T q \qquad (4.8)$$

If the scheduler detects that it cannot execute any actor before executing the actors the number of times specified in the firing vector q, then the SDF graph is deadlocked and no PASS exists [33]. Figure 4.13 shows an example of a deadlocked graph, where the graph can be fixed by adding an initial token on Arc 2 as shown in Figure 4.14, adding the token initializes the input value for Actor 3 allowing it to fire.

Arc 3 in Figure 4.15 is an example of a situation where tokens can accumulate and require an infinite buffer. This is because Actor 3 consumes one token from Arc 2 and one token from Arc 3 after each firing, but Actor 1 produces two tokens on Arc 3 after each firing, eventually causing the infinite buffer. The flowgraph can be fixed by changing the output rate for Actor 2, as shown in Figure 4.16. Now Actor 3 will have two input tokens, on Arcs 2 and 3, when if fires; it will fire twice to consume all the input tokens; Arcs 2 and 3 return to their initial state by not having outstanding tokens in any of its Arcs.

4.4.2.3 Boolean and Integer Data Flow

Boolean-controlled Data flow (BDF) [34] extends the SDF MoC by adding boolean controlled actors, namely *switch* and *select* actors that are configured via *control ports*. These ports accept boolean inputs, either *TRUE* (T) or *FALSE* (F), to select between two possible inputs for *select* actors and between two possible outputs for *switch* actors. An example of a BDF graph is shown in Figure 4.17, where Actor 2 can be used to detect a packet header and Actor 3 can

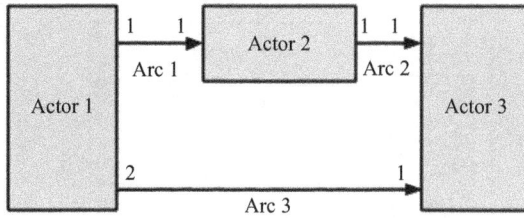

Figure 4.15   Inconsistent SDF graph.

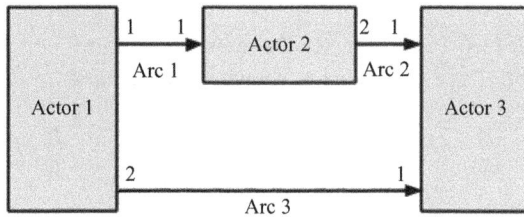

Figure 4.16   Consistent SDF graph.

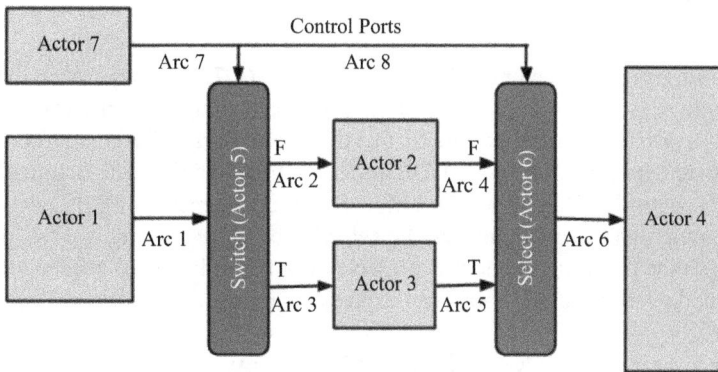

Figure 4.17   BDF graph example.

be used to process the body of a packet. The switch actor passes the tokens to the appropriate actor for processing and the select actor routes the data after processing.

It is possible to construct an SDF graph to perform the same functionality in Figure 4.17; but, since SDF does not support selective execution, when Actor 2 is processing a packet header Actor 5 needs to generate dummy data for Actor 3 to process in order to comply with SDF semantics. The SDF example is given in Figure 4.18.

Production and consumption rates in BDF are either static, as in SDF, or two-valued functions dependent on control input of an actor. The two-valued function includes the statistical quantity p_i which represents the long-term proportion of TRUE tokens over the control port of

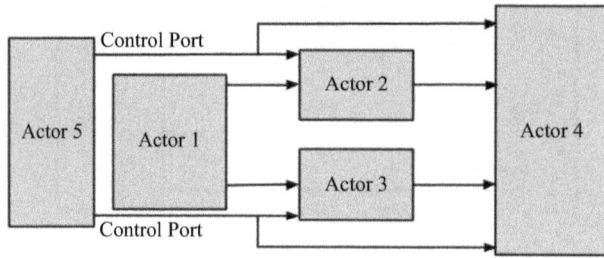

Figure 4.18   SDF graph control example.

a *switch* or *control* actor [34] and i indicates the referenced control port. The topology matrix for the graph in Figure 4.17 is constructed as shown in (4.9).

$$
\Gamma = \begin{bmatrix}
1 & 0 & 0 & 0 & -1 & 0 & 0 \\
0 & -1 & 0 & 0 & (1-p_1) & 0 & 0 \\
0 & 0 & -1 & 0 & p_1 & 0 & 0 \\
0 & 1 & 0 & 0 & 0 & -(1-p_2) & 0 \\
0 & 0 & 1 & 0 & 0 & -p_2 & 0 \\
0 & 0 & 0 & -1 & 0 & 1 & 0 \\
0 & 0 & 0 & 0 & -1 & 0 & 1 \\
0 & 0 & 0 & 0 & 0 & -1 & 1
\end{bmatrix}
\tag{4.9}
$$

In case a valid repetition vector can be calculated using (4.7), for any p_i value then the BDF graph is *strongly consistent*. But if a repetition vector exists only for some values p_i then the graph is *weakly consistent*. It is important to note that the authors of Reference 34 indicate that it is impossible to determine whether a BDF graph can be scheduled with bounded memory.

The Integer-controlled Data flow (IDF) MoC [35] is an extension of BDF allowing switch actors to have more than two outputs and the *select* actor to have more than two inputs. This extension allows the model to be more expressive such that it becomes possible to construct loops in this model where control loops can express the number of iterations.

4.4.2.4 Cyclo-static and Cyclo-dynamic Data Flow

The Cyclo-static Data flow (CSDF) [36] MoC allows applications to express conditional execution of actors by expressing *phases* of execution. Basically, actors can have different code modules, or phases, to be executed. Even though actors can have multiple phases, the final executed graph exhibits a periodic behavior. Figure 4.19 represents a CSDF with three phases, as indicated by the digit sequences, e.g., "0,1,1," where a 0 indicates the absence of tokens and a 1 indicates the existence of tokens. During the first phase, Actors 1, 3, and 4 fire, while in the second and third phases, Actors 1, 2, and 4 fire. As we can see the graph is periodic but different actors can execute during the various phases thus allowing conditional execution as in *BDF* but without the use of control ports and actors.

In this MoC, it is possible to determine graph consistency and bounded memory execution statically in the same manner as in SDF since graph periodicity is specified statically [36]. Note that in BDF the *proportion* of TRUE statements is observed while in CSDF the exact proportion of TRUE/FALSE statements is already defined over a period.

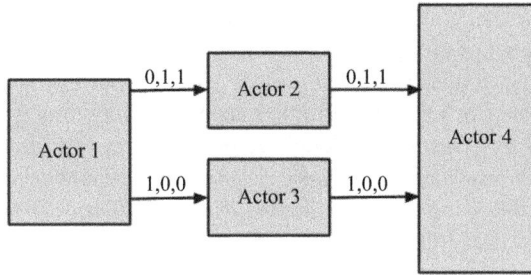

Figure 4.19   CSDF graph example.

Cyclo-dynamic Data Flow (CDDF) [37] is an extension of CSDF where the number actor phases and consumption/production rates are dynamic. The dynamic runtime behavior in this MoC can be impossible to prove graph consistency which is the same as with BDF and IDF.

4.4.2.5 Parametrized Synchronous Data Flow

Parametrized Synchronous Data flow (PSDF) [38] allows graph actors to be reconfigured during runtime. In Reference 38, the authors discuss the parameters as having associated domains, which are the set of all possible configurations for a particular parameter. Actors have an associated *maximum token transfer function* providing an upper bound on the maximum number of tokens they can produce which guarantees bounded memory execution [38]. Each arc also has an associated *maximum delay value* which determines the maximum number of delay tokens that can exists on an arc.

The notion of reconfiguration in PSDF graph uses *quiescent points* [39], where a graph can reconfigure upsampling and downsampling factors, essentially modifying the relative rates of nodes at particular points of graph execution. A quasi-static schedule, where token rates are described symbolically, can be evaluated once the graph reconfiguration parameters are defined during execution.

4.4.3 Process Algebra

The following section discusses Process Algebra. This is an appropriate class of MoCs to express a CE's reactive perspective of a radio description. We will discuss Communicating Sequential Processes is at length to illustrate the capabilities of this class of MoCs. While the other algebras mentioned in this section are not discussed in the same length, interested readers are encouraged to read further in the cited material.

Process Algebra can be defined as "the study of the behavior of parallel or distributed systems by algebraic means" [40]. Process behavior, composition, and input/output interfaces can be described by using the algebra. The benefit of relying on algebraic expressions in describing behavior is that systems lend themselves to verification [40]. From this perspective, process algebra provides semantics suitable for describing reactive systems and verifying them.

4.4.3.1 Communicating Sequential Processes

Communicating Sequential Processes (CSP) [41] is a process algebra which treats input/output operations as the basic primitives of computation. The following section introduce some of the basics of CSP; more details can be found in Reference 42.

Semantics: When defining processes, the process *alphabet*, αP, refers to the alphabet α of process P, which is the set of events used to describe the process. As an example, the alphabet of a vending machine, V, might be {*dollar, choose_drink1, choose_drink2, dispense_drink1, dispense_drink2*}. Users can interact with the machine by inputting a *dollar* and choosing either *drink1* or *drink2* by pressing the associated buttons. Pressing a button generates one of the following events: *choose_drink1* or *choose_drink2*. The machine then dispenses a drink by generating one of the following events *dispense_drink1* or *dispense_drink2*. The alphabet of the vending process, V, is as follows:

$$\alpha V = \{dollar, choose\_drink1, choose\_drink2, dispense\_drink1, dispense\_drink2\}$$

The process V is restricted in behavior to its alphabet; it is neither able to understand new events nor able to generate new response events.

An event, x, can be combined with a process P, using the prefix operator, $\rightarrow$, to generate a new process, for example:

$$x \rightarrow P \qquad\qquad (x \text{ then } P)$$

In the previous vending machine example, if a customer, C, keeps feeding a dollar, *dollar*, into the vending machine and keeps requesting *drink1* without picking up the drink and keeps doing so infinitely the prefix operator can describe this behavior recursively:

$$C = \mu X : \{dollar, choose\_drink1, dispense\_drink1\}$$
$$\cdot (dollar \rightarrow (choose\_drink1 \rightarrow (dispense\_drink1 \rightarrow C)))$$

For convenience, the alphabet from the process description will be dropped to simplify the expression as described in Reference 42. The new customer definition becomes:

$$C = dollar \rightarrow (choose\_drink1 \rightarrow (dispense\_drink1 \rightarrow C))$$

Choice: Customers are able to select drinks from the vending machine, so the process description for the vending machine needs to account for the customer choice. The choice operator, $|$, in $(x \rightarrow P | y \rightarrow Q)$, allows the choice between process P and process Q, depending on whether event x or event y occurs. In this context, the vending machine description becomes:

$$V = dollar \rightarrow ((choose\_drink1 \rightarrow dispense\_drink1 \rightarrow V)|$$
$$(choose\_drink2 \rightarrow dispense\_drink2 \rightarrow V))$$

Concurrency: For now, we have described the vending machine and customer as separate processes. But in a realistic situations, both processes interact with each other: a customer interacts with the vending machine to obtain the desired drink. The parallel composition of both processes can be described as:

$$P \| Q \qquad\qquad (P \text{ in parallel with } Q)$$

The parallel composition of processes can be synchronous, meaning that both processes must contain process definitions allowing the interaction of both processes. Some relevant concurrency laws are as follows:

$$(c \to P) \| (c \to Q) = (c \to (P \| Q))$$

$$(c \to P) \| (d \to Q) = STOP \qquad\qquad \text{if } c \neq d$$

STOP corresponds to stopping the process execution, as will be discussed later. The concurrency operator can be used to provide *choice*, or *interleave* capabilities to the concurrent process composition. The choice operator can be applied to the vending machine as follows:

$$X = (C \| V)$$

$$
\begin{aligned}
X = (&dollar \to choose\_drink1 \to dispense\_drink1 \to X) \| \\
     &(dollar \to ((choose\_drink1 \to dispense\_drink1 \to X) | \\
     &(choose\_drink2 \to dispense\_drink2 \to X)))
\end{aligned}
$$

$$
\begin{aligned}
X = (&dollar \to choose\_drink1 \to dispense\_drink1 \to X) \| \\
     &(dollar \to (choose\_drink1 \to dispense\_drink1 \to X) | \\
     &dollar \to (choose\_drink2 \to dispense\_drink2 \to X))
\end{aligned}
$$

$$
\begin{aligned}
X = (&dollar \to choose\_drink1 \to dispense\_drink1 \to X) \| \\
     &(dollar \to choose\_drink1 \to dispense\_drink1 \to X)
\end{aligned}
$$

$$X = (dollar \to choose\_drink1 \to dispense\_drink1 \to X)$$

In the previous example, we were able to describe the parallel processes C and V as a single processes X. This is possible because the two-process interaction can only yield as single behavior as derived in the example.

The choice operator can operate with two processes which do not share the same alphabet, where an *interleaving* of events takes place when events occur that are not shared between both processes. For example, let us assume a second vending machine, $V2$, which makes a *beeping* sound after each drink is selected. The associated process definition is as follows:

$$
\begin{aligned}
V2 = dollar \to (&(choose\_drink1 \to beep \to dispense\_drink1 \to V) | \\
                  &(choose\_drink2 \to beep \to dispense\_drink2 \to V))
\end{aligned}
$$

Let us also assume that a customer, *C2*, continuously chooses a drink from the vending machine, which has an infinite supply of drinks. Let us also assume that the customer *cheers* in anticipation of the drink being dispensed. The *C2* process definition is then:

$$C2 = (dollar \to choose\_drink1 \to cheer \to C2)$$

The concurrent composition of $V2$ and $C2$ is:

$$X = (C2 \| V2)$$

$$X = (dollar \to choose\_drink1 \to beep \to (dispense\_drink1 \to cheer \to X$$
$$| cheer \to dispense\_drink1 \to X))$$

Deadlock: A process with alphabet A which never interacts with any of the events in the alphabet A is defined as $STOP_A$. Looking back at our vending machine example and assuming that there is a customer, $C3$, who insists on choosing an unavailable drink, *drink3*, from the vending machine, and wants to do so an infinite number of times without picking up the drink. The customer process definition is:

$$C3 = (dollar \to choose\_drink3 \to dispense\_drink3 \to C3)$$

When the customer, $C3$, interacts with vending machine V, the following occurs:

$$(C3 \| V) = (dollar \to STOP)$$

The concurrent combination of both processes terminates in the $STOP$ process which indicates that a *deadlock* has occurred. The customer essentially tries to find a button to select *drink3*, but since the vending machine does not have an associated *choose_drink3* button execution halts and a deadlock occurs and there is no viable future interaction between both processes.

Communication: Communication between processes occurs over *channels*, where:

$$\alpha c(P) = \{ v | c.v \in \alpha P \}$$

represents the set of all messages, v, which process, P, can communicate over channel, c. A process, P, accepts input v over channel c and then goes back to behaving like P, is described as:

$$(c?v \to P) = P$$

A process outputting a value over channel c then behaves like P is described as:

$$(c!v \to P) = P$$

In CSP, communication channels are synchronous, meaning that they are blocking. A process receiving over a channel blocks until the associated process is ready to write over the channel. Also, a process writing over a channel blocks until a corresponding process is ready to receive over the same channel.

4.4.4 Calculus of Communicating Systems and Π-calculus

Calculus of Communicating Systems (CCS) [43] is an alternative process algebra worth noting. Many of the semantics discussed in this CSP section can be applied using CCS semantics, however a closer inspection of algebra semantics between both algebras would reveal differences not discussed in this chapter.

Another process algebra, Π-calculus [44], provides process mobility allowing dynamic reconfiguration between processes. From a CE perspective, Π-calculus can be used to describe runtime reconfigurations of radio flowgraphs.

4.5 Models-of-computation Use

In the next chapter, we will discuss an example application of MoCs in writing Cognitive Radio (CR) software, as discussed in Reference 28. The basic principle is that applying MoC to radio design makes it possible to provide new computational knobs and meters that a CE can continue to tweak and monitor, in addition to the other, existing, knobs and meters, all in order to create a radio configuration that meets its user's and operational needs.

Other examples of using MoCs to design radios in the literature are as follows. These do not discuss the CE-to-SDR mapping. In Reference 45, Berg *et al.* analyze the applicability of various models-of-computation for SDR applications. The authors discuss the following models: synchronous data flow, cyclo-stationary data flow, heterochronous data flow, cyclo-dynamic data flow, and parameterized synchronous data flow. They explore the issues and benefits of each model in the context of designing a Long Term Evolution (LTE) [46] receiver but do not discuss implementation. In Reference 47, Siyoum *et al.* discuss the use of *scenario aware dataflow* models for LTE systems. These models use Finite State Machines (FSMs) to define different computational *scenarios*, which are different operational modes that a radio can expect during its use [48] as for an LTE receiver. Each FSM state, or scenario, is then implemented using the Synchronous Data flow model. In Reference 49, Hsu *et al.* apply the mixed-mode vector-based data flow model to simulate the LTE physical layer. By using mixed static and dynamic data flow models, the authors are able to use static scheduling techniques used in static data flow models while making use of dynamic data flow models to express conditional aspects of the LTE physical layer. The authors present alternative models that can be used for implementing SDR systems which can complement the work presented in this dissertion.

In References 50 and 51, Zaki *et al.*, discuss implementing SDR flowgraphs using NVIDIA Graphic Processor Units (GPUs) [52]. GPUs have multiple processors that allows the parallel execution of algorithms. The authors developed GPU accelerated signal-processing functions in the GNU Radio software. In addition, the authors use an integer program to schedule signal processing functions on cores while taking into account communication times between processors, signal processing function execution times, and the number of processors in the GPU. In addition, the authors use the Data Flow Interchange Format (DIF) language [53], a language for defining data flow-based graphs. They also explore the effect of *vectorizing*, or scaling, of buffers on flowgraph performance.

4.6 Conclusion

In selecting computational hardware for SDR/CR applications, there is not necessarily a single platform and/or combination of computational devices that address an application. Having a particular platform that is readily available, familiarity with a particular set of computational devices, cost, time-to-market, etc. will influence a developer's choice of a platform. By understanding the underlying computational device architecture and the associated tradeoffs designers can better understand how to leverage and utilize the devices in their systems. The use of MoCs ultimately provides an appropriate framework to analyze, develop, and integrate application development. MoCs provide various design capabilities, guarantees, and programming complexity. In some applications, a combination of computational devices might be necessary, and a combination of computational models might also be necessary such as when bridging the reactive and real-time system perspective between CEs and SDRs.

Bibliography

[1] P. M. Heysters, "Coarse-grained Reconfigurable Processors – Flexibility Meets Efficiency," Ph.D. Dissertation, University of Twente, Enschede, Sep 2004.

[2] J. L. Hennessy and D. A. Patterson, *Computer Architecture, Fourth Edition: A Quantitative Approach*. San Francisco, CA: Morgan Kaufmann Publishers Inc., 2006.

[3] D. A. Patterson and J. L. Hennessy, *Computer Organization &Amp; Design: The Hardware/Software Interface*. San Francisco, CA: Morgan Kaufmann Publishers Inc., 2013.

[4] J. B. Dennis and D. P. Misunas, "A Preliminary Architecture for a Basic Data-flow Processor," in *Proceedings of the 2nd Annual Symposium on Computer Architecture*, ser. ISCA '75. New York, NY: ACM, 1975, pp. 126–132. doi: 10.1145/642089.642111

[5] I. Watson and J. Gurd, "A Practical Data Flow Computer," *Computer*, vol. 15, no. 2, 1982, pp. 51–57.

[6] P. Lapsley, J. Bier, E. A. Lee, and A. Shoham, *DSP Processor Fundamentals: Architectures and Features*, 1st ed. Wiley-IEEE Press, 1996. [Online]. Available: https://www.amazon.com/DSP-Processor-Fundamentals-Architectures-Features/dp/0780334051

[7] OpenEmbedded, "OpenEmbedded." [Online]. Available: http://www.openembedded.org/wiki/Main_Page

[8] Yocto, "Yocto Project." [Online]. Available: https://www.yoctoproject.org/

[9] Xilinx. [Online]. Available: http://www.xilinx.com/

[10] Altera. [Online]. Available: http://www.altera.com/

[11] Xilinx, "Xilinx Intellectual Property." [Online]. Available: http://www.xilinx.com/products/intellectual-property/

[12] Altera, "Altera Intellectual Property and Reference Designs." [Online]. Available: http://www.altera.com/products/ip/ipm-index.html

[13] OpenCores.org, "OpenCores." [Online]. Available: opencores.org

[14] Mathworks, "HDL Coder." [Online]. Available: http://www.mathworks.com/products/hdl-coder/

[15] Xilinx. [Online]. Available: http://www.xilinx.com/products/silicon-devices/soc/zynq-7000/index.htm

[16] Altera. [Online]. Available: http://www.altera.com/devices/fpga/cyclone-v-fpgas/hard-processor-system/cyv-soc-hps.html

[17] Xilinx. [Online]. Available: http://www.xilinx.com/tools/microblaze.htm

[18] Altera. [Online]. Available: http://www.altera.com/devices/processor/nios2/ni2-index.html

[19] Mathworks, "Simulink." [Online]. Available: http://www.mathworks.com/products/simulink/

[20] N. Instruments, "LabVIEW." [Online]. Available: http://www.ni.com/labview/

[21] P. Athanas, J. K. Bowen, T. G. Dunham, *et al.*, "Wires on Demand: Run-time Communication Synthesis for Reconfigurable Computing," Mar 2011, US Patent 7,902,866.

[22] D. Koch, *Partial Reconfiguration on FPGAs: Architectures, Tools and Applications*. Springer, 2012, vol. 153.

[23] S. Fahmy, J. Lotze, J. Noguera, L. Doyle, and R. Esser, "Generic Software Framework for Adaptive Applications on FPGAs," in *Field Programmable Custom Computing Machines, 2009. FCCM '09. 17th IEEE Symposium on*, April 2009, pp. 55–62.

[24] W.-K. Chen, *Memory, Microprocessor, and ASIC*. CRC Press, 2004. [Online]. Available: https://www.amazon.com/Memory-Microprocessor-ASIC-Applications-Engineering-ebook/dp/B000Q363FM

[25] Y. Lin, H. Lee, M. Woh, *et al.*, "SODA: A Low-power Architecture For Software Radio," in *Computer Architecture, 2006. ISCA '06. 33rd International Symposium on*, 2006, pp. 89–101.

[26] A. Benveniste and G. Berry, "The Synchronous Approach to Reactive and Real-time Systems," *Proceedings of the IEEE*, vol. 79, no. 9, 1991, pp. 1270–1282.

[27] C. L. Liu and J. W. Layland, "Scheduling Algorithms for Multiprogramming in a Hard-real-time Environment," *Journal of the ACM (JACM)*, vol. 20, no. 1, 1973, pp. 46–61.

[28] A. S. Fayez, "Design Space Decomposition for Cognitive and Software Defined Radios," Dissertation, Virginia Polytechnic Institute and State University, 2013.

[29] D. B. West, *Introduction to Graph Theory (2nd Edition)*. Prentice Hall, August 2000. [Online]. Available: http://www.amazon.ca/exec/obidos/redirect?tag=citeulike09-20&path=ASIN/0130144002

[30] E. A. Lee and D. G. Messerschmitt, "Synchronous Data Flow," *Proceedings of the IEEE*, vol. 75, no. 9, 1987, pp. 1235–1245.

[31] P. C. Treleaven, D. R. Brownbridge, and R. P. Hopkins, "Data-driven and Demand-driven Computer Architecture," *ACM Computing Surveys (CSUR)*, vol. 14, no. 1, 1982, pp. 93–143.

[32] G. Kahn and D. B. MacQueen, "Coroutines and Networks of Parallel Processes," in *IFIP Congress*, 1977, pp. 993–998.

[33] E. Lee and D. Messerschmitt, "Static Scheduling of Synchronous Data Flow Programs for Digital Signal Processing," *Computers, IEEE Transactions on*, vol. C-36, no. 1, 1987, pp. 24–35.

[34] J. T. Buck and E. A. Lee, "Scheduling Dynamic Dataflow Graphs with Bounded Memory using the Token Flow Model," in *Acoustics, Speech, and Signal Processing, 1993. ICASSP-93. 1993 IEEE International Conference on*, vol. 1. IEEE, 1993, pp. 429–432.

[35] J. T. Buck, "Static Scheduling and Code Generation from Dynamic Dataflow Graphs with Integer-valued Control Streams," in *Signals, Systems and Computers. Conference Record of the Twenty-Eighth Asilomar Conference on*, vol. 1. IEEE, 1994, pp. 508–513.

[36] M. Engels, G. Bilson, R. Lauwereins, and J. Peperstraete, "Cycle-Static Dataflow: Model and Implementation," in *Signals, Systems and Computers. Conference Record of the Twenty-Eighth Asilomar Conference on*, vol. 1. IEEE, 1994, pp. 503–507.

[37] P. Wauters, M. Engels, R. Lauwereins, and J. Peperstraete, "Cyclo-dynamic Dataflow," in *Parallel and Distributed Processing, 1996. PDP'96. Proceedings of the Fourth Euromicro Workshop on*. IEEE, 1996, pp. 319–326.

[38] B. Bhattacharya and S. S. Bhattacharyya, "Parameterized Dataflow Modeling for DSP Systems," *Signal Processing, IEEE Transactions on*, vol. 49, no. 10, 2001, pp. 2408–2421.

[39] S. Neuendorffer and E. Lee, "Hierarchical Reconfiguration of Dataflow Models," in *Formal Methods and Models for Co-Design, 2004. MEMOCODE '04. Proceedings. Second ACM and IEEE International Conference on*, 2004, pp. 179–188.

[40] J. C. Baeten, "A Brief History of Process Algebra," *Theoretical Computer Science*, vol. 335, no. 2, 2005, pp. 131–146.

[41] C. A. R. Hoare, "Communicating Sequential Processes," *Communications of the ACM*, vol. 21, no. 8, 1978, pp. 666–677.

[42] C. A. R. Hoare, *Communicating Sequential Processes*. Upper Saddle River, NJ: Prentice-Hall, Inc., 1985.

[43] R. Milner, *A Calculus of Communicating Systems*. New York, NY: Springer-Verlag, Inc., 1982.

[44] R. Milner, *Communicating and Mobile Systems The Pi Calculus*. Cambridge University Press, Cambridge, UK, 1999.

[45] H. Berg, C. Brunelli, and U. Lucking, "Analyzing Models of Computation for Software Defined Radio Applications," in *System-on-Chip, 2008. SOC 2008. International Symposium on*. IEEE, 2008, pp. 1–4.

[46] M.-S. David, M. Jose F, C.-P. Jorge, C. Daniel, G. Salvador, C. Narcís *et al.*, "On the Way Towards Fourth-generation Mobile: 3GPP LTE and LTE-advanced," *EURASIP Journal on Wireless Communications and Networking*, 2009.

[47] F. Siyoum, M. Geilen, O. Moreira, R. Nas, and H. Corporaal, "Analyzing Synchronous Dataflow Scenarios for Dynamic Software-defined Radio Applications," in *System on Chip (SoC), 2011 International Symposium on*. IEEE, 2011, pp. 14–21.

[48] B. D. Theelen, M. Geilen, T. Basten, J. Voeten, S. V. Gheorghita, and S. Stuijk, "A Scenario-aware Data Flow Model for Combined Long-run Average and Worst-case Performance Analysis," in *Formal Methods and Models for Co-Design, 2006. MEMOCODE'06. Proceedings. Fourth ACM and IEEE International Conference on*. IEEE, 2006, pp. 185–194.

[49] C.-J. Hsu, J. L. Pino, and F.-J. Hu, "A Mixed-mode Vector-based Dataflow Approach for Modeling and Simulating LTE Physical Layer," in *Design Automation Conference (DAC), 2010 47th ACM/IEEE*. IEEE, 2010, pp. 18–23.

[50] G. F. Zaki, W. Plishker, S. S. Bhattacharyya, C. Clancy, and J. Kuykendall, "Integration of Dataflow-Based Heterogeneous Multiprocessor Scheduling Techniques in GNU Radio," *Journal of Signal Processing Systems*, vol. 70, no. 2, 2013, pp. 177–191.

[51] G. Zaki, W. Plishker, T. Oshea, *et al.*, "Integration of Dataflow Optimization Techniques into a Software Radio Design Framework," in *Signals, Systems and Computers, 2009 Conference Record of the Forty-Third Asilomar Conference on*, 2009, pp. 243–247.

[52] NVIDIA, "CUDA GPUs." [Online]. Available: https://developer.nvidia.com/cuda-gpus

[53] C.-J. Hsu, M.-Y. Ko, and S. S. Bhattacharyya, "Software Synthesis from the Dataflow Interchange Format," in *Proceedings of the 2005 Workshop on Software and Compilers for Embedded Systems*, ser. SCOPES '05. New York, NY: ACM, 2005, pp. 37–49. doi: 10.1145/1140389.1140394

Integrating and Programming RF and Computational Platforms for Cognitive Radio

This chapter discusses Software Defined Radio (SDR) and radio frequency integrated circuit (RFIC) platform choices and their programming and system integration for Cognitive Radio (CR) applications. It describes platforms from computational and Radio Frequency (RF) perspectives, platform considerations when choosing a platform, how to approach programming such platforms, and presents some design examples.

5.1 SDR Platforms

A platform is defined as "A common hardware denominator that could be shared across multiple applications" [1]. There are three main types from which to choose for CR.

1. **RF Platform:** Includes the supporting hardware necessary for creating an SDR/CR. Such devices need to be coupled with a computational platform to allow the realization of a radio application.

2. **Computational Platform:** Includes a single or a combination of computational devices and need to be coupled with an appropriate programmable RF front-end to enable CR/SDR applications. The main consideration in integrating a computational platform with a configurable radio front-end is that the platform needs to possess an interface allowing it to communicate, control, and configure the front-end.

3. **Integrated SDR Platform:** Includes an integrated radio front-end with computational platform and do not require external supporting hardware.

RF platforms for SDR contain two major components, supporting hardware and an SDR-implementing software application, as shown in Figure 5.1.

- **Supporting Hardware**: The hardware needed for transmitting and receiving radio signals.
  - A programmable analog RF front-end with tunable RF parameters.
  - Analog-to-Digital Converter (ADC) to digitize the analog RF signals and a Digital-to-Analog Converter (DAC) to generate analog RF signals from digital data.

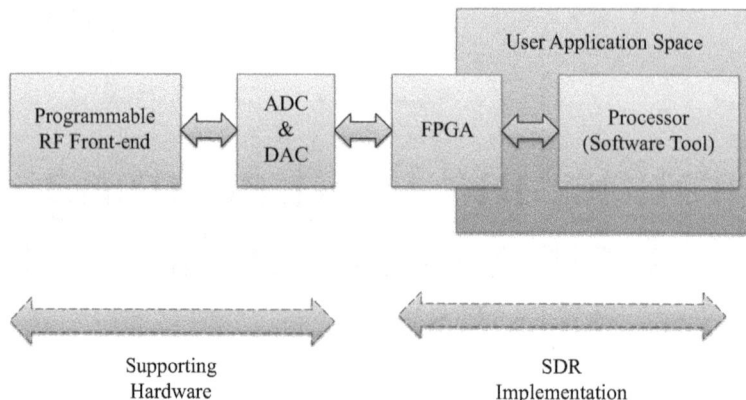

Figure 5.1   Description of an SDR system implementation.

- A device to perform decimation and downconversion in receivers and interpolation and upconversion for transmitters. Field-Programmable Gate Arrays (FPGAs) typically provide this functionality.

Note that both the RF and computational hardware systems may be available in a single integrated package, as discussed in Chapter 3.

- **Software Application**: Typically, one or a combination of General Purpose Processors (GPPs), Digital Signal Processors (DSPs), and FPGAs. The computational devices used can vary in cost, processing capability, and power consumption. Developers can choose to write their applications from scratch or they can use software development frameworks, like GNU Radio [2] and OSSIE [3]. Such frameworks allow developers to leverage existing communication function libraries and provide integration with supported programmable radio front-ends.

Developers can take the *classic* approach in targeting SDR platforms, using the individual chip maker's tool for each computational device. For GPPs, developers can elect to choose free GNU tools to compile programs [4], or choose to use vendor-specific compilers and/or Integrated Development Environments (IDEs). For Texas Instrument DSPs, the IDE Code Composer Studio (CoCS) [5] can be used and for Xilinx FPGAs Integrated Software Environment (ISE) [6], Embedded Development Kit (EDK) [7], or PlanAhead [8] can be used. Some platforms support *Model-based* design, where developers can leverage third-party software to simulate and express overall system behavior, such as Mathwork's Simulink [9], and generate SDR application system models. For example, Simulink hides the details behind running vendor-specific tools.

5.2 Choosing a Platform

This section provides some guidance for how a cognitive radio designer might go about selecting a platform and the approaches that can be taken to write software for it.

5.2.1 Choosing Between RF Alternatives

Many (or perhaps academic) CR systems are built with one of two radio platforms: the Universal Software Radio Platform (USRP) series from Ettus Research [10] tends to dominate in

PHY layer applications, and the Wireless Open-Access Research Platform (WARP) from Rice University [11] is often favored by network researchers. See Chapter 3 for more discussion.

Engineers' perceptions of SDRs and their approach to designing them differs somewhat according to their backgrounds and experience. Communication theorists may view a radio as a series of blocks performing ideal mathematical operations, perhaps unaware of issues like spurious response, dynamic range, bandwidth, etc., which do not arise in ideal cases. RF engineers may not initially appreciate that an SDR is really a computer system that must perform real-time calculations in a way that mimics the pipeline processing of analog radios, where a continuous sequence of signal values works its way through a chain. The difficulty of processing the current block of signal samples in time to be ready for the next block can be a challenge for SDRs running on a single-board computer.

Software radio offers capabilities for CR design, including potentially unlimited reconfigurability and the ability to build new components, but it also brings problems like overhead, resource consumption, and time delays associated with reconfiguration. These issues are nontrivial. For many practical applications, SDRs are inherently embedded real-time systems, requiring hardware platforms and tools and expertise that engineers focusing on applications of CR may well lack. Hardware and software issues can severely limit the performance of real SDR-based CRs, which, instead of having infinite configurability, when deployed may simply select from a relatively small set of preprogrammed modulations and modulation parameters. For many CR researchers, the net result may be a radio whose cost, reliability, and RF capabilities are comparable to those of an off the shelf RF Integrated Circuit (RFIC) or single board transceiver. Consequently, for some applications requiring low-cost and small-form factor CRs, there is now interest in CR designs that leverage WiFi and cellular chip sets and general purpose RFICs.

None of the above is meant to downplay the power and unlimited freedom that baseband processing of RF and microwave signals offer for the implementation of CR as well as its widespread use of SDR for many applications.

5.2.2 Processor Choices

Picking the right computational device(s) for an application can be complicated and multidimensional and a holistic approach needs to be taken in making the decision.

There is no easy way to specify the right processor or combination for a specific applications. Questions to be considered include: what platforms are readily available? What type of computing devices are available on them? What computing devices are application developers more comfortable with? Subjective matters can affect time to market, e.g., is using a familiar technology better than learning a new one? Is using a more capable but more expensive computing device justified? Will the processor be used in a streaming application where processing latency can be detrimental, or is it a non-streaming application where processing latency is acceptable?

It is also critical to look at whether a processor is able to perform the necessary computations, e.g., filtering, at acceptable throughput rates. This can be done by manual testing and/or reviewing processor benchmark results available in the literature. However, benchmark results do not provide a realistic characterization since they do not account for other software running concurrently, different memory access scenarios, the use of operating systems, or various other factors that can influence overall performance in actual systems.

If a processor has multiple cores, where multiple sub-processors can be used, it is important to note the software tools available to articulate and make use of concurrency while programming. The performance of the hardware interconnect can impact applications and the ease and/or difficulty of defining the use of multiple cores and data synchronization across them can be detrimental to developer's productivity.

If a heterogeneous computing platform is used, application development can be complicated by using different compilers and software development tools to program the various computing devices available. The latency and throughput of the interfaces connecting the computing devices are considerations, e.g., imagine a GPP and DSP available in a simple platform but with a slow interface between them. If developers need to implement a digital filter, the DSP might be able to implement the filter faster than the GPP. But if the data need to be moved by the GPP to and from the DSP, the slow interconnect interface might make the GPP/DSP-based filter slower than a GPP-based one.

5.2.3 Benchmarks

Benchmark results provide one means to compare performance between computing devices. The intention of benchmark suites is to have processors run representative tasks that can realistically run on actual systems. They provide a number of pre-written functions, e.g., FIR filtering, and collect information such as processing, time, and latency. Thus, it is possible to observe performance differences between processes under realistic workloads [12]. Since different applications can have drastically different workloads, there are many variations of benchmark suites. The following are some examples.

The Standard Performance Evaluation Corporation (SPEC) [13] focuses on CPU intensive benchmarks. Source code is distributed for specific applications, e.g., compression, compilation, and signal processing, and processors run the programs. The benchmark suite then aggregates performance measures and can be used to compare the various processors. The Embedded Microprocessor Benchmark Consortium (EEMBC) [14] provides benchmark suites for embedded processors in automotive, industrial, digital entertainment products (e.g. digital cameras, audio, graphics, etc.). The Berkley Design Technology, Inc. (BDTI) [15] offers benchmark suites that largely focus on embedded digital signal processing applications. It is important to note that benchmark results do not necessarily include off-chip memory access times and the effect of multiple concurrently running applications; we would expect these to be important factors in complete systems. Some application developers might be more interested in the performance of platforms comprising multiple computing devices, e.g., FPGAs, DSPs, and GPPs. Here, vendor benchmark results, for example, applications, might be the best resource.

5.2.4 Processor Interconnect

SDR and CR platforms can have different interfaces for transferring sampled RF data to the computing devices running the developed application. It is important to pick platforms that provide appropriate interfaces, at least in terms of the rate of samples that the platform can transfer to computer running the SDR/CR application, and connectivity needs, *i.e.*, what connection interface is needed and/or available. Here are some of the possible interfaces.

- **Network Interfaces:** Such as 1 Gigabit Ethernet and 10 Gigabit Ethernet.
- **Standard Bus Interfaces:** By standard, we mean that these interfaces are readily available and accessible using everyday computers. Such interfaces include USB 2.0, USB 3.0, and PCI Express.

Table 5.1   SDR platforms with their host interfaces and throughput.

Platform	Host interface	Host throughput	Duplex
SKIRL [21]	SPI	48 MHz	Full
USRP 1 [22]	USB 2.0	256 Mbps [23]	Half
USRP N200/N210 [24]	1.0 Gigabit Ethernet	800 Mbps [23]	Full
USRP B200/B210 [25]	USB 3.0	1.97 Gbps [23]	Full
USRP X300/X310 [26]	10 Gigabit Ethernet	6.4 Gbps [23]	Full
	PCI Express	6.4 Gbps [23]	Full
Nutaq Pico SDR [27]	10 Gigabit Ethernet	6.4 Gbps [28]	Full
	PCI Express	6.4 Gbps [28]	Full
Nutaq ZeptoSDR [29]	JESD204B	12.5 Gbps[a] [30]	Full
BEEcube [20]	Sting I/O	20 Gbps [31]	Full
Nutaq TitanMIMO [32]	MicroTCA.4	1.0 Tbps [33]	Full

[a]Performance measure provided by Xilinx for the JESD204B interface.

- **Custom Bus Interfaces:** By custom interfaces, we mean those that are not readily available on computing platforms. Examples include the JESD204B [16], a high-speed serial interface for data converters, with the specification developed by JEDEC [17]; MicroTCA [18] (Micro Telecommunications Computing Architecture) defined by the PCIMG open modular computing standards group [19]; and Sting I/O, the I/O interface used in BEEcube [20] platforms.
- **Peripheral Bus Interfaces:** By peripheral interfaces, we mean a computer data bus designed to support computer peripherals and components. Examples include RS-232, I^2C, and SPI.

In Table 5.1 we compare some of the SDR platforms available in terms of their host interface, the interface used to transfer data to where user applications are hosted; the theoretical throughput of the host interface; and whether the interface is half or full duplex.[1]

5.2.5 Other Considerations

Other considerations include: Form factor, e.g., is the radio supposed to be handheld? Power consumption, e.g., is the radio supposed to be battery operated? Price, does the system only require one platform or does it require multiple platforms to be purchased? It is important to note that benchmark results provided by manufacturers are typically for ideal best case scenarios; once developers design and run a complete solution, their power consumption and processor throughput might diverge from the ideal values provided by manufacturers.

5.3 Programming

This section explores different approaches that can be taken to program SDR and CR platforms. Approaches can be microscopic, such as the classic approach where users program each computing device directly, to macroscopic, as with the model-based approach where users

---

[1]Ettus provides the host sample rate in MS/s with a sample being a 16-bit I and a 16-bit Q value, so a sample is composed of 32-bits. The values were converted to Mbps by multiplying by 32.

develop applications using third-party tools that generate code for the individual computing devices. In addition, we will present that an approach we originally presented in Reference 34 to incorporate Models of Computations (MoCs) to expose computational knobs and meters for CR applications.

5.3.1 Classic Approach

By the *classic approach*, we mean that developers design their initial applications, simulate them, and physically program them using a non-integrated software suite. For example, users might model and develop their initial prototypes using Mathwork's Simulink [9]. Then they would port the system simulation to a GPP by writing a C++-based implementation. In case of targeting a platform composed of an FPGA in addition to a GPP, the user would need to write HDL code for the FPGA and the code necessary to transfer data between both computing devices using a suitable interface, e.g. a bus or memory mapped interface.

The challenge with such an approach is that the development environment, may require the initial modeling and implementations to be done separately. This means that the initial models need to be tested and verified, the final design needs to be tested and verified, and in addition the translation between the initial model and the final design must be correct. Also developers need to be experienced with the computing devices they are using, e.g., they have to know how to program a GPP and an FPGA, as in the previous example. However, by using the classic design approach, designers are able to maintain a lot of flexibility since they are programming each computing device directly with no intermediate software layers before translation or code generation.

5.3.1.1 Example Application

The Center for Wireless Telecommunications (CWT) Public Safety Cognitive Radio (PSCR) [35] is a radio built to address radio interoperability issues for public safety personnel. The PSCR is able to detect radios in the spectrum and classify their modulations, it can configure itself to communicate with the radios, and it can be used in *gateway* mode to bridge two incompatible radios on the fly. The original radio designed was done using GNU Radio and ran on a laptop/desktop-type GPP. The radio design was eventually transitioned to an embedded platform [36] to make it possible to commercialize the PSCR for a sub $2 000 price point.

The initial platform was BeagleBoard [37] which contains a Texas Instrument (TI) OMAP3530 [38], which combines an ARM Cortex-A8 embedded GPP [39] GPP and a TI C64x+ [40] fixed-point DSP. The BeagleBoard was connected to a first-generation Ettus Universal Software Radio Peripheral (USRP) [22] with a WBX RF daughterboard [41] 50–200 MHz transmit and receiver frequency range. We then switched the platform to an Ettus USRP E100 [42], which is an integrated platform combining a Gumstix Overo computational board [43] and a programmable RF front-end while still using the same WBX RF daughterboard. In the E100, the Gumstix Overo is connected to the programmable RF front-end FPGA via the TI General-Purpose Memory Controller (GPMC) bus.

We used OpenEmbedded (OE) [44] to cross-compile our code, compiling code on a regular computer so it can run on an embedded processor. The goal was to transition a GNU Radio and Python-based PSCR from a GPP to run on an embedded processor while using the available DSP to accelerate processing. The DSP code was written using TI's CoCS IDE and the GPP/DSP interface was done via the TI DSPLink library [45]. We then developed GNU Radio-based

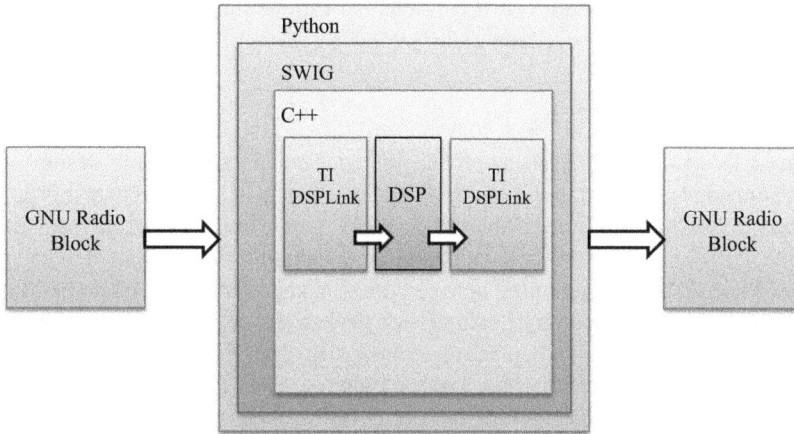

Figure 5.2   DSP-based GNU radio block abstraction.

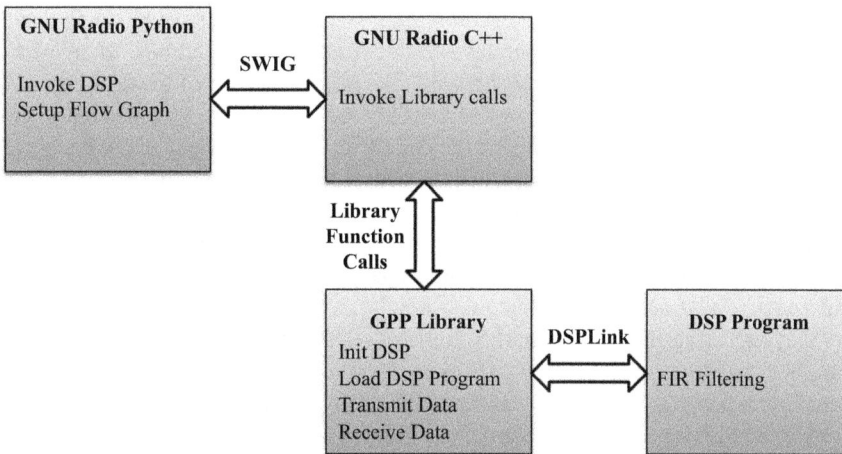

Figure 5.3   The programming flow between GNU radio, GPP library, and DSP executable.

blocks in C++ which perform function calls to the DSP through the DSPLink interface which we ultimately called as standard GNU Radio blocks using Python, by making use of the Simplified Wrapper and Interface Generator (SWIG) [46] wrapper which allows C++ to be called from Python. Figure 5.2 shows the encapsulation of DSP function calls as GNU Radio-based blocks. The interaction between the various software interfaces is show in Figure 5.3. To read more about the details of the development environment and GPP/DSP interfacing please refer to Reference 47.

Taking the *classic* approach to system development provides development flexibility at the expense of complexity. It requires developers to establish an initial system prototyping environment, in our example case the original laptop-based PSCR system. Then it is necessary to set up and learn to use the individual development tools, e.g. OE, DSPLink, and CoCS. Next it is necessary to integrate the overall development flow and verify the operation of the final system with the initial prototype. In the embedded PSCR model case, testing was done on a system level where we verified that it can identify various radios, interoperate (be able to bridge audio communications between radios).

5.3.2 Model-Based Design

Model-Based Design (MBD) is an approach intended to incorporate the design, analysis, implementation, and verification of systems [48]. By using an MBD approach, it is possible to test and verify a system model and be able to generate a system implementation from the model. This way the implemented system behavior would maintain the original system model that was tested and verified. The tight coupling between the simulation and implementation facilitates the use of Hardware-in-the-Loop (HIL) simulation [49], where developers can incorporate data from the actual RF front-end into their simulations or run part of their design, for example, in an FPGA while running the rest of their design in simulation. The model-based design allows the continuous refinement of a design going back and forth between design and implementation.

It is possible to create SDR and CR applications using the MBD approach using software such as Mathwork's Simulink [9] and National Instrument's LabVIEW [50]. Both development environments allow users to simulate and prototype designs using various Models of Computation which can also be used to generate, compile (or synthesize in case of FPGAs), download, and run on target platforms. In addition, it is possible to use HIL simulations, where part of the design can be running on an SDR platform while the remaining part of the design runs in the simulation environment. This design approach can reduce the design time needed to develop SDR and CR applications, and, by generating code from simulation models, it becomes possible for users without FPGA development experience, for example, to develop FPGA-based solutions when utilizing code generation. In addition, such tools can make it easier to handle data exchanges between computing devices, e.g., a platform that has both an FPGA and a DSP, by generating the code necessary to communicate data between them. However, an issue with MBD approach is that applications become directly tied to the stability and lifespan of the software they are using. For example, if Simulink or LabVIEW stop supporting a particular platform, then developers might need to migrate their applications to new platforms if they want to maintain compatibility with the latest MBD software tools.

In Reference 51, Haessig et al. implement a subset of a military SATCOM waveform. They compare the amount of engineering time required to implement the same waveform using a conventional classic approach and using an MBD approach, using Mathwork's Simulink [9] as the simulation environment, using Xilinx System Generator [52] to develop FPGA system components, and Mathwork's Real-time Workshop (RTW) to generate their application C code. The authors observed a 10:1 productivity improvement by using the MBD approach compared to the classic approach. They observed the greatest benefit by focusing on developing an accurate simulation, noting that debugging in a simulation environment is typically easier and less time consuming than debugging the same system running on FPGA from hand-written Hardware Description Language (HDL) programs. By debugging and refining the simulation, they were able to use the Simulink tools to generate a correct by design system.

5.3.3 Application of Models-of-Computation

In this section, we discuss the application of MoCs and the mapping between a Cognitive Engine (CE) and SDR perspective as an example of incorporating MoC and tying it to CR. This work is explored in greater detail in Reference 34.

5.3.3.1 Mapping of MoCs for Radio Applications

In Reference 34, the authors explored defining Differential Binary Phase-Shift Keying (DBPSK) radios in a CE in Communicating Sequential Processes (CSP), programmed using the Occam language [53,54] which is based on CSP semantics. Within the CSP model, the authors define *meta-data*, parameters from the reactive model that ultimately allow the generation of an Synchronous Dataflow (SDF)-based SDR implementation using GNU Radio.

Assuming a CR is capable of operating as a DBPSK transmitter and receiver, its process definition in CSP can be:

$$CE = (select\_TX \rightarrow TX)|(select\_RX \rightarrow RX)$$

where *select_TX* and *select_RX* are events corresponding to the CE selecting the DBPSK transmitter, *TX*, or receiver, *RX*, processes. Each radio flowgraph at this stage is an Occam program that instantiates the specific signal processing functionality necessary to implement the radio. For example, the DBPSK transmitter is defined from the following processes:

- **Parameter Generator (PG)**: This process is responsible for defining the configuration parameters, or *meta-data*, for the DBPSK transmitter.
- **Data Source (DS)**: Instantiates the data source for the radio.
- **Packet Encoder (PE)**: Creates packets by breaking up the input data stream into fixed length arrays and encapsulating them with packet headers and footers containing information such as packet length and checksum values.
- **DBPSK Modulator (DM)**: Instantiates a DBPSK modulator.
- **Channel Filter (CF)**: Defines an Root-Raised-Cosine Filter (RRC) filter.
- **Baseband Scale (BS)**: Scales the baseband signal before sending it over the RF interface.
- **RF Out (RO)**: Instantiates the output RF interface.

The CSP semantic for defining the transmitter chain is as follows:

$$TX = PG\|DS\|PE\|DM\|CF\|BS\|RO$$

After the Occam radio flowgraph definitions are interpreted, the framework knows what processes are instantiated, the defined meta-data parameters, and the needed IO channel definitions. At this stage, the framework is able to generate suitable SDF-based SDR flowgraphs for GNU Radio and Ptolemy [55], a software suite for modeling MoCs. For each target environment, there are associated *domain-mapping* tables. For example, a DBPSK modulator block has an associated block definition for GNU Radio and Ptolemy. The domain mapping tables include the exact block name, IO port names, sampling rates, and meta-data mapping for each of the target environments. Meta-data such as signal gain, RRC filter roll-off factor, RF transmit/receive, and center frequency are all passed to the designated block definitions. Figure 5.4 provides a summary of the model generation process.

5.3.3.2 Core Framework

The target SDR flowgraphs are implemented using the SDF MoC. In order to provide proof-of-concept applications and collect performance measurements, the framework is able to generate and run GNU Radio-based flowgraphs. A first generation USRP [22] is the radio interface used for over-the-air flowgraphs.

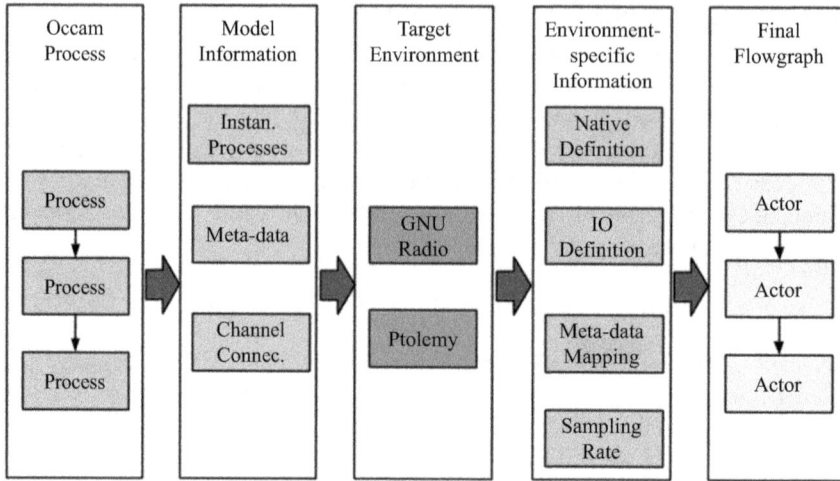

Figure 5.4   Model generation domain mapping process.

There are two different stages for implementing the radio flowgraphs. The first stage is based on the pure process definitions from the Occam files, where the instantiated processes, I/O connections, and meta-data are defined. During the second stage, the flowgraph definition interpreted previously is passed to GNU Radio and *expanded*. This means that if some blocks are hierarchical, composed of multiple sub-blocks, GNU Radio will replace the hierarchical blocks with the sub-blocks contained in it, hence *expanding* the flowgraph definition.

The core framework provides the following analysis and calculations:

- Topology matrix construction.
- Sampling rate consistency check.
- Repetition vector calculation.
- Performance measurement collection.

Figure 5.5 provides a summary of an example interaction between a CE and the framework which are discussed in detail throughout this chapter.

5.3.3.3 First Stage

During this stage, a topology matrix is constructed using information in the domain-mapping tables created by the model generator. The meta-data from the Occam file provides the exact sampling rates for each of the associated blocks. However, the topology matrix is constructed using the relative sampling rates and not absolute sampling rates, e.g., if the sampling rate of block 1 is 16 000 samples/second and the sampling rate of block 2 is 32 000 samples/second then the relative rate of block 1 is $\frac{16\,000}{16\,000} = 1$ and the relative rate for block 2 is $\frac{32\,000}{16\,000} = 2$. Basically, each sampling rate is divided by the greatest-common-divisor of the flowgraph's sampling rates.

After the first-stage topology matrix is constructed, a sampling rate consistency check is performed by calculating the rank of the topology matrix as presented in the previous chapter. The repetition vector is then calculated by constructing and solving a Mixed Integer Program (MIP) as will be discussed in more detail in a later section. Finally, the total memory needed to

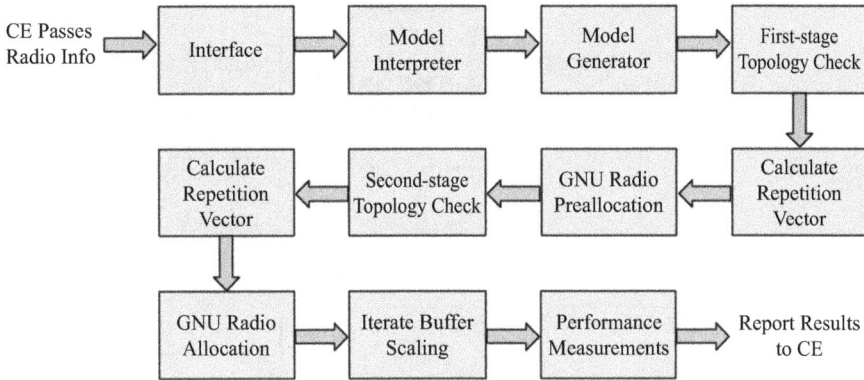

Figure 5.5 Sequence of events for CE/framework interaction.

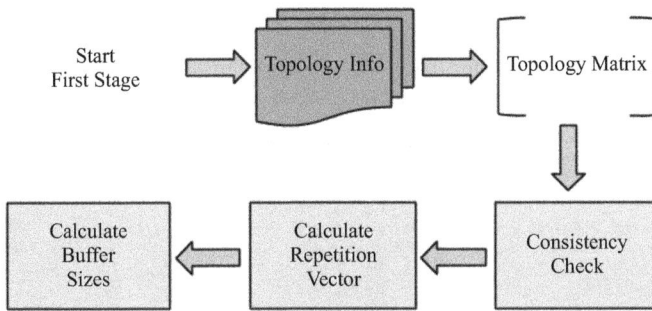

Figure 5.6 First-stage analysis.

implement the buffers is calculated by the framework. At this point, only the memory utilization can be calculated since throughput, latency, and reconfiguration times are measured from the actual implementation. Figure 5.6 summarizes the progression of events during the first stage of the core framework. At this stage, a CE will know if the flowgraph definition is consistent and the memory utilization for the current radio definition.

5.3.3.4 Second Stage

During the second stage, the framework queries GNU Radio for the *expanded* radio flow-graph. If any of the signal processing blocks defined during the first stage are hierarchical, the framework needs to redefine the topology matrix to account for the extra blocks and associated buffers. The repetition vector also needs to account for the extra blocks. A summary of the steps performed during the second stage is shown in Figure 5.7, which is followed by a more detailed explanation.

Assume that a CE selects a DBPSK link running over a Additive White Gaussian Noise (AWGN) channel. Figure 5.8 illustrates the framework's perspective of the flowgraph at the end of the first stage. The upward arrow and number 2 on top of the first RRC filter indicate an interpolation factor of 2 and the downward arrow and the number 2 in the second RRC filter indicate a decimation factor of 2. We can see that all blocks prior to the first RRC filter operate at a set sampling rate, assumed to be 1. Blocks between the first and second RRC filter

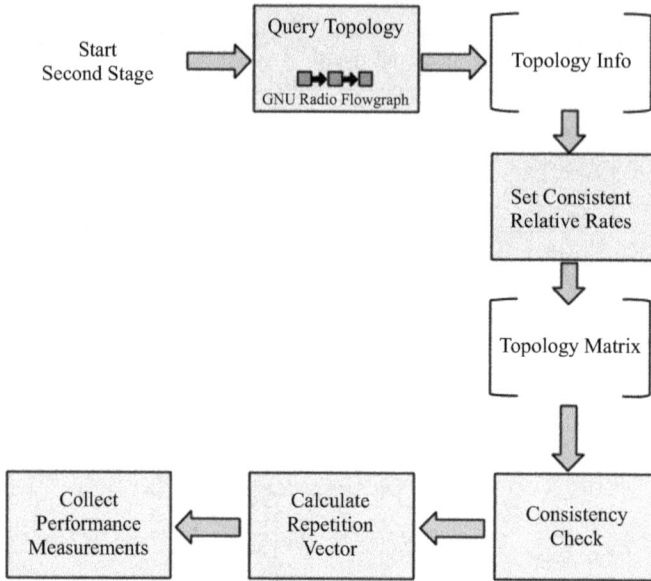

Figure 5.7 Second stage.

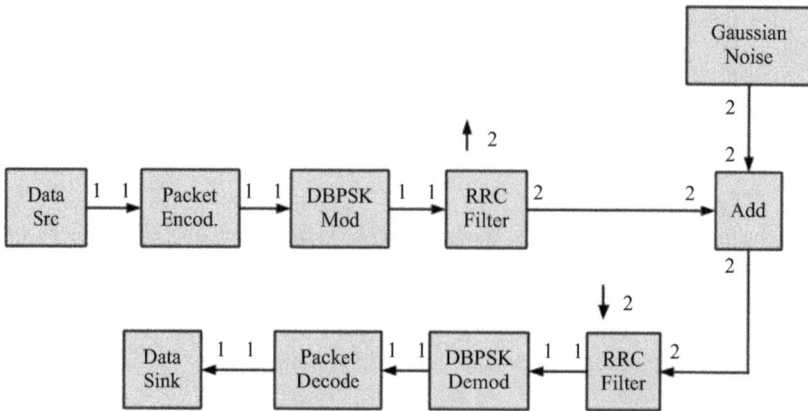

Figure 5.8 Flowgraph perspective after the first stage.

operate at a rate of 2. All blocks after the second RRC filter operate at a rate of 1 because the decimation in the filter. Note that, while the Gaussian noise generator does not specify any interpolation nor decimation, it should still operate at a sampling rate of 2 so the noise is generated at the same rate as the DBPSK radio signal.

The framework queries GNU Radio for the expanded flowgraph during the second stage, an example expanded flowgraph for Figure 5.8 is shown in Figure 5.9. The expanded flowgraph indicates that there are two gain blocks which are not in the original flowgraphs: a gain block after the first RRC filter and a second one before the second RRC filter. The assumption here is that the previous definition of the RRC filter is hierarchical and that the expanded GNU Radio flowgraph for the RRC filter includes a gain block. Also note that there is no relative

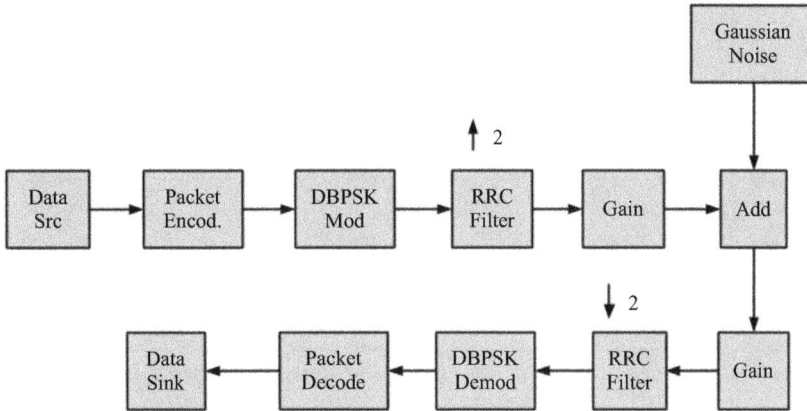

Figure 5.9 GNU radio expanded flowgraph.

rate information in the topology which is necessary to the construction of the second stage topology matrix. The only information available is the interpolation and decimation rates of the RRC filters.

A recursive algorithm addresses the issue of deducing the relative rates between the blocks to construct a topology matrix from the decimation and interpolation information in the graph. The algorithm assumes the following:

1. The flowgraph is a directed graph which is already implied by the use of the SDF MoC.

2. The relative rates at an input/output ports of a single arc are the same, e.g., if the output port of an arc is 1 then the input port is also 1. This assumption is made because interpolation/ decimation in GNU Radio is not implemented on a block's input port but physically in the block implementation.

3. The relative rates at an input/output port can only differ when a *stream-to-vector* or *vector-to-stream* blocks are used. In GNU Radio, a *stream-to-vector* block takes a stream of data and lumps them into a single vector, e.g., for performing FFT operations where a vector of 1 024 might be needed to perform the calculation. The *vector-to-stream* block performs the inverse operation. These blocks do not change the sampling rates.

4. All blocks performing interpolation and/or decimation should declare those factors explicitly.

5. Each arc traversed in the flowgraph has a Boolean value indicating whether or not it was visited.

6. The algorithm's base-case (terminating-case) occurs when a sink node is reached or if the algorithm reaches a previously visited node which does not require a relative rate correction.

Going back to the DBPSK link example, the algorithm traverses the flowgraph recursively starting from the source blocks. The source blocks in this case are the *Data Source* and *Gaussian Noise* blocks. Beginning at the *Data Source* block, the algorithm starts setting the relative rate as 1, then continues recursively traversing the flowgraph assigning a relative rate to each block as 2 after the RRC filter because of the interpolation factor. The relative rate will continue

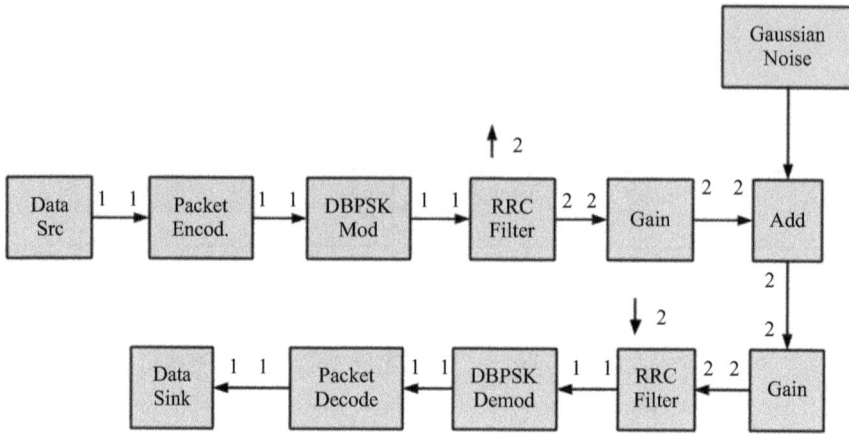

Figure 5.10 Setting flowgraph relative rates starting at data source block.

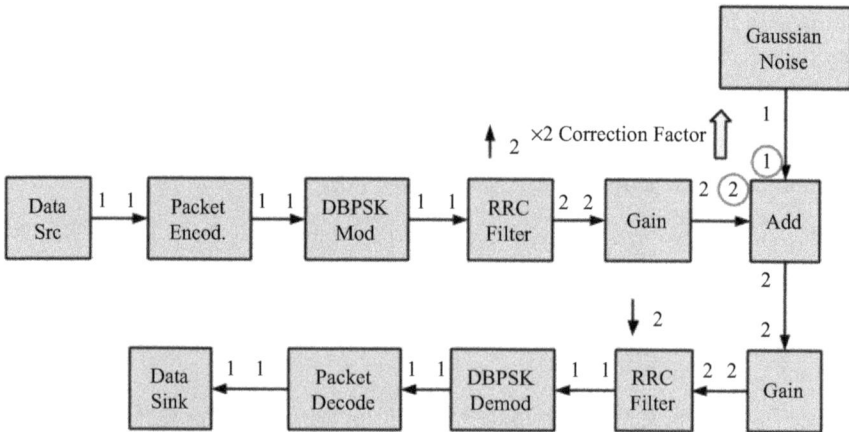

Figure 5.11 Flowgraph relative rate that must be fixed.

being set to 2 until the second RRC filter block is reached then the remaining blocks will have a relative rate of 1. Figure 5.10 shows the flowgraph overall relative rates at this stage.

The algorithm then starts setting the relative rates starting from the *Gaussian Noise* block. When it reaches the *add* block, it realizes that it already traversed this block. But when it compares the expected relative rate of 1 with the *add* block's current relative rate is 2, it realizes that it needs to increase the relative rate calculation for the path it just traversed, the *Gaussian Noise* source, by a factor of 2. This stage is shown in Figure 5.11. The algorithm then terminates, resulting in the expected flowgraph with the relative rates show in Figure 5.12.

Figure 5.13 provides another example of how the algorithm can handle setting the relative rates of flowgraphs. In the figure, the algorithm starts by setting the relative rates at *Block 1*, *Block 2*, *Block 5*, and *Block 6*. When the algorithm sets the relative rates from *Block 3*, it continues to *Block 4*, and reaches *Block 5* it detects an input rate mismatch on the two input ports of *Block 5* where one port has a relative rate of 3 and another as 2, as shown in Figure 5.14. Figure 5.15 then demonstrates how the greatest common multiple of the ports is set as the new

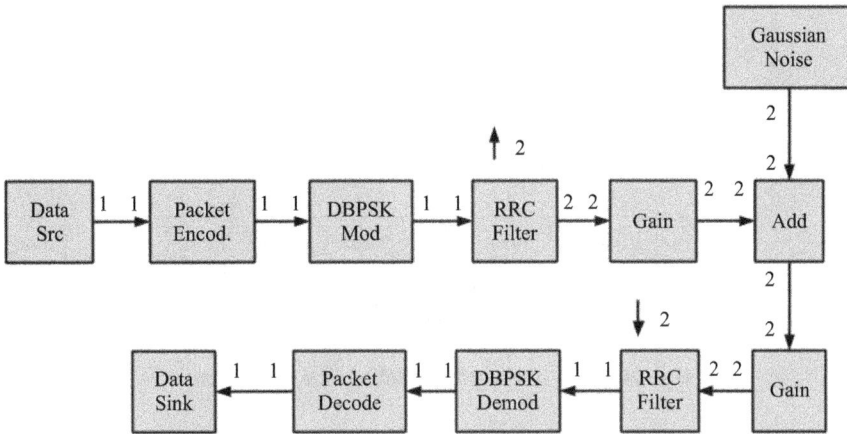

Figure 5.12 Final flowgraph with correct relative rates.

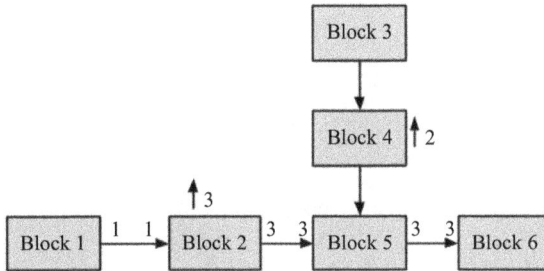

Figure 5.13 First iteration of setting relative rate starting at block 1.

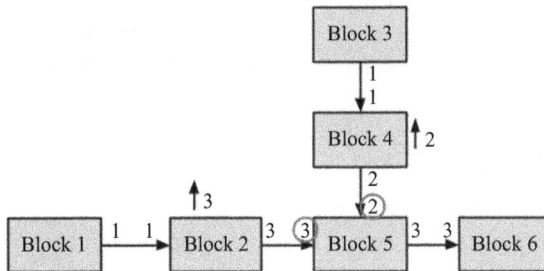

Figure 5.14 Second iteration of setting relative rate starting at block 3.

rate. While Figure 5.16 shows the final flowgraph with the final relative sampling rates. It is important to note that having different relative rates on the input ports of blocks is fine in terms of SDF semantics; however, in the context of the discussed radio applications it is not acceptable. If the ability to perform decimation and/or interpolation on the input ports is needed, then the application designer needs the ability to set the relative rates explicitly for each block. This is not inherently supported in GNU Radio and might not be applicable in the context of the presented radio flowgraphs.

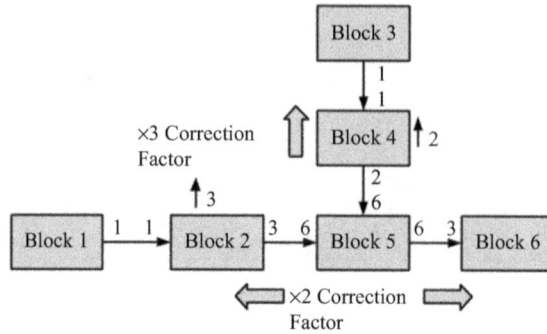

Figure 5.15   Flowgraph relative rate mismatch.

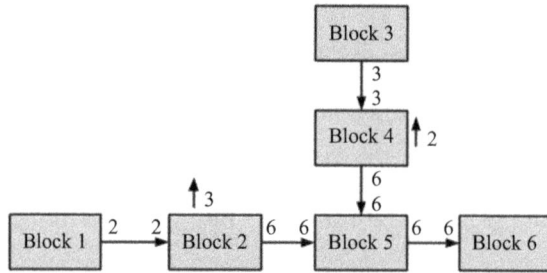

Figure 5.16   Flowgraph relative rate after algorithm is done.

5.3.3.5 Computational Knobs and Meters

Taking an MoC approach to mapping CR radio configurations to implementation it is possible to expose computational knobs and meters that can be leveraged by CEs. The work in Reference 34 demonstrates how scaling buffers between SDR radio components can be used as a computational knob and throughput, latency, and memory utilization as computational meters.

5.4 Concluding Remarks

In this chapter, we presented some considerations for choosing SDR/CR platforms and different approaches to programming them. There is no simple direct method for finding a platform for an application or to programming a particular application. It is a multifaceted problem where developers need to consider what processor(s) provide enough computational bandwidths for an application, whether an integrated platform should be used or should they combine separate computational and RF front-end platforms to suit their needs. Also, there are other considerations such as: does the project need increased flexibility by taking a conventional classic approach for programming platforms, if so, does the programming team have enough experience to program each individual computing device. If time to market is an important consideration and developers prefer to have a more integrated development environment where they can focus on system behavior and have the low-level code generated for them, then an MBD approach might be more suitable especially if they don't mind being dependent on the feature sets provided by the MBD software. If an MBD approach is taken, it is worth noting that there is always the risk of deprecated support for the platforms they're using or the specific version of the MBD tool being used. If this is an important consideration, then it might be worth engaging platform developers and software providers to know expected vendor support timeframe.

Bibliography

[1] A. Ferrari and A. Sangiovanni-Vincentelli, "System Design: Traditional Concepts and New Paradigms," in *Computer Design, 1999. (ICCD '99) International Conference on*, 1999, pp. 2–12.

[2] GNU Radio. [Online]. Available: http://gnuradio.org/redmine/wiki/gnuradio

[3] MPRG, "OSSIE." [Online]. Available: http://ossie.wireless.vt.edu/

[4] GCC, "GNU, the GNU Compiler Collection." [Online]. Available: https://gcc.gnu.org/

[5] T. Instrument, "Code Composer Studio." [Online]. Available: http://focus.ti.com/docs/toolsw/folders/print/ccstudio.html

[6] Xilinx. [Online]. Available: http://www.xilinx.com/products/design-tools/ise-design-suite/

[7] Xilinx. "Platform Studio and the Embedded Development Kit (EDK)." [Online]. Available: http://www.xilinx.com/tools/platform.htm

[8] Xilinx. "PlanAhead Design and Analysis Tool." [Online]. Available: http://www.xilinx.com/tools/planahead.htm

[9] Mathworks, "Simulink." [Online]. Available: http://www.mathworks.com/products/simulink/

[10] "Universal Software Radio Peripheral." [Online]. Available: http://www.ettus.com

[11] Rice University, "Rice WARP Board." [Online]. Available: http://warp.rice.edu/trac/

[12] L. K. John and L. Eeckhout, *Performance Evaluation and Benchmarking*. CRC Press, Boca Raton, FL, 2005.

[13] "Standard Performance Evaluation Corporation." [Online]. Available: http://www.spec.org/

[14] "The Embedded Microprocessor Benchmark Consortium." [Online]. Available: http://www.eembc.org/index.php

[15] "Berkley Design Technology Inc." [Online]. Available: http://www.bdti.com/

[16] "An Early Look at the JEDEC JESD204B Third-Generation High-Speed Serial Interface for Data Converters." [Online]. Available: http://www.eetimes.com/document.asp?doc_id=1279004

[17] "About JEDEC." [Online]. Available: http://www.jedec.org/about-jedec

[18] "MicroTCA Overview." [Online]. Available: http://picmg.org/openstandards/microtca/

[19] "PICMG Open Modular Computing Standards." [Online]. Available: http://wwwpicmg.org/

[20] "BEECube." [Online]. Available: www.beecube.com

[21] A. Young, "Unified Multi-Domain Decision Making: Cognitive Radio and Autonomous Vehicle Convergence," Ph.D. Dissertation, Virginia Polytechnic Institute and State University, Blacksburg, Virginia, December 2012.

[22] "Ettus Research USRP1 Bus Series," *Ettus Research*. [Online]. Available: https://www.ettus.com/content/files/07495_Ettus_USRP1_DS_Flyer_HR.pdf

[23] "USRP Bandwidth." [Online]. Available: http://www.ettus.com/kb/detail/usrp-bandwidth

[24] "USRP N Series." [Online]. Available: https://www.ettus.com/product/category/USRP-Networked-Series

[25] "USRP B Series." [Online]. Available: https://www.ettus.com/product/category/USRP-Bus-Series

[26] "USRP X Series." [Online]. Available: https://www.ettus.com/product/category/USRP-X-Series

[27] "Nutaq PicoSDR." [Online]. Available: http://nutaq.com/en/products/picosdr

[28] "Nutaq PicoSDR Datasheet." [Online]. Available: http://nutaq.com/sites/default/files/PicoSDR_V1.4_02_14_2014_web.pdf

[29] "Nutaq ZeptoSDR." [Online]. Available: http://www.nutaq.com/en/products/zeptosdr

[30] "Integrated Software-Defined Radio on Zynq®-7000 All Programmable SoC Design Seminar." [Online]. Available: http://www.zedboard.org/sites/default/files/ZYNQ_SDR_2014_slides_v1_0.pdf

[31] "Unique and Key Features of BEECube Technology." [Online]. Available: www.beecube.com/technology

[32] "Nutaq TitanMIMO." [Online]. Available: http://nutaq.com/en/products/titanmimo-4

[33] "TitanMIMO-4 Massive MIMO Testbed Product Sheet." [Online]. Available: http://nutaq.com/sites/default/files/datasheets/TitanMIMO_06_05_2014_web.pdf

[34] A. S. Fayez, "Design Space Decomposition for Cognitive and Software Defined Radios," Dissertation, Virginia Polytechnic Institute and State University, 2013.

[35] B. Le, F. A. Rodriguez, Q. Chen, B. P. Li, F. Ge, M. ElNainay, T. W. Rondeau, and C. W. Bostian, "A Public Safety Cognitive Radio Node," in *Proceedings of the 2007 SDR Forum Technical Conference, SDRF*, 2007.

[36] A. Fayez, Q. Chen, A. Young, N. Kaminski, H. Alavi, and C. Bostian, "An Embedded Platform Based Public Safety Cognitive Radio," *Wireless Innovation Forum Conference and Product Exposition*, January 2013.

[37] BeagleBoard, *BeagleBoard System Reference Manual Rev C4*, BeagleBoard, December 2009.

[38] T. Instrument, *OMAP3530/25 Application Processors*, Texas Instrument.

[39] ARM, "Cortex-A8 Processor." [Online]. Available: http://www.arm.com/products/processors/cortex-a/cortex-a8.php

[40] "C64x+," Texas Instrument. [Online]. Available: http://processors.wiki.ti.com/index.php/C64x%2B

[41] Ettus Research. [Online]. Available: https://www.ettus.com/product/details/WBX

[42] "Ettus Research USRP E100 Embedded Software Defined Radio," *Ettus Research*. [Online]. Available: www.ettus.com/downloads/USRP_E100_Series_temporary_datasheet.pdf

[43] Gumstix, *Gumstix Overo Product Page*, Gumstix. [Online]. Available: http://www.gumstix.com/store/catalog/index.php?cPath=27_33

[44] OpenEmbedded, "OpenEmbedded." [Online]. Available: http://www.openembedded.org/wiki/Main_Page

[45] "DSP/BIOS Link User Guide," Texas Instrument.

[46] "Working with Modules." [Online]. Available: http://www.swig.org/Doc1.3/Modules.html

[47] A. S. Fayez "Designing a Software Defined Radio to Run on a Heterogeneous Processor" Thesis Virginia Polytechnic Institute and State University, 2011.

[48] J. Jensen, D. Chang, and E. Lee, "A Model-based Design Methodology for Cyber-physical Systems," in *Wireless Communications and Mobile Computing Conference (IWCMC), 2011 7th International*, July 2011, pp. 1666–1671.

[49] M. Bacic, "On Hardware-in-the-loop Simulation," in *Decision and Control, 2005 and 2005 European Control Conference. CDC-ECC '05. 44th IEEE Conference on*, December 2005, pp. 3194–3198.

[50] N. Instruments, "LabVIEW." [Online]. Available: http://www.ni.com/labview/

[51] D. Haessig, J. Hwang, S. Gallagher, and M. Uhm, "Case-study of a Xilinx System Generator Design Flow for Rapid Development of SDR Waveforms," in *SDR Technical Conference*, 2005, pp. 14–18.

[52] Mathworks, "Xilinx System Generator and HDL Coder." [Online]. Available: http://www.mathworks.com/fpga-design/simulink-with-xilinx-system-generator-for-dsp.html

[53] A. W. Roscoe, "Denotational Semantics for Occam," in *Seminar on Concurrency*. Springer, London, UK, 1985, pp. 306–329.

[54] "Kroc." [Online]. Available: http://www.cs.kent.ac.uk/projects/ofa/kroc/

[55] "Ptolemy Project." [Online]. Available: http://ptolemy.berkeley.edu/ptolemyII/

CHAPTER 6

Cognitive Radio Evaluation[1]

6.1 Introduction

Evaluation is a rather important and deceptively complex topic. The central point of any evaluation is the determination of some object's goodness and herein lies the difficulty. The concept of goodness is ethereal at best; goodness means different things to different people or under different situations. In this way, evaluation takes on difficulties akin to quantum systems in which the particulars of the observer make a large difference in the observed phenomena.

As an instructive example of evaluation consider a household object, a fork perhaps. Let us briefly consider whether a fork is good. In the abstract, the question of "is a fork good?" is meaningless to the point of absurdity, clearly we need some information about how this fork is being used. Thus, we find that it is necessary to ask "is a fork good for eating soup?" instead of the more general question. However, it becomes quickly apparent that if we consider a different application, the goodness of our fork changes. For example, a salmon eater might evaluate a fork very differently than a soup eater. Additionally, it's worth noting that the possibilities increase exponentially as we add detail to the evaluation. For example, if we consider a small fork we have added size to the question, and the user's size must also be considered. That is, our small fork may be evaluated as good by a child but found lacking for an adult. Finally, note that we have avoided the question of preference to focus on utility. While this utilitarian approach may be appropriate for rigorous comparisons, preferences, and other more subjective factors can be extremely important.

Our brief consideration of a humble fork reveals several key principles about any evaluation. First, note that evaluation of an object depends heavily on the intended use (the mission) of the object. Of course, this is especially true when taking a utilitarian approach, where evaluation focuses entirely how useful something is for completing a task. Second, note that the complexity of evaluation depends on the number of factors involved. Evaluation, itself, presents a multidimensional problem where the meaning of goodness varies in several dimensions simultaneously according to complex relationships that depend on the details of the situation. Finally, note that those details include the inherently subjective preferences of the system's user for every real evaluation. A typical assumption for engineering disciplines is the omnipresence of objective approaches; however, as engineers chiefly design technology that impacts people this assumption is frequently baseless. Communications engineers, in particular, must handle this

---

[1]The material in this chapter is largely taken from Kaminski's thesis [1].

subjectivity in the form of quality of service, and, while this metric is typically handled through examination of a more objective measure, e.g. bit error rate, an evaluator must not forgot the end user. Thus, our evaluation thought experiment highlights evaluation as involving multiple dimensions which depend on the specifics of the problem at hand and the end user, as well as the object of evaluation itself.

The remainder of this chapter explores strategies for coping with these complexities, as they apply to the evaluation of cognitive radios (CRs). Note that evaluation itself is an evolving discipline and its application to the field of CR is certainly in a state of flux. That is, there is no single agreed upon method for determining the goodness of a cognitive radio, but rather each case must be examined separately. Thus, this chapter will focus on the foundational principles of evaluation systems, offering an example approach for insight into the application of fundamentals rather than provide a complete catalogue of evaluation methods.

6.2 Performance Evaluation Principles

Measuring CR performance is a complex and multilayered problem, further complicated by the inherently subjective nature of the task at hand, focused on enabling the comparison of goodness between different realizations of the CR concept. That is, the outcome of an evaluation provides a handle for describing the operation of CRs in a consistent manner across several possible variations in procedure. Furthermore, such an analysis must attach a value to CR agency that summarizes the worth of the methodology employed. Herein lies perhaps the most nuanced aspect of performance evaluation: a strict reliance on context. Fundamentally, analysis of the operation of a CR must be accomplished keeping in mind of the purpose of the CR to determine the value of the mechanisms in a useful manner. Therefore, a performance evaluation system must succinctly capture the operation of a CR in a consistent manner across several specimens while accessing the worthiness of the mechanisms employed for the task at hand.

In order to understand how such a goal may be accomplished, let us first decompose the core responsibilities of a goodness determining tool into the most universal principles of evaluation. Performance evaluation is a transform of information about the object being evaluated into some measure as discussed above. To achieve the consistency required for comparisons, this transform must operate in a repeatable manner, preferably without any bias away from the true goodness of the object. Moreover, to allow comparison between several discrete instantiations, this transform must not be overly sensitive to minor variations. Finally, as the transform will eventually provide information to guide some human decision, the evaluation process must clearly relate to the objectives of the user for the radio under test. These core responsibilities of evaluation allow us to define three fundamental properties that must be exhibited by any performance evaluation method: clarity, consistency, and robustness.

Achieving these properties in the performance evaluation of any non-deterministic system is an inherently complicated task. Evaluation requires a great deal of information, which is encoded in a variety of ways. We may conceptualize both background information and foreground information that is relevant to analysis of CR functionality. Foreground information directly encompasses radio operation, encoded as the actions of the agent under test. Background information provides the context in which this foreground information sits by capturing contextual information. This information is encoded as the results of sensors and the demands of users. Successful performance evaluation requires the joint processing of these two categories of information to achieve the desired understanding of radio operation.

6.3 Metrics and Factors for Cognitive Radio Evaluation

Performance evaluation is perhaps best discussed in terms of a performance function that maps from some n-dimensional domain to the real number line. This performance function captures the manner in which a CR operates and evaluates this numerically. The input to this function encompasses all of the factors that influence the value of CR operation, from stimuli to task difficulty, and the mechanisms of the performance function are specific to a particular CR. The output of this function is a metric that describes the value of the cognitive radio operation. Furthermore, note that the performance function for a CR may be assumed to be deterministic, but potentially chaotic. Also this performance is considered to be an inherent aspect of the cognitive radio under test rather than an external imposition to support evaluation; rather, the concept of the performance function is simply a more structured method for discussing the performance of the radio. Thus, the goal of performance evaluation may be restated as determinating an estimate for the typically complex and multidimensional performance function of a CR.

Understanding the factors that influence performance and form the domain of the performance function provides the means to understand the nature of performance functions and, subsequently, approaches to estimate them. As we will see, the domain of performance may be organized into a framework that allows for consideration of performance with regard to a subset of the total number of factors. Consideration of the resulting partial performance functions reduces the complexity and certainly the dimensionality of estimation and therefore evaluation. Moreover, partial performance functions yield a more understandable result as the metrics produced have a more certain meaning than a single overall figure for CR performance. In fact, consideration of performance as a vector of such metrics typically supplies the most useful description of overall performance. Therefore we begin by developing an organizational structure for the factors that influence CR performance as described by a collection of partial performance metrics.

As a means of organizing the factors of performance evaluation, let us consider various sets of factors that together form the domain of the performance function. Each such set encompasses a single aspect that influences the goodness of the operation of a cognitive radio, allowing us the means to select the factors most relevant for determinating any particular metric. Where appropriate we may also define a structure within factor sets to support the construction of indicators, such as distance between situations, that may be useful in comparisons. A grouping of factors into these sets may be readily determined by considering the questions most relevant to provide the information necessary for performance evaluation. For example, we may consider questions like "What is the radio trying to do?", "What can the radio do?", and "What is the radio doing?" as those that probe for the fundamental knowledge necessary for evaluation of radio operation. Each of the questions defines an independent manner in which the performance can be considered, or, loosely, a dimension of the topology over which the performance function is defined. Note that these questions encompass subquestions that examine a finer granularity of factors, e.g., "What is the radio trying to do?" includes questions about what goal is the radio pursuing and where the radio is trying to operate. Consideration of the possible answers to an individual question then allows us to construct a set of factors, possibly containing subsets as answers to subquestions, for the performance function.

The implications of each individual factor set on the final performance must be carefully determined. As an analogy, let us briefly consider competitive marathon running, where some tracks are known to be faster than others, even for a consistent group of runners. Clearly, there are several factors that contribute to time of an individual runner, which provides the metric for performance in this case, but the weight of each factor is certainly not always clear. Perhaps a runner consistently performs better on a particular track due to the altitude or perhaps a local restaurant serves the perfect fuel for the runner. Ideally, the contribution of each set to the performance would be directly evident and comparable to the contribution from any other set. However, in practice this is not the case and the weight of influence for any particular set may not be assumed. Organization of factors into sets does not necessarily imply any given relationship between them.

Here, we examine the factors for the performance evaluation of cognitive radio following this set organization as based on the high-level questions discussed above. Later in this chapter, we examine approaches for estimating the performance function over a particular collection of the sets discussed here.

6.3.1 Purpose

The first set of factors to consider are those connected to the question of what the radio is trying to do. This set is referred to as the set of purpose factors. The *purpose* set is primarily composed of two subsets connected to the subquestions of what goal a radio is pursuing and where, covering both the geographic location, operating frequency, and local radio frequency (RF) obstacles, the radio is trying to operate; consideration of these subsets is our focus here. We refer to these subsets as the mission and environment subsets respectively. The mission captures information related to the user objective for the CR and helps to define the concept of goodness for the radio. The environment information, on the other hand, captures the context for accomplishing this objective.

6.3.1.1 Mission

The mission subset captures the user's desires with regard to the operation of a CR. As such this is a somewhat special subset in that it provides the connection to the user and encompasses the subjectivity of performance. Due to the central importance of user desire, the specifics of mission information typically carry a large weight in performance evaluation.

The most straightforward structuring for the mission subset is a m-dimensional space where each dimension corresponds to a desired mission characteristic. For example, one may consider a mission space that includes throughput, latency, and power efficiency. Each point within the mission space then represents a collection of desired levels for the characteristics specified. Users may define regions within this mission space that represent desirable locations for operation. These regions then provide the target points for CR operation and may be considered attractors in the mission space. Further, note that the characteristics used to define this mission space will often also be used as the metrics for performance of the CR when considering the partial performance over the mission set. The distinction between mission characteristics and performance metrics is subtle, but conceptually significant; the characteristics of the mission space capture only the desired outcome of CR operation, while the metrics of performance provide a succinct description of the full operation of the CR. That is mission characteristics only provide a portion of the information captured by performance metrics in the general case.

The structure of mission spaces as discussed above requires a variety of mission spaces to discuss all of the possible missions for CR. Within this collection of possible mission spaces, there are several different levels of mission abstraction that can be considered. Here, we discuss abstraction in terms of the number of atomic dimensions required to describe a particular group of missions, taking atomic dimensions to be those that cannot be further subdivided. In this construction, higher levels of abstraction are, by necessity, combinations of lower levels. For example, an engineer working in a lab often considers low-level missions such as achieving a certain bit error rate (BER) or throughput threshold. A fire chief using a radio in the field, on the other hand, is more concerned about extinguishing a building fire safely. From the fire chief's point of view, the mission may be personal safety or speed while extinguishing the fire, which both feel a large impact from the use of radios. In the high-level missions for the fire chief, a good radio will have to achieve some combination of BER and throughput levels as well as some other high-level tasks like prioritizing data. When comparing CRs with divergent mission subsets it may be possible to compare them in terms of projections into the highest abstraction mission set shared by both.

Within a mission subset, the concept of distance enables comparison between different targets for CR operation. Since a mission space is geometric m-dimensional space of target points, the similarity between two missions, with the same dimensions, may readily be discussed in terms of the distance between the points or regions that represent the missions. This approach is advantageous for considering the application of a CR beyond a single intended purpose, as it enables estimation of performance for mission close to their design target. Note that Euclidean distance need not be the best measure applied for this purpose as other measures may better represent the trade-offs required for traversing mission space. This mechanism provides a quantitative method to examine the similarity in achieving various targets.

6.3.1.2 Environment

The environment subset holds the context for performance evaluation. Knowledge of the environment often directly reveals the reason for the actions of a CR, as the environment captures the various external stimuli that inform the cognitive radio's behavior. In this way, the environment often determines the variety of approaches exhibited by a CR during some situation. For example, a rather difficult environment (fast-fading, multi-path, many interferers) may force some radios to exhibit a broad range of skills to handle the environment. Alternatively, a difficult environment may simply overcome a radio that performs very well in a less extreme environment. Thus, the level of difficulty presented by an environment can greatly impact the behavior, and therefore the performance, of CRs in various ways. Understanding this impact of environment on radio performance is key to comparing performance across situations.

As the environment subset captures the external stimuli for a CR, the environment is most clearly expressed as an e-dimensional space, based on the perspective of the radio under test. For example, if the radio under test only understands RF information, the dimensions of the environment space could be the frequency, bandwidth, power, and timing of external signals. As the cognition of the radio grows to include more information, perhaps the physical location of the radio, these factors become directly important for evaluation as well. Expressing the environment in terms of the agent's understanding most clearly highlights the challenges faced by the cognitive element of the radio. Further note that this structure mirrors that of the mission subset and the concepts discussed there for abstraction and distance apply here.

6.3.2 Language

The second set of factors for CR evaluation are those connected to the question of what the radio can represent with its actions. This set of factors is the set of potential actions that

CR may take in pursuit of the objectives captured within this purpose set discussed above. We find many parallels between this set of factors for CR performance evaluation and language, leading to our examination of this set of factors in linguistic terms. The possibilities, for both acting and reasoning, available to a CR are defined by the knobs and meters employed by the cognitive element of the radio. Note that the important element within the scope of performance evaluation is which knobs and meters the CR uses rather than the values of the knobs and meters. This approach provides the means to discuss the options available to a CR.

Note that the concept of language for a CR is certainly complex and deeply impacts its performance. For these reasons, it is helpful to first consider some concepts from the study of human language. Human thought provides the only truly observable form of intelligence and as such provides the only truly stable foundation for the consideration any aspect of intelligence. As language is effectively the means by which intelligence is communicated (both internally and externally), consideration of language allows us to delimit the options available to a CR.[2]

6.3.2.1 Human Language

Modern linguistics largely agrees that human language is based on a highly parameterized structure, typically referred to as the Universal Grammar [2]. This structure describes every human language system in use and is based on the setting of various binary parameters. These parameters describe how phrases and mental language lexical items fit together. Mental language lexical items are the smallest units of self-contained meaning, and phrases are collections of these lexical items. Other parameters are then used to determine the pronunciation of these units of language.

This language structure provides the manner in which some pure thought or truth is encoded into a communicable form. Figure 6.1 shows the progress of this encoding in block diagram form. In this figure, the elements of the encoding process are organized into three tiers. The pure truth and Universal Grammar both occupy the top tier as Platonic ideals; both of these elements are the same for everyone. This top tier represents the foundation of the encoding. The next tier is the realization of language in a particular person. This realization is specific to one person and built up over time through experience. Such specificity does not mean that only one person understands this language, but rather that the small details of the language are unique to one person. This middle tier represents the encoding system used. Finally, the bottom tier gives the statement which is the expression of the pure truth by the realization of the pure Universal Grammar. This tier is distinguished as containing the result of a realization of the Universal Grammar acting on the pure truth. This bottom tier represents the product of the encoding. Note that no thought is ever expressed in pure form, but each communicated thought is a pure thought that has been encoded with some language. This encoding process allows communication, but results in each thought that is communicated undergoing some language-specific encoding.

We emphasize that this encoding of thoughts occurs both for thoughts that are destined for communication with external entities and those intended for internal communication. On a high level, intelligence is the consideration and manipulation of thoughts in a self-contained manner (typically contained in a single brain). Within the brain these thoughts are not generally

---

[2]Thanks to Dr. Joe Eska, English Department Chair at Virginia Tech, for providing background and guidance with this discussion of cognitive language.

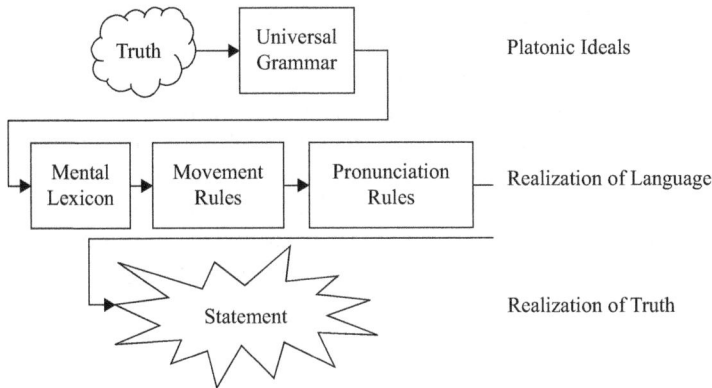

Figure 6.1 Encoding of some truth with language. ©2012 Nicholas J. Kaminski. Reprinted, with permission, from Reference 1.

operated on in their pure form, but are rather encapsulated in some language encoding first. Thus, the ability of the encoding used to describe concepts directly affects intelligence. These concepts of the impact of language on intelligence are further explored in the Sapir-Whorf Hypothesis (linguistic relativity) [3].

Extending these concepts to the performance evaluation, it is clear that language both affects which actions are externally reported and how the action itself is determined in the first place. Due to these effects, language, both internal and external language, must be considered during performance evaluation. Internal language encompasses the CR's understanding of the world and the items over which reasoning occurs. External language defines the expression of a CR to the external world. The scope of a CR's language effectively shapes the radio's ability to notice individual aspects of the scenario at hand and react to them appropriately. As such we refine the question of what can the radio do into the subquestions of what can the radio reason about and how can the radio act. We examine the subsets connected to these questions with the scope of internal and external language.

6.3.2.2 Internal Language

The internal language of the CR shapes the ability of the radio to notice and reason about situations. As such this is fundamentally related to the expressive power of the elements that the CR uses to represent the situations internally. Naturally, this representation is provided by the meters of the CR, whether sensed or internally calculated. The meters then provide the lexical items of the CR's internal language. The choice of meters determines the types of situations available to the attention of the CR. Furthermore, the granularity of each meter greatly impacts the expressive power of the meter; i.e., 2 potential values for sensed power are much less expressive than 20. Recall that truth must be encoded into language to make it communicable as a prerequisite for reasoning, and that this encoding distorts the truth into some representation based on the expressive power available in the language used. In the case of CR, we consider the expressive power of language to be primarily based on the lexical items of the language. The movement rules of a CR's language relate to the manner in which meters maybe manipulated for use in reasoning and are not tractably accessible. Pronunciation rules are largely insignificant for internal language and not considered here. We therefore considered the subset of internal language to consist of both a meter type and the potential values that the meter may assume.

Note that the set of language is not as easily structured as that of purpose. Within language, the notion of expressibility, which is central to determining the possibilities available to a CR, requires consideration in a manner that does not lead itself to structuring into a geometric space. However, this requirement does support the determination of a particular performance metric for the suitability of the language employed by a CR for a given scenario. In the case of internal language, the ratio of the observed entropy of meters to the maximum entropy for the set of meters employed provides a clear indication of the suitability of the employed set of meters for the situation at hand.

6.3.2.3 External Language

The external language of the CR shapes the statements made by a CR in the form of actions. In a manner analogous to the internal language, the lexical items of the external language are the knobs of the CR. In this case, the movement rules of the external language relate to the manner in which knobs may be applied; for example, throughput and bandwidth may not be set in a completely independent manner. The pronunciation rules relate to considerations of real hardware, which typically relate to time required to set values in this case. External language employs the same structuring and metric calculation mechanisms as internal language.

6.3.3 Actions

The final set of factors for CR are those connected to the question of what is the radio doing. Naturally, these are the actions of the CR. Conceptually, the action set is an independent set encompassing the actual impact of the CR on the environment. However, these actions represent some truth that must be encoded in some language to be useful for performance evaluation. In practice, this means that the set of actions depends on the other factor sets, especially that of language for expressing the manner in which actions are recorded or expressed by the language as discussed above. Alternatively, the actions may be encoded in some separate language, based on some separate language system, but this is typically unnecessary.

The set of actions therefore tend to follow the format employed by the set of language, specifically with regard to the subset of external language. The primary structural difference between these two sets is the need for the inclusion of a cost value for each action. Actions carry some cost in terms of energy, time, or computation that should be varied across each individual action. Some combination of these costs is the most typical metric associated with the action set of factors

6.4 Practical Evaluation Methods

As discussed above, performance evaluation is the search for understanding of the performance function of a given CR. Here, we discuss methods for conducting this search in a practical manner, beginning with our own framework for approaching this topic in order to offer a general strategy for the practical assessment of CRs. Figure 6.2 displays the structure of our framework for evaluation. As illustrated, evaluation consists of a sequence of four phases, each depending on the output of the prior phase, to arrive at an estimate of the performance function for a given CR. The first two of these are purely practical necessities, while the final two provide the core

```
Setup → Logging → Encoding → Interpolation
```

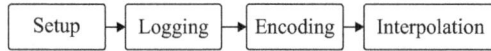

Figure 6.2   Structure of evaluation.

of the evaluation process. This structure provides a general framework to discuss the operation of performance evaluation and below we discuss a few options for each phases. Furthermore, this strategy is designed to be flexible in that the complete approach needs not be applied for every evaluation; rather, an evaluator is free to apply only the portion of the methodology required for their own assessment.

6.4.1 Setup

The setup phase of evaluation encompasses the mundane but extremely practical necessities of preparing the CR and environment for assessment. During this phase, much of the information that influences performance, as discussed in the previous section, will be determined, including environment information, mission information, and specification of the CR itself. Naturally, these elements must be carefully considered and noted for any success for assessment to occur.

6.4.2 Logging

Once the setup phase is complete, the logging phase can begin collecting data for evaluation. This phase encompasses the actual operation of the CR and the observing of its operation. The CR will work autonomously toward its goal, saving knob and meter values at some intervals. These records should be noted each time there is a significant change in either knob or meter values. This process results in a time series of the knob and meter values that is sampled non-uniformly and will be analyzed in subsequent evaluation phases. Recording continues until some end state is reached.

There are three general possibilities for end states of the logging phase of performance evaluation: (1) goal reached, (2) timeout, and (3) error. In some situations, the CR is working to a terminal goal that provides a clear end state. For example, a CR may be attempting to transfer a fixed amount of information. In these situations, the logging phase ends when the goal is reached. However, several missions do not imply such a terminal state and instead focus on continued operation within some bounds. For example, a CR may need to support continued communications with a given BER level. Naturally, for purely practical reasons an evaluator may want to provide a definite time limit for a particular assessment. In these situations, logging ends after some timeout occurs, which may be defined in terms of the total length of the assessment or in terms of the length of time in consistent operation. Finally, an evaluation may end simply because the CR crashes. This type of assessment termination is referred to as an error end state. These three end states encompass the general possibilities for CR performance logging and must be accounted for during an evaluation.

6.4.3 Encoding

The encoding phase begins the post processing that is central to performance evaluation. During this phase of the process, the samples of CR operation are encoded into a form that highlights various aspects of radio performance; that is, the observations are used to generate a set of metrics that together indicate the influence of the several factors discussed in the previous section. Here, we present a generic collection of such metrics that provide wide applicability for a broad range of CRs and operating scenarios. Encoding observations of radio behavior into this set of metrics provides several samples of the performance function, thus offering a route to estimation of this function.

In order to develop our general set of metrics, we will first examine the evaluation approach taken by Willink and Rutagemwa in Reference 4. These authors approach the performance evaluation from a network viewpoint, but much of the proposed approach is applicable well beyond that scope. Willink and Rutagemwa view a cognitive radio network as a complex system; that is, they notice that cognitive radio networks have some level of well-defined structure, while also exhibiting many variations [5]. Due to this complexity, "what is seen often depends on the size of the observer" [5], or perhaps more plainly "a more holistic approach to performance evaluation is required" [4], compared to the more typical targeted approach of considering separate components of a CR. Willink and Rutagemwa conjecture that the focus of performance evaluation should be on the ability of a cognitive radio network to deliver required services rather than a detailed characterization of the communication approach. To this end, they introduce six descriptors: effectiveness, survivability, efficiency, stability, security, and legality. Each of these descriptors measures an aspect of the general success of an arbitrary CR network with regard to an arbitrary task. These capacity descriptors provide a descriptive power that makes them well suited to serve as generic performance metrics.

The descriptors of Reference 4 are used to describe the various aspects of a cognitive radio network, such as, its effectiveness at accomplishing a given task or survivability in harsh environments. We apply these same descriptors here to capture a general assessment of a cognitive radio. Each individual descriptor provides a single metric that relates some specific as of CR operation, as driven by a particular group of factors. Taken together these metrics capture the overall performance of the cognitive radio for the samples collected during the logging phase. Table 6.1 provides brief definitions of each metric, based on the descriptors of Reference 4 as applied to more general performance evaluation.

Each metric is focused on a different aspect of CR operation. For clarity, we categorize these metrics into three groups based on the qualities they measure. The first group considers the impact of a CR operation on the continued performance of the radio. This continued performance group is made up of the survivability, efficiency, and stability metrics. The second group focuses on policy issues. This policy group simply contains the legality metric. The final group centers on higher level descriptions of performance. This high-level group, the security and effectiveness metrics, captures broad information that is heavily tied to the application

Table 6.1 Definitions of generic metrics.

Metric	Measured quality
Survivability	Impact of action on long-term operation of radio
Efficiency	Use of radio's resources in performing action
Stability	Likelihood of action causing unaccounted for state/oscillation
Legality	Potential of action to violate policy
Security	Vulnerability to malicious users caused by action
Effectiveness	Impact of action on current situation

of the CR. This organization of metrics extends their ability to describe CR operation and therefore relate an estimate of the performance function.

The continued performance group of metrics forms the foundation of performance evaluation; a radio must be operational to perform well. In this role, metrics within this group relate the degree to which a CR's operation may be sustained. For example, the survivability provides information related to damage done by an action. While this is seemingly not a concern for electronic systems, in truth, operations cause physical equipment to wear out and in many scenarios, such as long-term deployment, this effect can dominate the performance of a radio. Alternatively, the efficiency metric focuses on describing the resources consumed by a particular action. Note that within this context, resources are anything of limited quantity that is required for and consumed by the operation of a radio. The resources that limit the operation of a CR are most typically power, spectrum, and computational ability. Finally, the stability metric relates the potential of an action to destabilize a CR. This destabilization may cause oscillations or unwanted states in either the RF hardware or, more typically, in the cognitive engine (CE). Taken together these metrics describe a CR's ability to continue operating.

The policy group directly relates the legality of a CR's action. Note that the legality of a CR is often tied to the high-level purpose of the CR. For example, the CR of an emergency responder may be restricted from certain frequencies, unless in use during a disaster. More simply, legality of an action tends to vary with the location of the CR. As such the legality metric indicates whether the operation of a CR is legal in a particular context.

The final group of metrics relates high-level application consideration. The security metrics describe the vulnerability to some malicious user that results from a CR's operation. The type of vulnerability that is of most concern varies with the mission of the CR. For example, if a CR is used for some stealth-based purpose, vulnerability may equate to detectability. However, if the mission is instead to relay sensitive information, the security metric is more tied to the intelligibility of the information to some malicious observer. Also, for some missions, the security metric may not be of concern. The effectiveness metric provides the greedy performance of a CR. Specifically, this metric captures the progress toward a directly measurable goal. For example, if the goal of a mission involves transmitting some amount of data, successfully sending a large packet has high effectiveness. These final two metrics round out our generic set with strict focus on application considerations.

These metrics provide the conceptual framework for the generic consideration of the performance of a CR. Furthermore, the metrics discussed above may be used in the operation of performance evaluation. To accomplish this, we assign values between 0 and 1 to each metric, where 0 represents an absence of the measured quality and 1 represents full presence of that quality, for each sample collected during the logging phase. The result of this is a time series of metrics that describe the operation of a CR on the basis of its actions. Typically, this assignment is accomplished through the application of heuristic functions that quantify each metric. Note that application is not a means to remove subjectivity, which is inherent to performance evaluation; rather, the use of heuristic functions simply provides a measure of consistency to the process. In the next section, we present a brief example of performance evaluation including several heuristic functions.

6.4.4 Interpolation

The interpolation phase supports the other phases to provide a more complete picture of the performance of the CR under test. Within the context of performance evaluation, interpolation estimates performance metrics for knob and meter values between those sampled. Recall that the goal of performance evaluation is the estimation of the performance function of the CR. In pursuit of this, the setup phase determines the general parameters that affect the performance function, the logging phase collects raw samples from the function, and the encoding phase translates these raw samples into metrics with a defined and structured relation to performance. The interpolation phase offers a means to fill in the gaps between observed factor settings. Such an ability may be necessary in the situation that additional testing is impossible or cumbersome to perform. As such this phase offers additional utility to the evaluator, but is not always necessary.

There are certainly numerous techniques that may provide the interpolation of performance samples collected by the prior phases, however, artificial neural networks (ANNs) are particularly well suited to this task. First, ANNs display significant prowess in the general interpolation of functions [6]. In particular, Serra-Ricart and his co-authors show that ANNs provide one of the few approaches capable of interpolating vector functions. The authors of Reference 6 note three main properties of ANN-based interpolation, which they refer to as a neural network interpolator (NNI):

1. It allows complex multidimensional maps with minimal *a priori* information; that is, the functional form of f is not specified in advance.
2. The NNI treats the data in a global way, so it is an efficient method to deal with unbinned and/or sparse data.
3. The noise-resistant property of the neural network techniques gives the NNI robust behavior when one needs to deal with statistical fluctuations of real data.

Furthermore, Serra-Ricart *et al*. [6] also provide several guidelines for implementing ANN-based interpolation with these properties. For purely positive output, the authors suggest sigmoid activation functions as they model biological neurons well. The hyperbolic tangent maintains this shape over both positive and negative outputs, which provides for more general operation. The authors underscore the expressive ability of this type network, noting that in the ideal case a three-layer network with this style of activation function has the expressive ability to realize any continuous function. However, in practice, three-layer network, trained with back propagation face convergence issues in some cases. Thus, it is typically better to use one hidden layer for each important local maxima. Through the use of these guidelines, the authors of Reference 6 display the potential of ANNs for general interpolation.

Care must be taken, however, to ensure that ANNs provide interpolation rather than extrapolation. The distinction between the two is made in the training sets used to set the weights of the ANN [7]. ANNs used for interpolation are trained with examples that cover the full range of inputs that will be used in operation. Extrapolation ANNs, on the other hand, are trained based only on a portion of their functional domain. Barnard and Wessels show that interpolation-based ANNs perform well, while extrapolation-based ANNs do not [7]. Thus, an evaluator must collect sample points from all ranges of interest during the logging phase, rather than exploring new ranges during interpolation.

For use in the interpolation phase of performance evaluation, an ANN would be trained using the knobs and meters observed during the logging phase as inputs and the metrics produced by the encoding phases as outputs. Once trained, the evaluator may present theoretical knob and meter combinations to the ANN to determine the performance of particular actions in a given scenario. That is, ANNs provide a means to estimate the performance of a CR without having to observe the CR in a particular scenario. As such, the trained ANN represents an estimation of the performance function, over the range of samples observed in the logging phase.

In effect, the trained ANN becomes a model of the CR. Note that this modeling provides the means to overcome various limitations in the evaluation of the actual CR. These limitations may occur as practical limitations in terms of the time spent in evaluation or the conditions under which evaluation occurs. Alternatively, there may be limitations in terms of the knobs and meter values available to the CR. The model provided by the ANN offers a means to overcome these limitations.

6.4.5 Alternative Approaches to Evaluation

Despite the many associated challenges and importance of performance evaluation of CRs, there are few examples in the literature examining this topic. Most available work in this area targets specific applications of the cognitive radio concept, typically focusing on dynamic spectrum access (DSA); however, there are several notable examples of various methodologies for determining the characteristics of cognitive radios and identifying elements of the performance function. Zhao et al. provide a report-card-like system which allows some comparison of DSA techniques [8]. The approach taken in this work categorizes the factors that contribute to CR in the application of DSA and score the operation of the CR with regard to each. Dietrich *et al.* construct a cognitive model by probing a DSA CR with a psychometric approach which constructs a model of the CE to determine its potential to find useful solutions in Reference 9. Thompson *et al.* eschew detailed modeling of internal operation in favor of decoding the operation purely from observation of DSA CR behavior in Reference 10. Each of these works represents advancements in the characterization of CR performance, starting with the identification of important factors, to the modeling of the cognitive process, and finally to the determination of performance from pure observation. While the above works limit their focus to the application of DSA the process discussed above for general purpose CR evaluation owes a great deal to the advancements of these efforts.

6.5 Example Evaluation

In order to clarify the approach to performance evaluation discussed in Section 6.4, here we present a brief example of the first three phases of performance evaluation.

6.5.1 Setup Phase

As discussed previously, the setup phase focuses on the mundane necessities that determine many of the factors that influence performance. Within this example, we will use this phase to specify the mission and the CR itself. Each of these factors will subsequently influence the performance of the radio and our efforts to understand this performance.

Mission: For the purposes of this example, the mission of the CR is to simultaneously minimize the bandwidth, transmit power, and BER across a single radio link. The channel for this link is an additive white Gaussian noise (AWGN) channel, with a normally distributed average noise power featuring a mean of -15 dBm and variance of 2 dBm. For this mission, there are no security concerns, but transmissions above 10 dBm are deemed to be illegal. This mission is completed when the CR can determine knob settings that result in a link that requires no

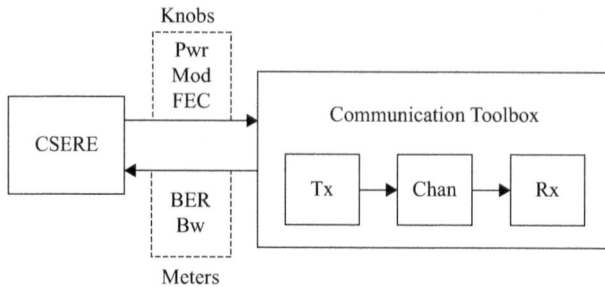

Figure 6.3   Example CR architecture. ©2012 Nicholas J. Kaminski. Reprinted, with permission, from Reference 1.

more than 10 MHz in bandwidth and 0 dBm in transmit power to achieve less than 10^{-10} BER. The CR is allowed to consume as much time as is necessary to reach this ending conditions.

Cognitive Radio: To avoid unnecessary complication, we use a CR that consists of a CE that is able to control both ends of the link through ideal control channels. However, the CE is given no information about the channel and must, instead, probe the channel for information. The CE has three available knobs: modulation, coding rate, and transmit power. Modulation is restricted to digital M-ary modulation, where M may be 2, 4, or 8. Available transmit power ranges from -10 dBm to 15 dBm. Forward error correction (FEC) is provided by BCH coding where the options are (15, 11), (15, 7), or (15, 5) codes. In addition to these knobs, the CE has the two meters of BER and occupied bandwidth. Since the CE has access to both ends of the link, the BER meter is simply the measured BER over the link. The bandwidth is calculated from the symbol rate and roll off factor ($\alpha = 0.5$) as shown in (6.1). The radios are stationary and the data rate at the input to the transmitter is fixed at 6 Mbps.

$$B_{occ} = R_s(1 + \alpha) \qquad (6.1)$$

Figure 6.3 provides the architecture of the CR used for this proof of concept. For this example, the entire CR is implemented in MATLAB; see Section 6.6. The radio portion of the CR employs MATLAB's communication toolbox, and the CE applies an implementation of Cognitive System Enabling Radio Evolution (CSERE), as discussed in Section 2.5.5. This configuration allows a simplified environment for evaluation. Note that from the perspective of performance evaluation, the optimization techniques employed by the CR do not matter; rather, here we are concerned only with estimating the performance function.

For the purposes of comparison, we evaluate two versions of the CR as described here. To highlight the abilities of the performance evaluation system, we intentionally hinder one of the two CRs by restricting it to only use the highest level of FEC protection. In the remainder of this example, we refer to the uninhibited CR as the free FEC CR and the hindered CR as the fixed FEC CR.

6.5.2 Logging Phase

As discussed in Section 6.4.2, logging consists of recording the actions of the CR. Within this example, the operation of the CR proceeds in a consistent sequence, consisting of sensing

Table 6.2 Free FEC cognitive radio knob and meter time series.

Action index	Modulation M	FEC level	Power (dBm)	Bandwidth (MHz)	BER
1	4	2	−10	9.6899414	2.6562500e−01
2	4	1	−10	6.1962891	6.9335938e−02
3	4	1	−10	6.1962891	1.5820312e−01
4	4	1	−10	6.1962891	1.4550781e−01
5	8	1	0	4.1308594	3.9062500e−03
6	8	1	0	4.1308594	5.8593750e−03
7	8	1	−5	4.1308594	1.5820312e−01
8	8	1	0	4.1308594	1.1718750e−02
9	8	1	0	4.1308594	1.8554688e−02
10	8	1	0	4.1308594	5.3710938e−02
11	8	1	0	4.1308594	8.0078125e−02
12	8	1	0	4.1308594	3.3203125e−02
13	8	1	0	4.1308594	1.7578125e−02
14	8	1	0	4.1308594	3.3203125e−02
15	8	1	−10	4.1308594	2.3535156e−01
16	8	1	−5	4.1308594	1.8847656e−01
17	8	1	−5	4.1308594	6.8359375e−02
18	8	1	−5	4.1308594	2.5097656e−01
19	8	1	0	4.1308594	4.8828125e−03
20	8	1	0	4.1308594	0

the environment, comparing the current situation to prior experience, optimizing knobs, and applying a solution. Once the solution has been applied, the selected knob and meters are recorded along with their result. Note that the environment noise is changing randomly, with statistics described above, which results in variable performance of any given knob and meter combination. At the start of evaluation, the CR simply has an empty knowledge base. As logging proceeds, we record the actions of both of our CRs in their pursuit of the mission specified above and report the results in Tables 6.2 and 6.3.

6.5.3 Encoding Phase

Recall from Section 6.4.3 that the encoding phases apply heuristics to translate the recorded knobs and meters into generic metrics of performance. Here, we present examples of such metrics and their results for describing the performance of the evaluated CRs.

Continued Performance Heuristics: This group contains the metrics of survivability, efficiency, and stability. Survivability provides a measure of the cognitive radio's self-damage. In this case, the primary feature that indicates whether self-damage has occurred is the end state of the logging phase. Of the two possible end states, failure provides the most information about survivability as it indicates that the CR did not survive its full operational period. As each action of the CR contributes to the final state of the CR, a failure state is the result of each action negatively affecting CR survivability. However, it is reasonable to assume that

Table 6.3 Fixed FEC cognitive radio knob and meter time series.

Action index	Modulation M	Power (dBm)	Bandwidth (MHz)	BER
1	8	5	9.0087891	0
2	8	5	9.0087891	0
3	8	15	9.0087891	8.7890625e−03
4	8	5	9.0087891	0
5	8	10	9.0087891	0
6	8	−10	9.0087891	1.6601562e−02
7	8	15	9.0087891	0
8	8	10	9.0087891	8.7890625e−03
9	8	5	9.0087891	0
10	8	0	9.0087891	0

the actions closer to the failure contributed more to the failure than to the remote actions. Equation (6.2) displays the heuristic used for survivability.

$$\text{Survivability} = \begin{cases} 0.5 - \dfrac{x}{2n} & \text{if end state is failure} \\ 1 & \text{otherwise} \end{cases} \tag{6.2}$$

In (6.2), x is the position of the knob and meter set in the times series and n is the position of the failure. This relation provides the desired behavior of producing a lower survivability for actions close to failure, while keeping the value between 0 and 1. In cases other than failure, survivability is simply set to 1 for each knob and meter set.

Efficiency provides a measure of the CR's utilization of necessary, but limited resources. In this case, the limited resource for the CR is spectrum. Marshall provides some guidance for developing a heuristic for this metric by convincingly arguing that consideration of bandwidth alone is not enough when determining use of spectrum [11]. He notes that consideration of bandwidth alone often leads to high order modulation schemes, to reduce bandwidth, and high transmit power to reduce BER. However, raising the transmit power to allow high order modulation causes signals to propagate further and interfere over a larger geographic area. Therefore, in order to avoid unduly rewarding CRs that trade width in spectrum for width in geography, determination of the efficiency metric must consider this trade off.

Based on the need for capturing the trade off highlighted by Marshall, we assign a value for bandwidth efficiency and then penalize this value based on power usage. Typically setting a value for bandwidth efficiency involves normalizing the spectral efficiency (data rate over bandwidth) by the Shannon bound, given in (6.3). This provides a comparison to how the CR performs with respect to the ideal (100% efficient) case. However, determination of the Shannon bound requires knowledge of the signal-to-noise ratio, which is typically unavailable. Thus, here an empirical upper bound on the signal-to-noise ratio, based on radio type, is used to provide an upper limit on the Shannon bound and thus efficiency. In this case, we use an empirical upper bound of 30 dB. Note that use of this upper bound reduces the value for efficiency, as a penalty for the inability of the CR to directly consider the signal-to-noise ratio. Once the bandwidth efficiency is determined the penalty for power efficiency must be applied.

As the power efficiency penalty represents geographical interference, a perfectly efficient signal, from a power perspective, uses no power and a totally inefficient signal uses infinite power. Thus, the power efficiency penalty is modeled by (6.4), where P is the power efficiency penalty and x is the power in watts. Once both the bandwidth efficiency and the power efficiency penalty have been determined, the efficiency is simply the bandwidth efficiency minus the power efficiency penalty.

$$\frac{C}{B} = \log\left(1 + \frac{S}{N}\right) \quad (6.3)$$

$$P = 1 - (1 + x^2)^{\left(\frac{1}{2}\right)} \quad (6.4)$$

Stability, the final metric of the continued performance group, captures the tendency of an action to cause an oscillation. While there are two major types of oscillation that affect CRs, RF and cognitive, our evaluation will only consider cognitive oscillations, as the radio is simulated. For this example, we will define a cognitive oscillation as any time the CR applies the same sequence of knobs repeatedly producing the same sequence of meters. There are two possible types of cognitive oscillations, recoverable and non-recoverable. Recoverable oscillations are those that CR eventually breaks, and non-recoverable oscillation continues for the remainder of operation. In either case, the action directly proceeding a state loop and actions within the loop all receive a 0 for stability. The action that breaks the loop receives a 1. All other actions receive a stability value of 1. Both types of loops are treated the same way, because the length of the loop directly provides the number of times an action is labeled unstable.

Policy Group Heuristics: The policy group of metrics, containing only the legality metric, is perhaps the most straightforward, in terms of heuristics. If a set of knobs and meters fall within the policy, as defined as part of the mission in the setup phase, the set receives a legality value of 1. Otherwise, the set receives a value of 0.

High-level Group Heuristics: The high-level metrics include the security and effectiveness metrics. As specified in the mission definition above that security is not a concern here. Recall that the effectiveness metric captures the greedy performance of the CR. In this case, since the objective is minimization of the bandwidth, power, and BER, the effectiveness metric is calculated as one minus the average of the normalized distance from the minimum possible value in these three areas, see (6.5). Note that this metric is often very similar to the scoring function used with the CR to make decisions and therefore often provides an estimate of how the CR views its own operation.

$$\text{Effectiveness} = 1 - \frac{1}{3}\left(\frac{bw - bw_{min}}{bw_{max} - bw_{min}} + \frac{pwr - pwr_{min}}{pwr_{max} - pwr_{min}} + \frac{ber - ber_{min}}{ber_{max} - ber_{min}}\right) \quad (6.5)$$

Results: Tables 6.4 and 6.5 display the results of the encoding of the samples collected during the logging phase, using the heuristics discussed above. Note that the metrics describe the evolution of the performance of the CR at each action point.

6.5.4 Interpolation

As discussed previously, the process of interpolation allows investigation of any gaps left by the encoding process; however, interpolation is not a perfect process. We propose the use of neural

Table 6.4   Free FEC cognitive radio metric values.

Action index	Survivability	Efficiency	Stability	Legality	Security	Effectiveness
1	1	0.415190225848	1	1	1	0.741984802834
2	1	0.649286835382	1	1	1	0.923706494736
3	1	0.649286835382	1	1	1	0.864461706736
4	1	0.649286835382	1	1	1	0.782925246736
5	1	0.973930745572	1	1	1	0.997395833333
6	1	0.973930745572	1	1	1	0.99609375
7	1	0.973930295572	1	1	1	0.894531253333
8	1	0.973930295572	1	1	1	0.9921875
9	1	0.973930295572	1	1	1	0.987630208
10	1	0.973930295572	1	1	1	0.964192708
11	1	0.973930295572	1	1	1	0.946614583333
12	1	0.973930295572	1	1	1	0.977864583333
13	1	0.973930295572	1	1	1	0.98828125
14	1	0.973930295572	1	1	1	0.977864583333
15	1	0.973930250572	1	1	1	0.84309896
16	1	0.973930295572	1	1	1	0.87434896
17	1	0.973930295572	1	1	1	0.954427083333
18	1	0.973930295572	1	1	1	0.832682293333
19	1	0.973930745572	1	1	1	0.996744791667
20	1	0.973930745572	1	1	1	1.0

Table 6.5   Fixed FEC cognitive radio metrics values.

Action index	Survivability	Efficiency	Stability	Legality	Security	Effectiveness
1	1	0.446587650021	1	1	1	1.0
2	1	0.446587650021	1	1	1	1.0
3	1	0.447082525095	1	1	1	0.660807291667
4	1	0.446587650021	1	1	1	1.0
5	1	0.446632648783	1	1	1	1.0
6	1	0.446582655033	1	1	1	0.988932292
7	1	0.447082525095	1	1	1	0.666666666667
8	1	0.446632648783	1	1	1	0.994140625
9	1	0.446587650021	1	1	1	1.0
10	1	0.446583150033	1	1	1	1.0

networks for this function less because of their accuracy and more because of their utility. The performance function to be interpolated is a complex mapping from several dimensions to several other dimensions, which causes the interpolation process to be complicated. The use of neural networks removes much of this complexity, but this convenience certainly carries a price in performance for the typical cause.

The current case is certainly typical in this regard. Within this example, we have already collected a number of samples that provide a view on how the cognitive radio performed. Let us suppose that we would like to investigate adding granularity to our cognitive radio. Thankfully, the data that we have already collected, combined with the process of interpolation, allow us to estimate the value of such additional granularity. Naturally, the accuracy of these estimations depends on several factors, which we will touch on below, but keeping this motivation in mind we now proceed with our example interpolation process.

For the sake of convenience, we have constructed our interpolator using Python, with code appearing in Section 6.6, although several options are available for this purpose. Our interpolator consists of three steps: data collection, network building, and point testing. For us, data collection simply consists of reading our data files using a get_data function. Network building is slightly more complex, involving the topology creation and training of the neural network. We have selected a fairly simple three-layer topology with a number of input nodes equal to the number of knobs and meters, 20 hidden nodes, and a number of output nodes equal to the number of generic metrics. This topology and the use of specifically 20 hidden nodes have been selected purely for their general application; certainly other topologies or even a different number of hidden nodes may improve performance and readers are directed to Reference 6 for a discussion of some of these factors. This topology is then trained with a back propagation algorithm for 1000 iterations using the collected knobs and meters as inputs and the observed generic metrics as outputs. We have used the PyBrain (http://pybrain.org/) library for network building. Once the training is complete, point testing simply involves presenting an interpolation point to the neural network and observing the output of the network. Recall, however, that for the output to have meaning, the presented point must be within the range of values represented in the collected knob and meter values. We use these three steps to interpolate data for both the free FEC CR and the fixed FEC CR, building a separate network for each.

Since the goal of our interpolation within this example is to examine increased granularity of new knobs to our cognitive radio we have selected test points with knob values between those in our collected set. Specifically, Tables 6.6 and 6.7 provide the test points used here. For both CRs, the first test point is simply the first action index observed above. In all other cases, the test points only contain knob and meter values within the ranges of the data used to train the neural networks. These test points represent the what-if cases for which we would like to have data, but have been excluded from the prior steps of evaluation due to some limitation. Note that there is no inherent dependence of the meters on the knob settings within these points, meaning that a more complete evaluation would require interpolating over a range of combinations and employing some external modeling to select the knob and meter combinations that seem most likely. Here, we simply examine a few points to focus on the interpolation itself. Once these points are presented to the appropriate neural network, the outputs shown in Tables 6.8 and 6.9 are produced.

Examining the results of the interpolation, the neural network is clearly able to achieve sufficiently accurate results that provide an understanding of the CR performance. Certainly inaccuracies exist, in that the observed results are not perfect matched for the first test point and that the metric values are not always within their defined limits. These inaccuracies result in part to the general topology choices discussed above, in part from the low number of training points, in part to the number of input and output dimensions, and in part to lack of pre-processing steps to make the training set more uniform. Methods exist to address each source of inaccuracy, but for our purposes these are not always necessary. As the results stand,

Table 6.6   Free FEC cognitive radio test points.

Index	Modulation M	FEC Level	Power (dBm)	Bandwidth (MHz)	BER
1	4	2	−10	9.6899414	2.6562500e−01
2	4	2	−7	9.6899414	2.6562500e−01
3	4	1.5	−10	6.1962891	6.9335938e−02

Table 6.7   Fixed FEC cognitive radio knob and meter time series.

Index	Modulation M	Power (dBm)	Bandwidth (MHz)	BER
1	8	5	9.0087891	0
2	8	7	9.0087891	0
3	8	−7	9.0087891	0

Table 6.8   Free FEC cognitive radio test results.

Index	Survivability	Efficiency	Stability	Legality	Security	Effectiveness
1	1.00332177	0.40740049	1.00731902	0.99903312	1.00346138	0.72507531
2	1.17855051	0.35378462	0.89142488	1.16429437	1.13702116	0.73253552
3	1.01853154	0.54197682	1.09370102	1.00226387	1.06763197	0.72472557

Table 6.9   Fixed FEC cognitive radio test results.

Index	Survivability	Efficiency	Stability	Legality	Security	Effectiveness
1	1.00332177	0.44556155	1.00085791	1.00008335	1.00115457	1.00317395
2	0.99343451	0.47696623	0.97870119	0.93763579	0.97183651	1.16477784
3	1.70310894	0.6568242	0.79911388	0.61207257	0.78450127	0.30273553

they provide an indication of the proposed additional knob values on the performance of the cognitive radio. For example, based on these results, we can expect allowing the free FEC CR a 1.5 rate FEC would yield improved efficiency with negligible impact on effectiveness over the 2 rate FEC case. Naturally, these results do not include the relationship between knobs and meters, but even still they provide an idea of performance beyond cases directly test during evaluation.

6.6 Example Code

6.6.1 Free FEC Cognitive Radio

This section provides the Matlab code of the evaluated cognitive radios (CRs).

6.6.1.1 Controller

```
function cserel()

pop= opt();
n_pop = size(pop,1);

n_knobs = 3;


count_out = 50;
ber_thres = 0;
bw_thres = 10e6;
pwr_thres = 0;

cur_ber = 99;
cur_bw = 99;
cur_pwr = 99;

good = 0;
counter = 0;
while not(good);

    aug_pop = pop;
    for p=1:1:n_pop
      act = pop(p,:);
      a = size(act);
      ber = kb(act);
      aug_pop(p,n_knobs+1)=ber;
    end


    [max_score, best]=rank(aug_pop);

    [bw, ber] = radio(best);

    kb(best,ber);


    cur_ber = ber;
    cur_pwr = best(n_knobs);
```

```
    cur_bw = bw;

    record = [best bw ber];
    save ('cserel.dat','record','-ASCII','-append')
    counter = counter +1


    if cur_ber <= ber_thres
      if cur_pwr <= pwr_thres
        if cur_bw <= bw_thres
disp('answer found')
disp(counter)
good = 1;
          end
        end
      end

if counter > count_out
    good = 1;
    disp('count out)
end


if not(good)
    pop=opt(best);
end
```

6.6.1.2 Optimizer

```
if nargin<1
    seed = -1;
end


n_pop = 15;
n_knobs = 3;
knob_lev = [3 3 6];
knobs = [2 4 8 0 0 0;
 1 2 3 0 0 0;
-10 -5 0 5 10 15];

pop = zeros(n_pop, n_knobs);


if seed == -1
  % Random Pop
  for p=1:1:n_pop
```

```
   for q=1:1:n_knobs
     top = knob_lev(q);
     idx = randi([1 top],1);
     pop(p,q) = knobs(q,idx);
   end
  end
else
  for p=1:1:n_pop
    direct = n_knobs-mod(p,n_knobs);
    for q=1:1:n_knobs
       if q==direct
pop(p,q)=seed(q);
       else
top = knob_lev(q);
idx = randi ([1 top],1);
pop(p,q) = knobs(q,idx);
         end
       end
     end
  end
```

6.6.1.3 Ranker

```
function [max_score, best]=rank(aug_pop)

max_score = -99
max_idx = -1;

n_pop = size(aug_pop,1);


for p=1:1:n_pop
    mod_sc = aug_pop(p,1);
    rate_sc = aug_pop(p,2);
    pwr_sc = aug_pop(p,3);
    ber_sc = aug_pop(p,4);

    mod_sc = log2(mod_sc);
    rate_sc = -1*rate_sc+4;
    pwr_sc = 4+ pwr_sc/(-5);
    ber_sc = round(-11*ber_sc+6);
    score = sum([mod_sc rate_sc pwr_sc ber_sc]);
    if score > max_score
       max_score = score;
       max_idx = p;
    end
  end


best = aug_pop(max_idx,1:3);
```

6.6.1.4 Radio

```
function [bw,ber] = radio(act)

M = act(1);
rate = act(2);
pwr = act(3);

% Source
x = randi([1,0],1024,1)

%Encoder
switch rate
  case 1
    n = 15
    k = 11;
  case 2
    n = 15;
    k = 7;
  case 3
    n = 15;
    k = 5;
end
len_x = size(x,1);
num_pad = k_mod(len_x,k);
code = [x;zeros(num_pad,1)];
enc = fec.bchenc(n,k);
code = encode(enc,code);

len_code = size(code,1);
eff = len_x/len_code;

%Modulation
mod_len = log2(M);
mod_pad = mod_len - mod(len_code, mod_len);
code=[code;zeros(mod_pad,1)];
h = modem.dpskmod('M',M,'InputType','bit');
y = modulate(h,code);

% Channel
n_pwr = -15+2randn(1);
snr = pwr - n_pwr;
rx = awgn(y,snr);

% Demodulation
h = modem.dpskdemod('M',M, 'OutputType','bit');
```

```
z = demodulate(h,rx);

z = z(1:len_code);

dec = fec.bchdec(n,k);
uncode = decode(dec,z);

uncode = uncode(1:len_x);

bw = 1/log2(M)*1/eff*6e6*(1.5);

bad_count = nnz(xor(x,uncode));
ber = bad_count/len_x;
```

6.6.1.5 Knowledge Base

```
function ber=kb(act,adber)

if nargin<2
  adber = -1;
end

filename = 'kb.dat';
act_size = size(act,2);
row = -1;
% Load Database
if exist(filename)
    dat_base=load(filename,'ASCII');

    act_base = dat_base(:,1:act_size);
    base_size = size(act_base,1);
    % Find Row
    finder = repmat(act, base_size,1);
    finder = act_base-finder;
    finder = sum(finder,2);
    row = find(finder==0);
    if size(row,1)==0
      row = -1;
    end
  else
    dat_base = -1;
  end

  if adber~ = -1
    %In
    sav_act = act;
```

```
    sav_act(act_size+1)=adber;
    if dat_base ==-1;
      dat_base = [sav_act];
    else
      if row == -1
        dat_base = [dat_base;sav_act];
      else
        old_adber = dat_base(row, act_size+1);
        new_adber = mean ([old_adber,adber]);
        dat_base(row,act_size+1)=new_adber;
        end
      end
      save(filename, 'dat_base','ASCII');
      ber =-1;
    else
     %out
     if dat_base == -1
       ber = 0.25;
     else
      if row == -1
         ber = 0.25;
      else
         ber = dat_base(row,act_size+1);
      end
    end
  end
end
```

6.6.2 Fixed FEC Cognitive Radio

This section provides the Matlab code of the evaluated CRs.

6.6.2.1 Controller

```
function csere2()

pop = opt2();
n_pop = size(pop,1);

n_knobs = 2;

count_out = 50;
ber_thres = 0;
bw_thres = 10e6;
pwr_thres = 0;

cur_ber = 99;
```

```
cur_bw = 99;
cur_pwr = 99;

good = 0;
counter = 0;
while not(good);

  aug_pop = pop;
  for p=1:1:n_pop
    act = pop(p,:);
    a = size(act);
    ber = kb(act):
    aug_pop(p,n_knobs+1)=ber;
  end

  [max_score, best]=rank2(aug_pop);

  [bw,ber] = radio2(best);

  kb(best,ber);

  cur_ber = ber;
  cur_pwr = best(n_knobs);
  cur_bw = bw;

  record = [best bw ber];
  save('csere2.dat','record','-ASCII','-append')
  counter = counter +1;

  if cur_ber <= ber_thres
    if cur_pwr <= pwr_thres
      if cur_bw <= bw_thres
disp('answer found')
disp(counter)
good = 1;
        end
      end
    end

    if counter > count_out
      good =1;
      disp('count out')
    end


    if not(good)
```

```
      pop=opt2(best);
    end
  end
```

6.6.2.2 Optimizer

```
function pop=opt2(seed)

if nargin<1
    seed = -1
end

n_pop = 1;
n_knobs = 2;
knob_lev = [3 6];
knobs = [2 4 8 0 0 0;
  -10 -5 0 5 10 15];

pop = zeros(n_pop, n_knobs);

if seed == -1
  % Random Pop
  for p=1:1:n_pop
    for q=1:1:_knobs
      top = knob_lev(q);
      idx = randi([1 top],1);
      pop(p,q) = knobs(q,idx);
    end
  end
else
  for p=1:1:n_pop
    direct = n_knobs-mod(p,n_knobs);
    for q=1:1:n_knobs
      if q==direct
pop(p,q)=seed(q);
      else
top = knob_lev(q);
idx = randi ([1 top],1);
pop(p,q) = knobs(q,idx);
      end
    end
  end
end
```

6.6.2.3 Ranker

```
function [max_score, best]=rank2(aug_pop)
```

```
max_score = -99;
max_idx = -1;

n_pop = size(aug_pop,1);

for p=1:1:n_pop
  mod_sc = aug_pop(p,1);
  pwr_sc = aug_pop(p,2);
  ber_sc = aug_pop(p,3);

  mod_sc = 2*log2(mod_sc);
  pwr_sc = 4 +pwr_sc/(-5);
  ber_sc = round(-11*ber_sc+6);
  score = sum([mod_sc pwr_sc ber_sc]);
  if score > max_sscore
    max_idx = p;
  end
end

best = aug_pop(max_idx,1:3);
```

6.6.2.4 Radio

```
function [bw, ber] = radio2(act)

M = act(1)
rate = 3;
pwr = act(3);

% Source
x = randi([0,1],1024,1);

% Encoder
switch rate
  case 1
   n = 15;
   k = 11;
  case 2
   n = 15;
   k = 7;
  case 3
   n = 15;
   k = 5;
  end
  len_x = size(x,1);
  num_pad = k-mod(len_x,k);
  code = [x;zeros(num_pad,1)];
```

```
enc = fec.bchenc(n,k);
code = encode (enc,code);

len_code = size(code,1);
eff = len_x/len_code;

% Modulation
mod_len = log2(M);
mod_pad = mod_len - mod(len_code,mod_len);
code=[code;zeros(mod_pad,1)];
h = modem.dpskmod('M',M,'InputType','bit');
y = modulate(h,code);

% Channel
n_pwr = -15+2randn91);
snr = pwr - n_pwr;
rx = awgn(y,snr);

%Demodulation
h = modem.dpskdemod('M',M,'OutputType','bit');
z = demodulate(h,rx);

z = z(1:len_code);

dec = fec.bchdec(n,k);
uncode = decode(dec,z);

uncode = uncode(1:len_x);

bw = 1/log2(M)*1/eff*6e6*(1.5);

bad_count = nnz(xor(x,uncode));
ber = bad_count/len_x;
```

6.6.2.5 Knowledge Base
Same as the free forward error correction (FEC) CR.

6.6.3 Interpolation Code

This section provides the Python code used for interpolation.

```
from pybrain.datasets import SupervisedDataSet
from pybrain.tools.shortcuts import buildNetwork
from pybrain.supervised.trainers import BackpropTrainer

def get_network(in_data, out_data):
```

```python
    dataset = SupervisedDataSet(len(in_data[0]), len(out_data[0]))

    for idx in range(len(in_data)):
        dataset.addSample(in_data[idx], out_data[idx])

    nn = buildNetwork(dataset.indim, 20, dataset.outdim)
    trainer = BackpropTrainer(nn, learningrate=0.01, momentum=0.5,
            verbose=True)
    trainer.trainOnDataset(dataset, 1000)

    return nn

def get_data(tag):
    reader = lambda x: [[float(l) for l in k.strip().split(',')[1:]]
            for k in x.readlines()]

    in_data = []
    with open(tag+'_in.dat') as f:
        in_data = reader(f)

    out_data = []
    with open(tag+'_out.dat') as f:
        out_data = reader(f)

    return in_data, out_data

def main():
    fixed_data = get_data('fixed')
    fixed_net = get_network(*fixed_data)

    free_test = []
    fixed_test = [[8,5, 9.0087891, 0], [8, 7, 9.00887891, 0]]

    fixed_out = []
    for i in fixed_test:
        tmp = fixed_net.activate(i)
        print tmp

        # fixed_out.append(fixed_net.activate(i))

if __name__ == '__main__':
    # a =get_data('fixed')
    # print a

    main()
```

6.7 Conclusion

Herein we have tangled with the challenging topic of performance evaluation as it relates to CRs. The primary feature of this process is the estimation of the performance function which is inherent to the operation of a CR and determines the reactions of the CR to any given scenario. Through estimation of this function, an evaluator determines the information necessary to make a judgment about the goodness of the CR's operation. We have discussed the major factors and metrics that influence this function and provided a framework for their consideration.

Bibliography

[1] N. J. Kaminski, "Performance Evaluation of Cognitive Radios," Master's Thesis, Virginia Tech, Blacksburg, VA, 2012.

[2] N. Chomsky, *Reflections on Language*. Pantheon, New York, NY, 1975.

[3] H. Hoijer, "The Sapir-Whorf Hypothesis," *Language in Culture*, pp. 92–105, 1954.

[4] T. Willink and H. Rutagemwa, "Framework for Performance Evaluation of Cognitive Radio Networks in Heterogeneous Environments," in *Electrical and Computer Engineering, 2009. CCECE '09. Canadian Conference on*, May 2009, pp. 199–203.

[5] N. Goldenfeld, "Simple Lessons from Complexity," *Science*, vol. 284, no. 5411, 1999, pp. 87–89. doi: 10.1126/science.284.5411.87

[6] M. Serra-Ricart, J. Trapero, J. E. Beckman, L. Garrido, and V. Gaitan, "Multidimensional interpolation using artificial neural networks: Application to an H I cloud in Perseus," *The Astronomical Journal*, vol. 109, 1995, p. 312. doi: 10.1086/117274

[7] E. Barnard and L. F. A. Wessels, "Extrapolation and interpolation in neural network classifiers," *IEEE Control Systems*, vol. 12, no. 5, 1992, pp. 50–53. doi: 10.1109/37.158898

[8] Y. Zhao, S. Mao, J. O. Neel, and J. Reed, "Performance Evaluation of Cognitive Radios: Metrics, Utility Functions, and Methodology," *Proceedings of the IEEE*, vol. 97, no. 4, 2009.

[9] C. Dietrich, E. Wolfe, and G. Vanhoy, "Cognitive Radio Testing Using Psychometric Approaches: Applicability and Proof of Concept Study," *Analog Integrated Circuits and Signal Processing*, vol. 73, no. 2, 2012, pp. 627–636.

[10] J. Thompson, K. Hopkinson, and M. Silvius, "A Test Methodology For Evaluating Cognitive Radio Systems," *IEEE Transactions on Wireless Communications*, pp. 1–1, 2015. doi: 10.1109/twc.2015.2452268

[11] P. Marshall, *Quantitive Analysis of Cognitive Radio and Network Performance*. Artech House, Norwood, MA, 2010.

CHAPTER 7

Cognitive Radio Design for Networking

7.1 Networks of Cognitive Radios Versus Cognitive Networks

Recall from Chapter 1 that a cognitive radio (CR) is a transceiver with three distinct features: (1) awareness, (2) adaptability, and (3) augmentation. By awareness, we mean that a CR is cognizant of its environment, purpose, capabilities, limitations, and policies, as well as those of its peers. Adaptability refers to a CR's ability to configure and reconfigure itself as needed. Finally, augmentation describes a CR's ability to improve its operation over the course of time by applying past experience in some manner. Note that the extent of each of these features may vary between instantiations of the CR concept, but each is present in CRs in general.

A cognitive network may be viewed as the extension of the concept of a cognitive radio to encompass an entire network. In the first definition of the term "cognitive network" available in the literature, Thomas et al. note that:

> A cognitive network has a cognitive process that can perceive current network conditions, and then plan, decide and act on those conditions. The network can learn from these adaptations and use them to make future decisions, all while taking into account end-to-end goals [1].

The parallels between cognitive radios and cognitive networks are clear from this definition. Thomas specifically highlights perception of current conditions which directly corresponds to the awareness we have discussed for cognitive radio. Similarly, the ability to plan, decide, and act relates to the adaptability of a CR, with additional stressing of the intelligence necessary to determine appropriate actions. Both cognitive networks and cognitive radios employ prior experience to improve current and future operation. Finally, note that Thomas specifically highlights the end-to-end goals of a cognitive network, which directly parallels the mission focus of a cognitive radio, extended to a network wide scale.

This focus on end-to-end functionality is especially important in the discussion of cognitive networks as it serves both to demonstrate the commonalities of cognitive radios and cognitive networks and to differentiate the two concepts. End-to-end operation incorporates all parts of a network to achieve some high-level goal set forth by the users of a network. In the case that the network embodies the attributes discussed above and is cognitive, this drive to fulfill the desires of the user directly echoes the motivation of CRs. Simultaneously, the difference of including the entire network draws a stark distinction from CRs. As discussed so far, CRs operate as individuals, aware of their peers, but ultimately discretized in terms of operation and goals. Cognitive networks, instead, operate as a single entity, distributed across heterogeneous elements.

Cognitive networks are perhaps best discussed as super-organisms, made up of several individual elements totally committed to the collective goals and operation of the whole. Social insects provide the most ready example of the concept super-organisms, in which individual ants or bees operate in a manner completely directed by the goals of the whole. This relation to social insects is so strong that such compositions of individual elements totally driven by the collective are commonly referred to as swarm behavior or a hive mind. Hive mind is a particularly good moniker as it highlights the notion that decisions occur at the level of the collective rather than the level of the individual. Note that this does not preclude centralized decision making; rather, the notion of collective cognition simply states that the operations of cognition, the awareness, adaptation, and augmentation discussed above, occur in a manner that considers the collective rather than a single element.

The positioning of cognition is then the key feature that distinguishes a cognitive network from a network of cognitive radios. Within the former, cognition occurs in a manner that spans the entire network. That is, the cognitive element is aware of the whole network, may adapt the collective, and benefits from the composite experience. In a network of cognitive radios, cognition is localized to the individual transceivers. Here, there are several cognitive processes involved, with each only aware of its own phenomena, able to adapt only the local aspects of operation, and benefiting from only its own experience. In this way, cognitive networks transcend networks of cognitive radios by demonstrating a higher level of cognitive operation. Maddeningly, however, networks of cognitive radios are able to support this higher level of cognition, leading to scenarios where a single network is a cognitive network composed as a network of cognitive radios. Again, though, the key factor defining a cognitive network is the collective cognition that operates across the network.

Certainly, the concept of collective cognition that underpins cognitive networks is fairly challenging and perhaps seemingly purely philosophical. In fact, this concept is of increasing importance to telecommunications as networks increase in complication, heterogeneity, size, and importance. Cognition that considers the collective may soon provide the only means of successfully operating networks. Once again, note that this sort of cognition does not necessarily imply centralization or decentralization of decision making and instead represents a broad range of possibilities. Therefore, we will briefly discuss some of the most important aspects of networks exhibit collective cognition here to advance this increasingly important concept.

7.2 Cognitive Network Goals

As discussed above, cognitive networks employ collective cognition to achieve end-to-end goals. While this concept may seem far-fetched at first, we are beginning to see this approach reflected in several commercially deployed networks and it is certainly a major feature in any discussion of future networks. Specifically, the recent focus on self-organization approaches, encompassing self-healing, self-configuration, or self-optimization, may be considered an early realization of the concept of a cognitive network. Furthermore, networks focused on the ideas around spectrum sharing or even infrastructure sharing may rely on the collective cognition of cognitive networks. Here, we will briefly discuss how these examples relate to the concept of cognitive networks.

The LTE notion of self-organizing networks (SONs) currently represents the most mature instantiation of the cognitive network concept. This concept, developed on the basis of several prior works in the area of intelligent and dynamic networks, outlines some of the most tangible uses of the collective cognition provided with cognitive networks. The technical report [2] provides a complete definition of SONs within the scope of LTE. In this report, the 3GPP categorizes SON approaches into self-configuration, self-optimization, and self-healing largely on the basis of the goals of the collective cognition employed. In fact, each of these categories is almost entirely defined in terms of the end-to-end effects of techniques from a given category; implementation is left entirely to network operators. Such a mature concept for the use of cognition for the benefit of networks, defined largely in terms of the goals of cognition, serves as the ideal starting point for our discussion of cognitive network goals.

The first category of SON approaches, self-configuration, comprises the setup phase of an LTE network. Specifically, much of the material related to this phase discusses the setup and configuration of an LTE base station, termed as an evolved node B (eNB). While this seemingly lacks the end-to-end focus that defines the collective cognition of cognitive networks, the eNB being configured is merely the central entity in the end-to-end process. Certainly, the rest of the network must be involved in the establishment of a new eNB: already established eNBs must contribute to the configuration of inter-base station interfaces, the operations and management (OAM) system must determine appropriate settings for the new eNB, and the automatic neighbor relation (ANR) actions must occur. Ultimately, the goal of self-configuration approaches is the expansion of a network through the activation of an additional eNB, which requires contributions from the entire network to understand the current state of network operations, create interfaces for the new eNB, apply appropriate settings, and update knowledge of the network state for later use. In this way, the elements of awareness, adaptation, and augmentation that define cognition span all the elements in the network to realize the configuration of additional network elements.

The next category of SON techniques, self-optimization, broadly includes automated methods to optimize the on-going operation of an LTE network. This category includes the techniques and approaches most commonly associated with cognitive networks, including optimization of coverage, capacity, handovers, interference, and load-balancing. Each objective of the optimization that occurs within this category requires the consideration and interaction of the network as whole. The highly interconnected nature of complete network deployments necessarily complicates self-organization approaches by extending the impact of alterations at single entities. In fact, the coupling between elements within a network provides an important consideration for any collective cognition operation, defining both the possible outcomes of a given variation and the degree of difficulty in reaching that outcome, in terms of the number of entities that must take action. This consideration is especially crucial for self-optimization techniques which must be designed in a manner that automatically navigates the coupling of network elements to improve the operation of a network, while avoiding any outages. For the sake of this discussion, we will avoid overly detailed analysis of network improvement and simply accept enhancement of a network-wide metric or set of metrics as improvement. Further, note that there are a multitude of possible methods for achieving such improvements, largely composed of extensions of the approaches discussed in Chapter 2, with most fitting the general framework for cognition discussed herein.

The final category of SON techniques, self-healing, concerns methods for recovering from network failures. Network failures are an unfortunate reality of any large-scale network and doubly so for complicated networks such as those supporting LTE cellular systems.

These specters of reality are present in numerable guises ranging from hardware failures, to miss-configurations or installations, and even as far as state oscillations or indeterminacy. Furthermore, self-healing operations present a challenge beyond those of self-optimization due to the typically unpredictable timing of failures and even the difficulty in identifying the presence of a failure. Therefore, we note that self-healing approaches rely on the awareness, adaptation, and augmentation of cognition for both the identification (we hope at a future point in time) and the resolution of failures in the network, whatever form they may take. Moreover, the resolution of failures will often resemble the operation of either self-configuration or self-optimization operations, in setting up backup eNBs or readjusting load balance or coverage across a network.

The SON concept within LTE therefore provides a convenient framework for considering potential goals of cognitive networks. First, the automated setup of a network provides a category of goals, encompassing examples such as the creation of interfaces. Second, optimization of network operation encompasses several goals, including load balancing or interference management. Third, identification and resolution of failures summarizes a range of goals from traffic re-routing to eNB re-activation. In each case, these goals require the use of awareness of the current state of the entire network, adaptation of network elements in a holistic manner, and augmentation of abilities through application of prior experience. As such, these goals from the SON concept of LTE capture the goals of a cognitive network from the perspective of current cellular systems.

Looking beyond current cellular systems, we can conceptualize additional goals for collective network cognition. Paralleling spectrum sharing for cognitive radios in the form of dynamic spectrum access (DSA), resource sharing in networks is increasingly a topic of interest. Note that in the context of networks resource sharing goes beyond simply sharing spectrum to include infrastructure or even service sharing. Infrastructure sharing encompasses the use of network elements, typically base stations, by several operators and service sharing goes beyond even this to include the use of network services, such as computation, by several operators. Each variety of sharing allows network providers to operate more efficiently by distributing resource costs across several parities; however, achieving resource sharing in networks often requires the dynamic adjustment of network operation to support multiple user in a manner that maintains isolation between the various users of a network. While such automatic adjustment of network operation seems similar to the self-optimization goals discussed above, note that the purposes of the adjustments are rather different. Instead of attempting to optimize performance in terms of traditional measures of network operation, such as coverage or capacity, a new measure, based on the external goals, guides network adaptation. Such a model for guiding the operation of networks allows cognitive networks to support as of yet unknown usage models for networks. In fact, in the space of resource sharing for networks, the functionality provided by the automatic dynamic tuning of cognitive networks is the very same necessary to support the decoupling of physical networks from the services, allowing the creation of virtual networks tailored to the needs of a single service [3].

7.3 Interaction Methods for Cognitive Radios

Regardless of the particular approach used to achieve the goals of a cognitive network, achieving the goals of a cognitive network requires some mechanism for coordinating the elements of the network. In fact, this mechanism and its use often determine the difference between a network of cognitive radios and a cognitive network, as the interaction method determines the ability of the cognitive radios to consider end-to-end objectives. With this in mind, we provide a framework for such interaction mechanisms to coordinate the actions of a cognitive network based on work first proposed in Reference 4.

7.3.1 Social Language

Any collection of individuals that exists together in the same environment must compete and/or cooperate to utilize some limited pool of resources. Typically, cooperation yields the most beneficial utilization of resources for the collection as a whole, which drives the formation of societies to support this cooperation. Societies, themselves, rely on communications mechanisms to maintain cohesion and cooperation within a group. Specifically, communications provide the means to align the goals of individual members and share the relevant data for coordinating members of the society. As such, communication mechanisms underpin the existence of societies for several organisms, from ants and bees to human beings. Through the mechanisms of communication, these groups are able to successfully cooperate and effectively make use of limited resources to survive.

A group of intelligent, flexible cognitive radios has many parallels with natural organisms that employ various communication mechanisms to underpin society operation. Certainly, the artificial creation of a CR has not achieved the complexity or autonomy of even a bee, yet CRs share many of the fundamental attributes that benefit from social coordination. In particular, CRs adapt to their environment to achieve a goal based on the cooperation of several individuals and the utilization of limited resources. Therefore, CRs in general and especially cognitive networks demand the development of artificial societies based on communication mechanisms to allow for coordination without greatly diverting limited resources away from the communication of application data, which is the primary purpose of CRs and cognitive networks.

We will limit the development of artificial societies to the creation of efficient communication mechanisms that support the coordination of individuals, excluding the more subtle aspects often exhibited by natural societies. In order to create such mechanisms, we will examine communication within natural societies to extract the fundamental properties of their coordination mechanisms. Based on this analysis, we will then explore the development of similar mechanisms to support frequent, efficient CR interactions to support the coordination of individual radios.

Natural Social Language: In sociology and linguistics, the term social language is used to refer to the union of direct language and the contextual information that surrounds the explicit message [5]. The addition of contextual information removes much of the ambiguity that is otherwise present in language, eliminating the need for highlight-specific explicit communication or other forms of clarifying communication. Such contextual information is added to a message simply by including consideration of available knowledge of the environment and the societal rules of the communicating parties [6]. In this way, the union of direct and contextual information provides an efficient society-specific means of communication. For our purpose, we will take social language to be a form of communication that relies on the combination of direct messages and contextual information, heavily influenced by societal rules.

Nature provides many examples of social language as defined above. Bees perform a waggle dance to indicate the locale of an object of interest, which is generally either a new nesting site or a good food source. In this dance, a bee travels a straight path waggling its body for some duration before making a non-waggling return to its starting point in order to repeat the process [7]. As a result of the rules of bee society, this dance contains a great deal of information. The direction traveled by the dancing bee during the waggling portion of the dance corresponds to the orientation of the object of interest relative to the sun, with up typically being directly in line with the sun. The duration of the waggling phase indicates the distance to the advertised object. Perhaps most interestingly, the intensity of the waggle reveals the bee's excitement about the object. The nature of the object is indicated entirely by the current broad goals of the bee society. During normal bee life within the hive, the dance indicates the location of the food, since the collection of sustenance is the dominating goal. Once bees have exited the hive to form a swarm, the dance communicates information about new nest locations, adapting the purpose of the dance to serve the swarm's goal of finding a new home. Additionally, this dance serves as the basis for a voting mechanism for the bees, with each dancer attempting to elicit acolytes to the support of a particular strategy. We further discuss the importance of this voting aspect later in the context of group learning. Thus, the waggle dance is a bee social language that provides a context-reliant means for the bees to communicate without ambiguity.

Social languages are evident within human society as well, although due to the increased complexity of human society compared to that of bees, a variety of social languages are employed by any given collection of individuals. One of the most straightforward examples of this is pantomime in which gestures that mimic shared experiences within a given group are used to successfully communicate some message. In this case, the direct message of some gesture is augmented with the social experience of the individuals involved. Extending this concept slightly to more broadly include all body language, the influence of societal rules quickly becomes clear. For example, in the United States, a direct message of a thumb extended upward from a closed fist typically indicates approval based on the augmentation from the rules of the local society. The same gesture in Germany may simply indicate the numeral 1 and in parts of West Africa carries a pejorative meaning [8]. Further, we can decode an even more nuanced meaning when considering environmental information alongside societal rules. In this case, the simple single gesture of a thumb up takes on the meaning of a request for transportation or a positive critical rating on American highways or in American movie theaters, respectively. Note that so far we have only considered broad societal rules, leaving aside those that may exist in small groups. Beyond gestures, humans may employ facial expressions or verbal language as the direct messages of a social language. Catchall words such as "thing," "stuff," or "deal" are often used in a manner that depends on the context and particular societal rules. The complexity of human society and the ingenuity of human beings allows for the fluid and dynamic creation of social languages to fill the needs of the moment, resulting in a myriad of actualizations of the concept.

The prevalence of social language in natural societies may be explained by the ability for a group to reach a common understanding of both individual abilities and society goals. According to celebrated scholar Jürgen Habermas, successful societies are based on communicative action, which has the sole purpose of achieving mutual understanding between members [9]. In fact, Habermas claims that reasoning is "thinking codified in language" and that rationality rests in how agents acquire and use knowledge rather than the possession of any particular information [9]. This suggests that social language provides the mechanism necessary for a collection of agents to transform from merely a group to a thinking society. Extending the understanding of individual ability to contribute to society wide goals allows for the society to take on a cognition of its own.

Societies based on communicative action benefit from consensual coordination, i.e., members freely agreeing on goals and plans of action [9]. This contrast with the so-called strategic action-based groups in which members focus only on their own goals. Coexistence in a strategic action society is inherently contentious in nature; individuals are forced to create threats or bribes to coerce peers into agreeable behavior. This contention ends up achieving a situation that is the least bad for all individuals involved, typically with the cost of coercion as a detraction factor. Consensual coordination, on the hand, allows for achieving situations that are the most beneficial to society members by aligning the individuals in a pursuit of goals. However, such a consensus typically may only be reached by members predisposed to coordination, with the means to achieve common understanding of individual abilities and goals.

Utilizing predisposition to coordination is a key benefit purpose of social language. In these way, social languages are distinct from more general language in that they provide a society-specific mode of communications enabling the efficient coordination of the group's actions. As such social languages are not concerned with conveying the fullest expression of emotion, but rather to relate some basic information immediately useful within a society together. As such, social languages aim to transfer only the information most necessary for coordination to allow for frequent use without great penalty. The bee waggle dance is an excellent example of this concept [10]. Honey bees have established a method of broadcasting pertinent information about food or new nesting sites based on actions suited to bees. Certainly, this concept is extended in more complex societies, but the focus on providing an efficient mechanism of communication remains even in human society. Fundamentally, social language provides highly stylized and compressed means of communication based on the combination of a direct message with contextual information, largely from societal rules, especially useful for coordinating the actions of individuals.

Cognitive Radio Social Language: Ultimately, the purpose of a CR is the transport of application data, which requires multiple radios to work together on some level. The challenge within such groups of CRs is that while these devices are designed to enable efficient communication with other entities, they themselves do not yet have an efficient method of communicating with each other. The problem is rooted in the variety and amount of data that CRs typically share for accomplishing their goals. The standard method for solving this problem is some form of centralized coordination [11, 12]. However, such approaches reduce the scalability of the network and increase associated overhead. Several natural societies provide clear evidence that such centralization is not required to achieve successful societies; Doerr points out that "there is no 'master fish'" in a school of fish [13]. Instead, these societies tend to rely on efficient stylized forms of communication that are particular to their group, e.g., the waggle dance of bees [10], and the pheromone trails of ants [14]. This "social transmission of information" is necessary to hold many natural societies together without the need of a central authority or data exchange [15]. Cognitive radios and networks require a social language to shed the weight of centralization in order to more efficiently accomplish their goals.

In the case of cognitive radios, social language supports coordination by allowing information transfer through understanding a peer's actions based on knowledge of societal rules. Each CR makes decisions about how to control the communications under its own supervision in order to establish and maintain necessary communication links with its peers based on its understanding of its situation. These decisions are then translated into actions based on the societal rules of the radios involved and realized through currently available radio hardware. In this way, the CR's actions always encode its understanding of its situation; the use of social language in these scenarios simply requires decoding of this understanding and applying it to subsequent decisions with a CR. Spreading understanding across the entire society in this way then allows peers to achieve the mutual understanding discussed by Habermas in Reference 9.

Decoding the information that drives a CR's actions requires a common basis of societal rules that drive the determination of actions. This is most easily accomplished when nodes share a common method for determining their actions. In this situation, each node can gather information from its peers that is useful for determining how it should proceed in order to fit into a society. Each node is able to understand the situation of its peers and determine its next action based on an estimate of its peers' actions in order to achieve coordinated and interference-free operation.

The use of social language in CRs provides convergence of nodes without requiring the exchange of future plans or detailed sensing information. Each radio in a group makes local decisions about its own action and performs this action in a manner that allows for the easy decoding of the understanding that lead to using one action instead of another. Each other node then simply observes the information conveying details of the performer's action and uses this information to determine its own action. This results in a direct connection between all the actions of every radio in a society without the need for a central control. Through proper use of this connectivity, the society can converge to an useful operating point in an emergent fashion.

Inherently, this connectivity extends the situational awareness of each node in the society. Since each action encodes the reason for using that particular action and each reason is based, at least partially, on other observed action, observing only a few actions can provide enough information to make far-reaching estimates. While this extended awareness allows for the coordination of several nodes, it must be managed to avoid unwanted feedback loops of influence. That is, if all nodes strongly base their actions on other nodes' actions, any action (whether correct or not) will guide the future actions of nodes that observe it. In a realistic scenario, nodes will take improper actions for a variety of reasons, and the web of awareness must be designed in a manner that prevents false convergence from an action from rippling too far.

The extended situational awareness provided through the action-based communication of social language is well suited to the co-ordination of CRs, since this mechanism for coordination does not detract from their primary purpose, i.e., the communication of application data. The society specialization and focus on efficiency of a social language make the cognitive engine (CE), the reasoning center of a CR, the ideal component to handle the encoding and decoding of actions. The CE already determines which actions a CR should take based upon the built-in interaction methods. Adding the ability for the actions themselves to carry coordination information achieves the benefits of social language without detracting resources from the transfer of application data.

Recall that Mitola posited cognitive radios as a combination of artificially constructed intelligence and flexible radios to enable new uses of communications. Mitola borrowed the concept of an observe, orient, decide, and act (OODA) loop from Boyd [16] to describe the operation of CRs. This loop is a cycle of observation, orientation, decision, and action, as shown in Figure 7.1. Radios operating in accordance with the OODA loop execute tasks associated with each stage of the cycle. First radios observe their environment, taking in data. Next radios orient themselves according to the information collected, which consists of comprehension of the data through determination of metrics and/or filtering. Once comprehension is carried out,

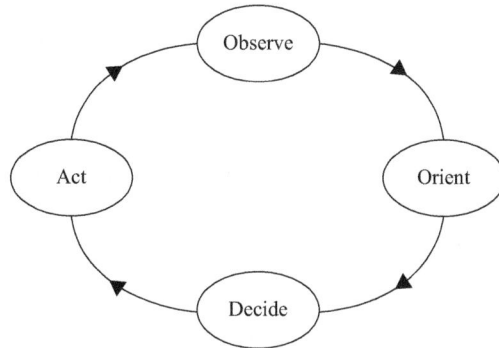

Figure 7.1   OODA loop. © 2014 Nicholas J. Kaminski. Reprinted, with permission, from Reference 4.

a decision about what a radio should do, based on the understanding gained, is made. The radio than takes action based on its decision.

Social language serves to augment the steps of the OODA loops. A social language provides radios with the ability to observe previously unobservable information about their peers. Once these observations are made, radios can orient themselves by decoding the information contained in the social language messages. This information can then be used to decide and act. For example, as a bee searches for food sources, the bee follows an OODA loop augmented with social language. The bee observes a waggle dance by one of its peers. The bee determines the goodness of the food source by comprehending the observed amount of waggle. The bee then decides whether or not to visit this source and acts on that decision. Thus, social language is used to encode and transfer information that can aid the operation of peers.

7.4 Components of Interaction

Fundamentally, there are two parts to the successful interaction: observability and understanding. Any social language developed for use within a cognitive network must be based on sending messages that may be observed by other members of the society. Following this observation by the receiving node, the message must be understood in a useful fashion by that node. In order to accomplish this, the message itself must efficiently be attached to some action that is both available to the communicator and observable by the intended recipients. Furthermore, the recipients must be imbued with the necessary capabilities to decode the required meaning. In our case, we will further design a homogeneous understanding across all receiving entities within the society; although non-homogeneous understanding is certainly possible, it greatly complicates the use of a social language in a designed system. Here, we explore these components of the interactions within the context of social language, however, these components are indeed fundamental to any communicative interaction.

7.4.1 Observability

Observability of a social language has two complementary factors: the lucidity of the sender and the discernment of the receiver. Sender lucidity captures the ease of identifying a message from the sender as important. Receiver discernment, on the other hand, concerns the ability of the receiver to identify the details of a message. Note that in both the case of the sender and that of the receiver, we are examining factors associated with identification of a message; the observability component of a social language is primarily concerned with the detection of the direct message employed. Certainly, the factors of lucidity and discernment are intertwined to determine the observability of a particular message and deficiency in one aspect may be rectified with strength in the other. Social languages require some degree of matching between these factors, although the balance between the two may be adjusted to suit the needs of the situation.

The concept of observability greatly depends on the details of the actualization of the social language. Naturally, the domain in which a message is sent affects the mechanisms required for observation. For example, a fish may be sending messages through its position in space. This requires receivers to be able to discern the important features of the position, i.e., the relative position of two fish may be critically important but the absolute position of a fish may not matter at all. Additionally, the complexity of information to be communicated determines the degree of detail required from observation. In the fish scenario, the dominant social language is concerned with simply communicating a fish's position. The sending fish need only exist in the physical space and the receiving fish must determine a relative distance to an accuracy of approximately its own width. Bees, on the other hand, wish to communicate the location, through distance and heading, and goodness of an object through their waggle dance. This requires the sender to encode several details into a lucid dance and the receiver to discern the details of heading, length of dance, and intensity of waggle. Note that the domain of the bee's social language is still the physical space that the bee occupies, but the bee also incorporates the dynamics of this space to attach many more details than in the case of the fish. The details that a social language is attempting to convey and the direct actions for conveying messages ultimately determine the requirements on the agents sending and receiving messages.

Note that the environment of the direct action itself imposes certain challenges to the observability of a social language. Consider the bee performing a waggle dance, a form of communication used both inside a hive and on the surface of a swarm which has left a hive to start a new colony. When this dance is performed within a hive, there is not artificial lighting, so visual observation of the dance is unlikely. Rather the receiving bees must be in physical contact with the dancing bee or in range of the electric field generated by the dance [17]. As bees do not generate especially intense electric fields and electric fields decay as the range from the source increases, those that wish to receive the signal must be within some maximum range. Additionally, note that physical realities of the space influence the utility of various potential attachment points for information; for example, since the electric field decays with distance, attaching information to the exact value of field intensity is foolhardy. This notion already impacts the design of communication schemes in the case of radios. Radios would have a very difficult time communicating information by observing each other's absolute signal power levels, as the signal power level decays with distance. Instead, radios may employ amplitude modulation based on relative power levels within a signal to communicate. Understanding the variation of a message transmission in time and space is therefore an important factor for social language, just as it is for any formation of communication.

Finally, the concept of interference is as important for social language as it is for any communication scheme. A sender cannot be considered lucid in the case that its messages consistently overlap with some other sender. In fact, such a sender damages the lucidity of all those with which it interferes. Again, as with any other communication this does not defeat the ability to transfer useful data; it simply requires more sophisticated techniques for successful communication [18]. Typically, sophisticated techniques for managing communication through interference are not available for social language, as the concept of social language aims to deliver a low-cost solution for transferring only pertinent coordination information. That is, the fundamental low-complexity of the communications schemes employed for social languages amplify the impact of interference, typically to the point of resulting in total data loss. This inevitable loss is typically handled through reliance on the web of situational awareness reinforcing useful messages through the actions of others, e.g., bees that miss an original dancer's message may catch a repeated performance from a mutual peer. Depending on the methods used for implementation of a social language, message fidelity could be quantified in a number of ways, just as any other communication method. Ultimately, many of the techniques for designing and analyzing more tradition communication mechanisms have application to the observability of social language as well.

7.4.2 Understanding

Understanding is the act of determining the useful information encoded within some observed message. Recall that actions inherently encode some amount of data; this is the principle upon which all communication is built. Further, recall that social language centers on the encoding information within a direct message through the application of contextual information. These encoded data typically hold only a small amount of information and are typically only relevant for a short period of time. These features suggest that the encoding and decoding of a social language should be as inexpensive, in both computational effort and time, as possible. Moreover, the type of information communicated is designed to coordinate the decision-making processes of several individuals. The combination of this and inexpensive encoding methods often allows data to directly decode and use within the decision making without separate explicit stages. The specific goals and attributes of a social language serve to differentiate these mechanisms from more general purpose communications in that social language stresses society specialization and cohesion through compact communications over generalized utility. As such, extracting the useful information from social language messages in order to achieve understanding employs specialized features.

Since the object communicated by social languages is some form of coordination data, societal rules often simply dictate deliberate specialized actions in response to important stimuli. Decoding messages then requires that receiving nodes need only match the action observed to the associated societal rule to determine why a peer took some action and gain the benefits of this communication. For example, a fish might observe a peer move closer to itself. In this situation, the fish must consider the societal rules governing such an action within this context to determine why the peer took this action; perhaps the peer encountered an obstacle or perhaps the peer wishes to mate. This principle holds in more complex social languages, as well. For example, bees need only to review the societal rules of a waggle dance in order to decode the reason behind the particulars of a peer's dance to determine the location of an object of interest. Therefore, matching societal rules to the observed action allows a receiver to determine the driving stimulus of the communicative actions of social language and respond appropriately.

Note that the information transferred by social language is for internal consumption of the agents. While it is conceptually helpful to think of the agents as decoding a particular message by asking themselves why a peer took a particular action, this process is not always necessary. In the fish example, it is unlikely that one fish pauses to consider why one of its peers is darting toward it sharply. Rather the fish automatically reacts to the rapidity and direction of the movement by moving away from the peer who has encountered an obstacle. This reaction represents both the understanding of a message and the application of the information gained. Separating the process into encoding, decoding, and acting is really only useful for early examination of a particular social language, where the separation highlights the basis for an interaction and the expressibility of a particular message. In practice, social language is more concerned with encouraging appropriate reactions to behavior than the actual transfer of bits.

7.5 Analyzing Interactions

Quantifying the information encoded into the actions of a CR provides insight into the potential for attaching meaning to this information to support a social language for CRs. As stated above, this meaning is imbued through a structure of societal rules that relate some stimulus to a particular action, which encodes that stimulus as the action selected. Examining the amount of stimulus information that actions of CRs contain following this encoding process provides insight into the differentiability of the various stimuli under the rules employed by the CRs. That is, examining of the residual stimuli information available from the actions of various CR approaches provides direction in structure societal rules to support a social language and therefore light weight coordination of CRs. Thankfully, the information flow analysis techniques introduced in Chapter 2 provide the tools necessary for this examination.

Recall from Section 2.6 that the CE encodes both built-in and environment information into the actions that a CR takes. Understanding the relative contributions from these sources of the information contained in an action, rather than simply the amount, is of primary importance to social language. Since the goal of designing social language for CRs is to communicate coordination information through the actions of the radios, understanding the source of the information contained in the actions provides insight into which CEs are well-suited for social language communications. Note that achieving this understanding requires examination of the roles of the information used by a CE in the selection of an action. Recall that, in the example of the coin, the number of available values determines the number of bits required to describe the value and therefore the amount of information in a coin flip. In the same way, the number of options available for a given action determines the amount of information contained in that action. Since the pool of options for any action is determined by the built-in information of a CE, this sort of information determines the amount of information an action may contain. However, the influence of built-in information does not end there. Rather the process for selecting an action may depend on some internally generated information, such as an internal random number generator. In this case, the value of the random number generator contributes to the selection of the action. Thus, the built-in information related to the number generator is in some way reflected in the action selected. In a similar manner, environment information is typically used in some way to help select actions and is therefore represented in the selected action.

Given this understanding of how various sources of information contribute to the information in an action, a modification of the process may be made to allow for social language communications. Recall that the built-in information is intrinsic to the selection process employed by the CE, both determining the amount of information contained in an action and contributing to the selection of the action directly. As such, the built-in information results from the decisions of the designer of the CE related to the action selection process. The environment information, on the other hand, represents the insertion of external information into the action determining process. This information can be observed by the CR and incorporated into the selection of a particular action. The built in information may be viewed as the information related to the encoding process and the environment information may be looked on as the message that is encoded. That is, the built-in information contains the societal rules that shape reactions to various stimuli and the environmental information contains the stimuli. Within the goal of communicating the stimuli experienced by one radio to another through a social language, we are then interested in determining the proportion of environment information represented in a selected action under the influence of various approaches to designing the CE, each with a different amount and use of built-in information.

The uncertainty coefficient, introduced as part of the information flow analysis of Section 2.6, offers the perfect tools for analyzing the potential for social language use. Recall that this metric gives the portion of bits in some signal A that may be predicted using knowledge about a second signal B. Conceptually, this can be considered as the proportion of information in A that comes from B. This measure is limited to values between 0 and 1, providing a clear measure of the accuracy with which signal A (the reason for selecting a particular action) can be predicted given knowledge of signal B (the action in question). This measure therefore gives the capacity of an action to carry social language information.

7.5.1 An Example Analysis

Herein, we expand the simplified test scenario of Section 2.6.1 to examine the utility of social language based on the measures introduced above. In the course of this examination, we will determine the proportions of information in an action from environment information and built-in information for the three example cognitive radios presented earlier. Recall that, each CR chooses an action based solely on a measurement of the noise present in the environment. These actions are cataloged along with the noise measures that proceed them. Once the cataloging is concluded, we will employ the tools presented above on the recorded actions and noise to calculate the proportional influence of each type of information. The CR takes action by selecting a power level and modulation scheme. The noise is assumed to purely additive white Gaussian noise (AWGN).

Within this analysis, we explore the impact of various CE approaches on the use of social language purely by examining the proportion of action information from environment information. Recall that the exact nature of the environment information is not important; rather, we are more interested in determining the degree to which some stimuli may be encoded with a given set of societal rules. The analysis here provides an indication of the CE methods that are best suited to encoding and decoding social language information. To accomplish this we quantify the degree to which environment information is represented within an action in terms of the ability to predict the environment information upon observing the action; that is, we employ the uncertainty coefficient. Note that this analysis simply indicates trends for a selection of CE techniques and does not exhaustively cover the nuances of optimal CR strategies; further analysis would likely be needed for a particular implementation.

Before we begin our formal analysis, a conceptual analysis of the CE approaches examined here is already revealing. Considering that we are quantifying the degree to which environment information in the form of a noise measurement is represented in the selected action of our CRs, we can identify broad characteristics of this relationship simply by examining the CE strategies.

Table 7.1 Total action results under 100 000 tests.

CE under test	Action entropy	Mutual information	Uncertainty coefficient
Random	2.708010	0.023096	0.008529
GA	1.610803	0.166126	0.103133
Table	2.085064	2.085064	1.000000

Table 7.2 Power sub-action results under 100 000 tests.

CE under test	Action entropy	Mutual information	Uncertainty coefficient
Random	1.609423	0.006733	0.004183
GA	0.565302	0.117135	0.207208
Table	1.099828	1.099828	1.000000

Table 7.3 Modulation sub-action results under 100 000 tests.

CE under test	Action entropy	Mutual information	Uncertainty coefficient
Random	1.098610	0.003165	0.002881
GA	1.046020	0.031977	0.030570
Table	1.097648	1.097648	1.000000

In the case of the Random CE, we note that there is no connection between the noise measurement and the selected action. Therefore, we immediately expect that the proportion of information in the action from the environment information to be very close to 0. On the other hand, the Rule Based CE has a fairly simple and direct connect between noise measurement and action, suggesting a strong connection between the action information and the environment information. Finally the genetic algorithm (GA) CE employs the most complicated strategy, including some internal random number generation, which indicates that this CE will fall between the other two extremes. Understanding the approaches employed in the CEs in terms of their use of information provides a great deal of insight into their potential for social language use.

7.5.2 Analysis Results

The expanded results of the analysis discussed previously are given in Tables 7.1 and 7.2. The tables each relate results concerned with a particular type of action. Table 7.1 covers the total action, consisting of both the power setting and the modulation selection. Table 7.2 focuses on only the power setting portion of the action. Table 7.3 focuses on only the modulation selection portion of the action. Examining the total action and each of the sub-actions separately reveals aspects of the influence of environment information on the action that are otherwise obscured. Each table shows the total information contained in its action type (in the form of the action entropy), the mutual information between the action and the noise, and the uncertainty coefficient relating the action and the noise. Each of these figures are calculated from the recorded noise values and actions. In this testing, the uncertainty coefficient is calculated by normalizing the mutual information between the action and the noise by the information contained in the action.

Recall that we are employing the uncertainty coefficient to quantify the proportion of information in the action that comes from the environment information, given knowledge of available action options. Specifically, we are here using this metric to quantify the purity of the stimuli representation within the action as a result of the application of societal rules. As expected, the most straight forward relationships exist for the random CE and the rule-based CE. In the case of the random CE, the environment information was not used at all in the determination of the action and therefore not represented in the selected action, as expected. Note that the value is not exactly 0 due to the finite nature of the analysis conducted here. This, unsurprisingly, indicates that societal rules encouraging random behavior are not particularly useful for supporting the communication of stimuli through a social language. Contrast this to the rule-based CE, in which the action is completely determined by the environment information. In this particular case, the environment information dominates the information represented in the selected action because mutually exclusive rules are employed. Such a strong dominance makes stimuli communication simple; nodes need to simply share rules to understand which stimuli a peer experiences based on that peer's actions. For more complicated rule sets, the connection between environment and action information would likely be slightly weaker, complicating the use of social language slightly. This analysis therefore indicates that the uncertainty coefficient provides a measure of the degree of difficulty for understanding the stimuli experienced by a node via an action based social language.

In light of this, the GA CE provides an intermediary case in between the other two CE extremes. The results for this CE clearly indicate that the action information is a mixture of both the environment information and the built-in information. That is the action produced does not have a direct and simple mapping to a single stimuli and instead a more complex decoding of the message is required. Recall that the GA randomly generates and evolves a collection of candidate solutions based on a fitness landscape that involves the environment information as well as other information. Thus, the selected action is the product of the random number generation, the random evolutions, and the other calculated attributes in addition to the calculation that involved the environment information. All of these built-in sources of information contribute to the final selection of the action, as reflected in a low uncertainty coefficient. Such a complicated set of societal rules suggest that additional contextual information is necessary to decode the stimuli from an observed action.

In fact, a low uncertainty coefficient signals the need for more complex social language implementations due to the need for additional decoding information. The variety in social language complexity based on the need for additional information is clear when considering ants and humans, both of which utilize social language to underpin their societies. Ants are less intelligent than humans; they tend to use comparatively simple societal rules for determining appropriate actions. The ant action-based social language therefore relies on an action selection method where the message of the language completely determines the action taken, mirroring our CE that employs simple and mutually exclusive rules. Decoding their social language is therefore very simple; one must just test for the intensity of particular chemicals. Humans, on the other hand, are highly intelligent and tend to consider several sources of information to make any given decision. As discussed previously human body language, for example, depends on a great deal of additional contextual information, requiring a potential decoder to have access to several sources of information. This need for additional information is reflected in the tendency of a person to understand the body language of a person who they are familiar with, but won't necessarily pick up the nuances of a stranger's message. Certainly, some connection between environment and action information is necessary, but through the application of additional contextual information social language may relate stimuli even when the environment does not dominate the information within an action, additional information. In this way, the uncertainly coefficient provides a quantification of the degree of difficulty in decoding environment information. In the case of more complex social languages, which employ low uncertainty coefficient direct messages, typically captured in a model of the transmitter, is required.

Note that the proportion of action information representing external information may not be the same for all actions, despite a common action determining method. Examining the uncertainty coefficient for the individualized actions reveals a telling point. Note that the coefficient is much larger for the power sub-action than for the modulation sub-action. This suggests that this particular GA more strongly relies on the environment information to select a power level than the modulation setting. This situation can result from the specific implementation of the algorithm, such as a bias in altering power more than modulation during evolution, or the fitness functions used for the algorithm. This indicates that decoding the environment information from the power of the selected action requires less additional information than decoding the same information from the modulation. Note that the lower uncertainty coefficient of the total action versus the power sub-action does imply that decoding environment information is harder simply because the modulation is known. Rather it signifies that the environment information is reflected in a smaller proportion of the action in the case of the total action than in the case of the power sub-action.

The results presented here provide a number of suggestions for the design of an action-based social languages for CRs. Chief among these is the notion that rule-based determination of actions to be used for social language will require the smallest amount of additional information for decoding. Rule-based social languages simply require all members of a society to have the same rule set to decode a particular action, whereas a social language based on a GA would require both a common knowledge of the fitness function to be used as well as communication of the states of internal random number generators. Additionally, the results suggest that in non-rule based social languages, some parts of the message could require more information to decode than others. This suggests that multi-layered social languages, where certain members of a society receive more detailed information based on their greater knowledge, are possible.

Through this information theory analysis, we have gained insight into the sort of decision process that leads to the best understanding of the reasons behind a radio's behavior, which provides the basis for action-based social languages. Our analysis has clearly shown that rule-based methods, with mutually exclusive rules, provide the most direct connection between a radio's actions and its reasons. The behavior of radios controlled with such rule-based methods point unequivocally to particular input stimuli. While this method does not necessarily allow for behavior optimization at the individual radio level, it does provided the most straightforward approach for an initial development of a CR society. As is discussed below, employing social language allows for learning on the society-wide or network level.

7.6 Group Learning

Societies of individuals with connected awareness exhibit the ability to learn as a group. In this societal learning, the collective transforms itself in a manner that stores the information to be learned. In this scenario, the information is stored in between the members themselves, in their interactions. The state of the society as whole represents the storage of information learned through interacting with the environment. Storing information in this manner allows the collective whole to learn information that is not necessarily directly apparent to any individual member, but influences the behavior of all members. Such societal learning removes dependence on any particular agent, preserving performance of the group in the face of the entry or exit of any members. In this way, an entire society learns as though it was a single organism.

Such super-organism learning is actually fairly common. Seeley describes the concept as it applies to swarms of bees which are seeking a new nesting site:

> ... there is no central decider who posses synoptic knowledge or exceptional intelligence and directs everyone else to the best course of action. Instead in both swarms and brains, the decision-making process is broadly diffused among and ensemble of relatively simple information-processing units, each of which possesses only a tiny fraction of the total pool of information used to make a collective judgment. ... These similarities point to general principles for building a sophisticated cognitive unit out of far simpler parts [7].

Social language provides the connectivity necessary for the process of societal learning. As the society interacts with its surroundings, the strength of the connection between assorted nodes varies based on their communications through social language. For example, bees in a swarm looking for a new home interact with peers through the waggle dance. This interaction influences other bees' decision-making process, which then results in bees either agreeing with the initial bee and reinforcing that bee's declarations or disagreeing and offering alternatives. In this way, the language of the waggle dance provides the conduit that connects the individual reasoning of the bees. The bees use this conduit to reinforce or subdue the firings of other bees in a manner analogous to the action of neurons in a brain.

The connections between the individuals in a swarm result in a transformation of the whole system. Bees travel out from the swarm and bring back information about the goodness of a potential site. Their campaigning draws other bees to independently assess the site. The result of this is an emergent learning of appropriate nesting sites at the swarm level. No individual bee tallies up the number of its peers that vote for a particular site, rather the voting process simply influences bees to investigate individually. Determination of the swarm's overall assessment of a site must be made on the swarm level by determining the proportion of bees interested in a particular location. In this way, the number of bees that a swarm devotes to investigating a location stores information about the swarm's evaluation of that site. Once a critical mass of bees displays great interest in a new nesting site, the swarm, as a whole, has determined that the new site would make an appropriate home. Selection of a new site signals that the swarm has learned the local landscape of potential homes and made a decision by weighing its options before choosing the one best suited to its needs.

Artificial neural networks provide a technique for learning-based entirely on varying the strength of influence between neurons that each determine their own action through the consideration of all connected peers [19]. Societies perform the same process in a slightly more sophisticated manner. The interactions between members of a society is not typically as simple as the accumulation and repeating of signals, as is usually the case in artificial neural networks. Rather the interaction between peers in a social group is more dependent on the details of the interaction and the society. In fact, the nature of these interactions makes societies extremely complex to model. However, there is no denying that the fundamental principles that underpin artificial neural networks are exhibited by societies.

Societal learning, just as social language, is focused on advancing the state of the society, not on the explicit storage and retrieval of arbitrary information. Much like social language the processes of societal learning are simply the actions and interactions of the society members. Each member influences the actions of those around it, until the society learns how best to interact with its environment. During this learning phase, the society as a whole probes its environment to discover the pertinent information it desires and stores information in the interactions of its members. For example, bees may discover a new nesting site not fit for their needs. Rather than having to continually re-examine this site, the bees collectively remember its unfit qualities through the interactions of bees directing new searches elsewhere.

The society must be taken as a whole to witness societal learning. Examining the action of a single member, bee or radio, does not reveal what the society knows. Rather it is through the examination of how the group influences its member that the learning is revealed. As bees discover prime nesting sites, the group influences several members to examine these sites, revealing that the society has learned both the location of a good side and the local fitness landscape as it pertains to nesting sites. The bees make their final observation by optimization over the nesting fitness landscape of the terrain in their range. To do this, they must first learn this landscape; no outside force provides it to them. When the whole group makes the decision about a new nesting site, it is clear that the society has learned the necessary information about its local fitness landscape.

7.7 Building a Cognitive Network with Social Language

7.7.1 MAC Layer Considerations

Certainly, the ultimate goal of a cognitive network is to transfer application data. In fact, this goal is so central to a cognitive network that it parallels the goal of survival for bee hives; it ultimately drives all other goals of the network, even if it is more typically broken down into sub goals. To achieve success, each member of the network must cooperate in pursuit of the transfer of application data. Thus, a fundamental requirement of successful networks is the organization of individuals to minimize interference between members. That is, radios must first organize their spectrum access before application data may be transferred. Achieving such an organization in support of efficient data transfer will serve as the focus of the example cognitive network discussed here.

The role of the medium access control (MAC) in a radio system "is to coordinate the access to the channel so that information gets through from a source to a destination" [20]. The MAC is responsible for providing the organization of individuals, which is fundamental to the operation of a cognitive (or indeed any) network. Additionally, the MAC implicitly enables the parallel application of several techniques in the development of a radio system. That is, the abstraction of the stack model allows the MAC layer solutions to apply to a variety of physical layer solutions.

To support cohesion of our network, while simultaneously achieving the critical organization of radios, we will apply a rule-based system for guiding MAC layer actions here. As discussed above, a rule-based approach allows for the most straight forward use of a social language to coordinate radios within a cognitive network. Additionally, allowing the parallel operation of techniques at different layers of the stack requires that no single layer consumes excessive computational resources, and rule-based systems avoid the need for computationally intense searching as is common in many optimization techniques. Importantly, the field of robotics has demonstrated that a rule-based technique, termed behavior-based design, is well suited to developing multi-agent systems [21].

7.7.2 Behavior-based Design and Social Language

Behavior-based design employs several goal-oriented behaviors in order to pursue multi-objective goals [22]. This approach more closely resembles the natural brain's ability to juggle several objectives than do most other artificial intelligence approaches [23]. Each individual behavior constitutes a rule that affects the overall behavior of an agent. While these rules are not necessarily mutually exclusive, they offer the straight forward connection between stimuli and action that is essential to social language. Mataric has notably used this method of multi-agent system design to demonstrate the emergence of beneficial behaviors similar to those needed in cognitive networks [21, 25–26].

Mataric employed several robots in a two-dimensional plane to develop principles for using behavior-based design to achieve desirable emergent behavior [21]. The robots relied on a basic behavior set consisting of safe-wandering, following, dispersion, aggregation, and homing in order to accomplish tasks like flocking and foraging. This basic behaviors represent a minimal set of control rules that allow for the "structuring, synthesizing, and analyzing" of a system's overall behavior [21]. Two methods for combining these basic behaviors allowed Mataric to achieve the emergence of more complex behavior to fit the tasks at hand.

While Mataric makes a large contribution to multi-agent system design, her work is not a complete solution. Specifically, Mataric does not investigate the connection of peers through a social language to yield an information processing array. Additionally, all of Mataric's work focuses on robots. The differences between robot and radio domains are significant. Insofar as CRs and robots both represent intelligent agents with common classes of problems, general approaches may serve both domains. However, the parallels between the challenges faced by robots and radios make a behavior-based approach to building multi-agent systems in either domain appealing, but the differences between the same challenges require such an approach to be fitted to each domain individually.

The behavior-based approach offers several advantages to a cognitive network. Since each member of the network has the same behaviors, they all react to stimuli in a similar way. This allows each individual in a society to decode social language messages without maintaining a separate rule set for each peer. Additionally, the commonality means that peers will react predictably. The rule-based nature of the approach maintains the straight-forward connection between action and message, easing the complexity of social language systems. However, this connection is degraded when the interaction of several behaviors must be considered; thus social languages are best suited to communicating through actions that are influenced by a smaller number of behaviors. Overall, the behavior-based approach to multi-agent system design provides the means necessary for the development agents with complementary behaviors.

7.7.3 Tasks and Behaviors

The first step of applying the behavior-based approach to building our cognitive network is to determine the important tasks of the network. Recall that the ultimate goal of the network is the transfer of application data, and we will pursue this at the MAC layer. Based on this, we decompose the primary tasks of the network into organization of spectrum access in time (time flocking) and organization of spectrum access in frequency (frequency flocking).

We will accomplish the tasks of time flocking and frequency flocking by combining several basic behaviors, aimed at organizing the transmissions consists of radios in time and frequency. Each of these basic behaviors further decomposes the problem into atomic units that may be directly addressed by uniform strategies. We make no claim of optimality for the particular strategies discussed here; rather, the strategies demonstrate attributes conducive to combination in cognitive networks. In particular, the strategies applied to the example network focus on achieving their goals in a simple manner, based on a small number of information sources, in order to examine the utilization of social language to coordinate several agents in an efficient and emergent manner.

The basic behaviors of radios within our network are supported by a social language designed to allow them efficiently to communicate only the most relevant information for coordinating access to the spectrum. This coordination requires some communication of intent, as a radio's position in time and frequency is neither constrained to continuity in movement nor available for observation without active transmission. Each radio needs to inform others of its intentions in accessing the spectrum; social language provides the mechanism to do this.

7.7.4 Hardware Considerations and Implementation

For our example network, a periodic beacon transmission from each radio serves as the primary means of social language communication. This beacon serves to increase the observability of social language messages within our network in order to broadcast situational awareness throughout the network. As such, the beacon's transmission is representative of a sending radio's spectrum access and is affected by the various MAC behaviors of the radio. Each behavior affects the beacon in one of the orthogonal dimensions of time and frequency. The frequency of the beacon is set according to the frequency that the sending radio intends to use. A beacon is sent within the time slot that the sending radio intends to use, once every T seconds where T is referred to as the beacon period. The length of time slots depends on the negotiation of radios through the social language mechanism based on their behaviors, as discussed below. The beacon signal consists of a single packet with dummy pay load and a beacon flag in the header. The beacon does not require addressing information for either intended recipients or senders. This mechanism allows the radios to communicate all the information they need to coordinate themselves in time and frequency.

For the purposes of this example, our implementation hardware is the combination of a BeagleBoard-xM computational platform [27] and a Hope-RF RFM22B radio front end [28] named SKIRL, first proposed for CR by Young [29]. The system offers a flexible radio frequency integrated circuit (RFIC) supported by computational power comparable to the average smart phone. Young points out that the use of an RFIC instead of the more common software-defined radio approach improves the radio performance and decreases cost by removing the reliance on powerful field programmable gate arrays (FPGAs). While this does come at the cost of reduced options at the physical layer, such a system is ideal for the work conducted here. The SKIRL platform provides the hardware and minimal software for examining mechanisms for coordinating cognitive networks. Table 7.4 summarizes its capabilities.

The BeagleBoard-xM runs a minimalist form of the Ubuntu Linux operating system, which allows for direct interaction with the system through Ethernet. The computational system interacts with the radio front end by setting a collection of registers and a Python module

Table 7.4   Summary of hardware capabilities.

Frequency Range	240–930 MHz
Modulations	Frequency shift keying (FSK),
	Gaussian frequency shift keying (GFSK),
	On–off keying (OOK)
Maximum Power Output	13 dBm
Minimum Receiver Sensitivity	−121 dBm
Sensing Bandwidth	100 kHz

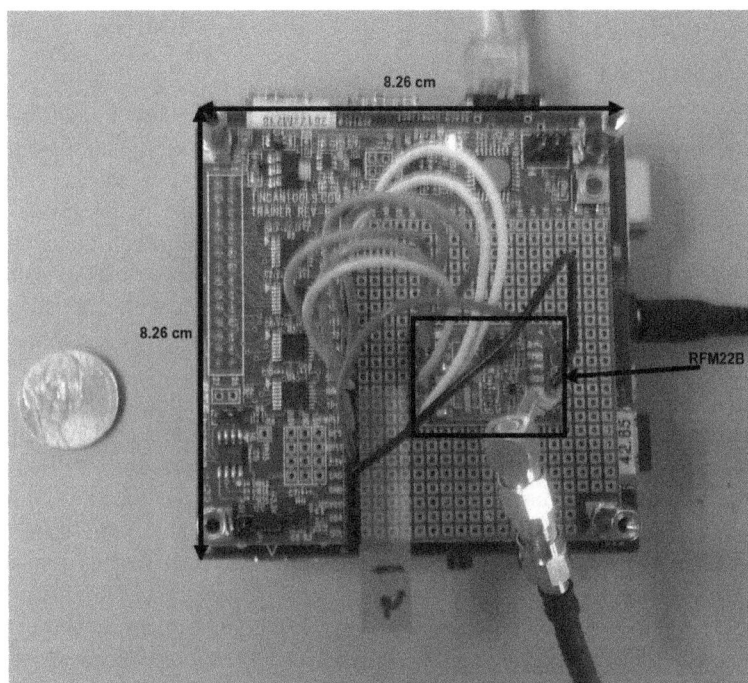

Figure 7.2   Picture of SKIRL platform. © 2014 Nicholas J. Kaminski. Reprinted, with permission, from Reference 4.

exists to streamline the process. The platform can access and sense only a single channel at a time, and it requires a finite reconfiguration time for any change. Sensing information takes the form of received signal strength indicator (RSSI) measures for the current channel. The details of controlling the SKIRL platform are discussed in Young's dissertation [30]. Figure 7.2 provides a picture of the SKIRL platform with a US nickel for scale. The entire platform costs approximately US $250.00.

The SKIRL platform is optimized for use in the 433 and 915 MHz bands; this example operates around 433 MHz, using 21 channels with center frequencies between 430.25 and 437.75 MHz. The centers of adjacent channels are spaced 750 kHz apart. Since the example focuses on MAC layer control, the same physical layer parameters (GFSK modulation with a bit rate of 4.8 kbps and a transmitter power of 17 dBm) are used for all transmissions. These physical layer parameters could be optimized by cognitive techniques independently of the techniques examined here.

Each transmission consisted of 64-byte packets. The first 14 bytes constitute the header, which includes the packet type, a local time stamp, the sender, and the intended recipient of the packet. Packets may be beacons, application data, or acknowledgment packets. The time stamp information is not used by receiving radios and is present simply for debugging purposes. The remaining 50 bytes are application data. No error checking is applied.

7.7.5 Implementing Behaviors in Software and Hardware

7.7.5.1 Time Flocking

Accomplishing the task of time flocking requires developing behaviors to address each goal associated with the temporal position of a transmission goal and combining them. Here, each temporal goal of radios centers on determining an appropriate time window for transmitting application data. These transmissions must not interfere with other radios inside or outside of the network. Interference within the network can be handled through the cooperation of neighbors, arising from communication and complementary behaviors. This interference is managed primarily with the time cooperation behavior. Interference with external radios must be handled in a different manner. This interference is controlled chiefly through the use of the time avoidance behavior.

The purpose of time avoidance is to eliminate interference. Here, we make no assumptions about the capabilities of external radios that may create interference. Thus, this behavior is designed to control interference through sensing and short transmissions. Sensing prior to transmitting prevents a radio from starting a transmission during the operation of some other radio, while short transmissions reduce the possibility of some other radio starting a transmission during the radio's operation. Here, this is accomplished by enforcing a sensing period prior to any transmission and limiting transmissions to single short packets. This strategy provides the desired interference elimination at the cost of throughput, a necessary trade off to accomplish the desired dynamic behavior. Additionally, while this approach is most general, it may be unnecessarily restrictive when information about external entities is readily available. The lack of such information limits the options for interference control and makes avoiding interference the most straightforward approach.

SKIRL platform sensing is RSSI based. Prior to any transmission, our radios monitor the channels for signals above a specified RSSI, corresponding to -22 dBm of received power, threshold for a specified sense time (0.1 second). In this case, these parameters were empirically determined. If a signal with RSSI above the threshold is sensed, the system immediately concludes that the channel is busy and prevents transmission; otherwise, it waits the full sensing time before declaring the channel clear.

The next behavior determines an appropriate position and size for application data transmission. The size of a transmission refers to the length of the time slot in which such a transmission is allowed and not necessarily the duration of time that the channel is actually in use, which depends on the amount of data to be sent and the physical parameters used. This definition for size implies a time division multiple access (TDMA) approach to time management, which has the benefit of clearly specifying who is authorized to use the channel.

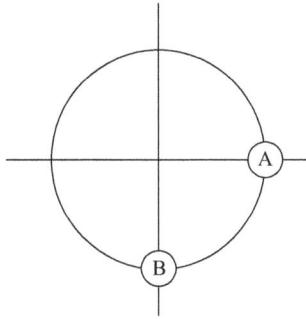

Figure 7.3   Visualization of radio phase. © 2014 Nicholas J. Kaminski. Reprinted, with permission, from Reference 4.

Using and Extending the DESYNC Algorithm: The time cooperation behavior applies the beacon-based social language to spread the transmissions of radios in time. This behavior is based on the DESYNC algorithm [31] which adjusts the beacon period of each radio for the maximum possible distance in time between adjacent beacon firings and defines slots based on beacon firing. The radios in a society negotiate appropriate positions and sizes for transmission slots through a combination of the time cooperation behavior and the beacon social language.

First, consider the time spreading of beacons, termed desynchronization in Reference 31. Initially, consider a single radio in order to understand the mechanism for splitting time into rounds. When started, our radio is given a value for its beacon period, T, of 10 seconds, for example. This means that every 10 seconds, from the activation of the radio onward, this radio will fire a beacon. Each firing signals the end of a time round of the radio. All radios in a group share the same value for T. The percentage progress of a radio through a round is referred to as the current phase $\phi(t)$ of the radio. Note that this term is borrowed from the study of pulse coupled oscillators (PCOs) [32] and that phase is a dimensionless quantity. For example, if the radio has 5 seconds until the next firing, it has a phase of 0.5 and if the radio has 2.5 seconds until the next firing, it has a phase of 0.75. Phase may be calculated as $(t_{current} - t_{start})/T$, where $t_{current}$ is the current time as given by a local running clock and t_{start} is the time when a round began. When radios reach a phase of 1, they fire a beacon and restart their timer, indicating a reset of the phase to 0.

Figure 7.3 illustrates the phase of radios. Phase advances in the clockwise direction. This circular depiction of phase provides a visual representation of a moment in time. Note that the DESYNC algorithm splits time into sequential periods; a single period is sufficient to depict any moment in time. In this diagram, the 12 o'clock position represents the beginning of a time period; this is the position that corresponds to both a phase of 1 and a phase of 0. As time passes, radios move around the circle in a clockwise direction. The radios first reach position A which corresponds to a phase of 0.25 and then proceed to position B, or a phase of 0.5. Note that each diagram only depicts a single moment in time and displays the various phases of the radios shown. Thus, Figure 7.3 shows two radios, A and B, just at the moment when radio A has a phase of 0.25 and radio B has a phase of 0.5.

Given this system for understanding the phase of a radio within a given time period, let us now consider the difference of phase between two radios. The internal clocks of radios are *not* synchronized, so no radio has access to the phase information of another radio. Instead, radios

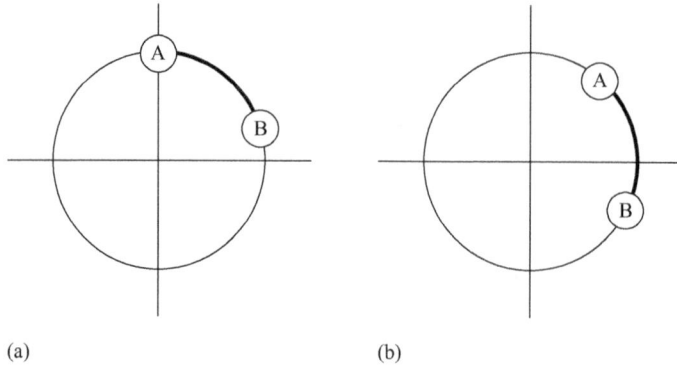

Figure 7.4   Phase difference mechanics. © 2014 Nicholas J. Kaminski. Reprinted, with permission, from Reference 4. (a) Radio phases just after A fires. (b) Radio phase at some later time.

have access to the phase difference between themselves and their peers. Imagine two radios A and B, each firing beacons every T seconds. Assume that radio A fires first. Radio B can determine the phase difference between itself and radio A, $\Delta_{BA}(t)$, at that instant simply by noting its own phase. This is because $\Delta_{BA}(t)$ is given as the difference $\phi_B(t) - \phi_A(t)$; when radio A fires its phase is reset to 0. Note that strictly the phase difference varies with time due to adjustments and clock drift, but clock drift is typically small and phase differences are fairly static between deliberate adjustments.

Figure 7.4 illustrates the principles of determining phase difference. Figure 7.4(a) shows the situation just after A fires and Figure 7.4(b) depicts a point later in time. Note that at each of these time points the phase difference is the same. However, consider the process of radio B determining this difference at each point. Just after radio A fires, radio B has a wealth of knowledge. First, since radio A just fired radio B knows that A will be resetting its phase to 0 and radio B knows its own phase. This allows radio B to easily calculate the phase difference between itself and radio A. At a later point, on the other hand, radio B does not have all necessary information. Specifically, while radio B knows its own phase, it does not have access to radio A's phase. Thus, radio B can easily determine the phase difference between itself and radio A just after radio A fires, but not at other times.

Given the concepts of phase and phase difference, along with a mechanism for a node to determine phase difference through observation, we may now discuss the mechanism for determining a new phase. For a group of n radios, the phase of the ith radio is given by $\phi_i(t)$. All radios are assumed to fire once every T seconds. The term round will refer to a cycle in which every radio fires once. For notational convenience, let us assume that radios are numbered such that $i < j$ implies that radio i fires after radio j in a given round. Finally, the phase difference $\Delta_i(t)$ will refer to the difference $\phi_i(t) - \phi_{i-1}(t)$. The concept of phase neighbors provides the final ingredient for determining a new phase for a radio. Note that any radio i is phase adjacent to both radio $i - 1$ *mod n* (its next neighbor) and radio $i + 1$ *mod n* (its previous phase neighbor). Recall from prior discussion that any radio i has access to both $\Delta_i(t)$ and $\Delta_{i+1}(t)$ through observation of its peers. Then the determination of the new phase simply requires applying the local goal of maximizing the time distance between adjacent firings to the local situation. Specifically, radio i wants its next firing to be the mid-point of its

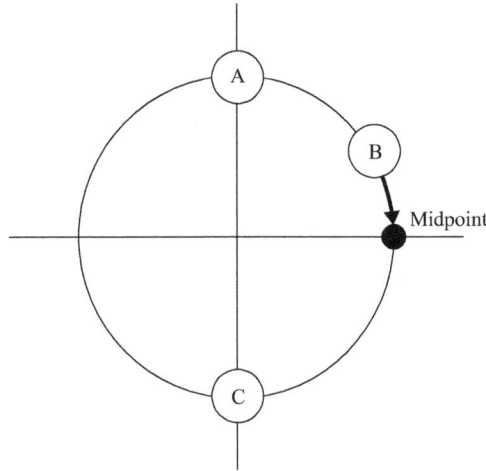

Figure 7.5   Midpoint of phase. © 2014 Nicholas J. Kaminski. Reprinted, with permission, from Reference 4.

two phase neighbors. Equation (7.1) shows how this midpoint may be determined with local information.

$$\phi_{i\,mid}(t) = \frac{1}{2}\left(\phi_{i+1}(t) + \phi_{i-1}(t)\right) \qquad (7.1)$$

$$= \frac{1}{2}\left[(\phi_{i+1}(t) - \phi_i(t)) - (\phi_i - \phi_{i-1}(t))\right] + \phi_i(t) \qquad (7.2)$$

$$= \phi_i(t) + \frac{1}{2}\left(\Delta_{i+1}(t) - \Delta_i(t)\right) \qquad (7.3)$$

Figure 7.5 illustrates the benefits of adjusting one radio's phase to the midpoint of its phase neighbors. In this figure, radio B has just determined the phase difference between itself and radio A, after previously determining the phase difference between itself and radio C. From this information radio B knows that it is closer in phase to radio A than radio C. Radio B then uses the process discussed above to determine the midpoint between radios A and C to use as its new target phase. This midpoint provides the greatest distance between radio B and its phase neighbors, without changing the firing order.

Given a new phase goal, achieving that phase for the next round is a matter of altering beacon timing. Recall that radios fire beacons once every T seconds, which corresponds to always firing with a phase of 1. When a radio needs to adjust its phase, it simply adjusts the time before firing its next beacon to be T' instead of T. If $T' < T$, the radio reduces its phase, and if $T' < T$, the radio increases its phase. After a phase adjustment is made, radios return to waiting T seconds to fire the next beacon. The adjustment period T' is easily determined from a target phase ϕ as ϕT.

Note that the authors of Reference 31 suggest tempering each phase adjustment to help convergence. Since phase adjustments affect the next beacon fired, but are determined from the most recent observations, adjustments inherently contain a delay of approximately T seconds between the determination of a new phase and acting on that determination. This delay can degrade overall performance since all radios are able to adjust their phase in a given round.

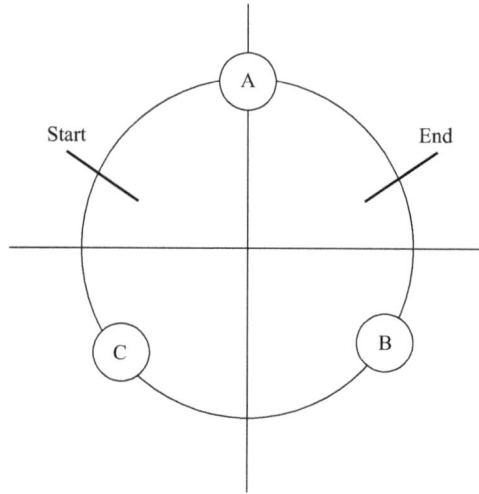

Figure 7.6   Depiction of slots as radio A fires. © 2014 Nicholas J. Kaminski. Reprinted, with permission, from Reference 4.

To avoid such degradation, the authors of Reference 31 suggest setting target phases as shown in (7.4). In this equation, ϕ_{target} is the new, tempered target phase, $\phi_{current}$ is the most recently used phase, and ϕ_{mid} is the mid-point of phase neighbors as discussed above. The parameter α is the tempering factor with value between 0 and 1 and typically close to 1.

$$\phi_{target}(t) = (1 - \alpha)\phi_{current}(t) + \alpha\phi_{mid}(t). \tag{7.4}$$

Phase adjustments allow radios to spread throughout a time period and provide the basis for determining slots. Radios simply start their slots at the midpoint of their prior neighbor's phase and their own firing. Slots are then ended at the midpoint of their next neighbor's phase and their own firing. Note this organization of slots is designed to keep a radio's firing within its own slot to prevent its beacon from interfering with another radio's application transmissions. The process for determining the target phase appears in Reference 4.

Figure 7.6 depicts the transmission slot of radio A. Note that the slot start point is at the midpoint of radio A's firing and radio C's phase and the end is at the midpoint of radio A's firing and radio B's phase. Additionally, note that the slot contains the firing of radio A. These properties allow the radios to split the time into slots based on the firings of peers.

The authors of Reference 31 suggest a mechanism for a new radio to join an existing network of radios whereby an interrupt packet signals the owner of the current slot that the new radio would like to join the network. This interrupt packet adds a packet to the algorithm with the goal ensuring that new beacons will be heard in an existing network.

Modifying the DESYNC Algorithm for Use in a CR Society: The above summarizes the DESYNC algorithm as presented in Reference 31, with the added term of social language. This algorithm provides a means to organized radios in time. However, the DESYNC algorithm is not fit for direct use as a behavior in a cognitive network. This deficiency arises because DESYNC does not include mechanisms to cope with the interactions that arise from multiple behaviors. Since both the time avoidance and time cooperation behaviors control the timing of transmissions, the interactions between these behaviors must be considered. In fact, as will be shown later, these minor interactions have a large impact on the overall system behavior.

For an immediate example of the role of behavior interaction, note that, when the avoidance behavior prevents the sending of a timing beacon that radio becomes invisible to its peers, effectively exiting the system. Unfortunately, the DESYNC interrupt packet that would be used for the radio to rejoin the system could then prevent additional beacons. This could lead to a cycle of forcing radios to exit the network and greatly damage the cooperation. The issue, in this case, arises from the additional sensing enforced by the avoidance behavior. Note that the increased focus on sensing also removes the need for the interrupt packet; if radios spend additional time sensing, missing the beacon of a new radio is less likely. Thus, interrupt beacons can safely be removed from the system. To prevent the overlapping of slots that can arise from this scenario, radios reset their slots if a beacon is received during their slot.

7.7.5.2 Frequency Flocking

Frequency flocking is the frequency complement to time flocking. It is a task aimed at dynamically organizing a cognitive network in frequency. Recall that time flocking controls both the position and length of radio transmission slots in time. Frequency flocking only controls the position of the radios in frequency. This is because the spectral size of transmission depends on physical layer characteristics, which are beyond the control of MAC layer solutions.

Just as in time flocking, the frequency goals of a cognitive network have two major components. First, the radios within a network need a means of finding one another. Ideally, the entire society would operate on the same frequency, which allows for the social language to reach, and therefore help coordinate, the largest possible number of radios. Second, the society needs a mechanism to find new open frequencies when a particular channel becomes congested, either through the presence of too many radios or from the appearance of an outside radio system. Each of these sub-goals requires the radios to acquire knowledge of peer positions in frequency and the state of a channel, as being either clear or congested. Additionally, radios should individually adapt to their spectral environment over time.

Frequency aggregation provides the mechanisms necessary to keep a network together in frequency. Grouping peer radios together allows the time behaviors to coordinate the radios and make effective use of the spectrum. Given the limitations of the SKIRL platform, peer recognition through sensing alone is not available. Instead, the system relies on the beacon-based social language for peer identification. Members of our cognitive network are able to recognize peers through reception of beacons. In this case, recognition through language requires radios both to be on the same channel at the time that one radio fires. Thus, the frequency aggregation behavior is designed around accomplishing this.

Frequency aggregation employs a frequency search pattern and a channel dwell time to accomplish its goals. The frequency search pattern is constructed to provide radios a defined rendezvous channel while still allowing radios to find peers on other channels. The radios always search the channel at the center of the specified range first and expand out from there. Specifically, the radios will search the center channel, then one channel above the center, then the center channel again, then one channel below the center, then the center again, then two above the center, and so on. A dwell time holds the searching radio on each channel for one beacon period to avoid changing channels before a peer sends a beacon to make its presence known. Note that this dwell time assumes that each radio has a different phase at any given time.

The final factor of the frequency aggregation behavior is its activation conditions. These specify when a radio should begin searching for peers. For this purpose, each radio keeps track of active peers on its current channel. This is accomplished by adding radios to a list when a beacon is received from them. Radios are removed from the list if two periods pass without receiving a beacon. When the list is empty, the radio assumes that it is alone in the channel. If a beacon is received from a radio already in the list, the count for that radio is reset. This prevents the radios from falsely determining an empty channel if beacons have been prevented due to time avoidance, but does not overly delay appropriate reactions to empty channels.

Frequency dispersion allows radios to leave channels that are too congested to use. Recall that the time avoidance behavior prevents radio transmissions if a channel is busy. In isolation, this could have the effect of a radio waiting an indeterminate time to send a beacon in a highly congested channel. Clearly, this is not ideal. Thus, the frequency dispersion method selects a new channel for use when a channel becomes overly congested. The system simply moves to the next channel when frequency dispersion is activated.

Changing frequency has a cost. If radios change frequency at the first sign of interference, societies would rapidly become fragmented. Additionally, reconfiguration is not an instant operation, and further delays the transmission of application data. Therefore, instead of changing frequencies at the first prevented transmission, radios wait until two consecutive beacons have been prevented. For our purposes, this method allows radios to effectively differentiate truly congested channels from minor interference.

Frequency learning allows the radios to improve their channel selection over time. This behavior supports the other frequency behaviors by adjusting the probability of selecting a particular channel. Each channel is assigned a visitation probability, p, that is incremented and decremented based on a radio's individual experiences. This visitation probability is the probability that a particular channel is actually visited when selected by one of the other frequency behaviors. Stated, differently $1-p$ is the probability that a channel is skipped and the radio instead moves to the next channel it would otherwise visit. The visitation probability for a channel is reduced by d when radios determine that channel is congested and employ dispersion to leave that channel. The visitation probability is incremented by a small amount i after n beacons to allow radios to forget about stale congestion information. Naturally, decrementing and incrementing must be accomplished in a manner that restricts p to the interval $[0,1]$.

Herein lies a problem. While frequency learning allows radios to adapt to their operating environment over time, this adaptation comes at the cost of deterministic radio operation. That is, a radio stochastically determines whether a new channel is visited based upon its current value for p. The amount of determinism that remains depends on the parameters d, i, and n. The values of these parameters must depend on the hardware's capabilities to handle non-deterministic behavior. Specifically, the sensing capabilities of the hardware determine how well peers motions in frequency may be determined in cases when they cannot be predicted. As discussed, the sensing capabilities of the SKIRL platform are limited. Thus, any non-determinism must be reduced as much as possible.

Thankfully, non-determinism can be effectively removed by setting d and i to 1 and n to 3. This parameter combination has the effect of completely ignoring channels that have been determined to be congested for three beacon rounds before forgetting about the congestion, three beacon periods later. This steers radios away from congested channels for three beacon rounds, which provides a balance between adaptability to the environment and management of non-determinism.

7.7.6 Network Evaluation

The evaluation of multi-agent systems is a non-trivial problem with complexities arising from mutual dependencies among agents, multiple objectives that must be handled simultaneously, and unanticipated interactions between strategies within individual agents. Consequently, roboticists have focused on the evaluation of behavior-based systems through implementation, the only available method to accurately capture all the relevant interactions that occur in realistic environments. Mataric provides a framework for the implementation-based evaluation of behavior-based systems. While this framework stresses the importance to exposing agents to realistic noise and interference, it also notes that the environments for such testing must be well controlled in order to understand the operation of the system. Additionally, Mataric suggests that individual behaviors should be confirmed as operating according to their designed purpose, as much as possible, before any combinations are assessed [21]. However, in some cases this strict isolation cannot be maintained, as one behavior is used to activate another, and the behaviors must be considered together. Here, we focus on some interesting scenarios starting with the smallest groups of behaviors possible before considering more complete sets.

The non-synchronized local clocks of the radios present a problem to the examination of system behaviors. Determining the time difference between the actions of radios requires a common time base. In order to determine this common time, shared events are necessary. Specifically, when a radio transmits a beacon, other radios receiving the beacon provide events that are shared between radios. We calculate the time difference between every two radios for each beacon transmitted and received. These differences are then averaged for each radio pair in a given test. This provides the means to shift all events to a common time base without altering their relative spacing.

For this example, we will examine a network of four radios, sharing a common implementation, each running separate instances of the behaviors and communication methods discussed above. The radios are identified by unique code name (fuhr, gillies, joliat, or howe) and graphical symbols (see Table 7.5). Testing occurred in the 400 MHz band.

7.7.7 The Entry Scenario

The entry scenario examines the ability of the system to allow new radios to join an existing group. To test this scenario, we first allow three radios, namely fuhr, gillies, and joliat, to organize themselves in time and then starting a fourth radio, howe, at some later time so that it may join the other radios and carve out a place for itself. Each radio fires (transmits) 23 packets, each with a beacon period of 30 seconds, prior to shutting down. This number of packets allows observation of the pertinent radio behavior, but limits the data to reasonable levels for visualization.

Table 7.5 Radio symbols.

Fuhr	X
Gillies	Square
Joliat	Triangle
Howe	Circle

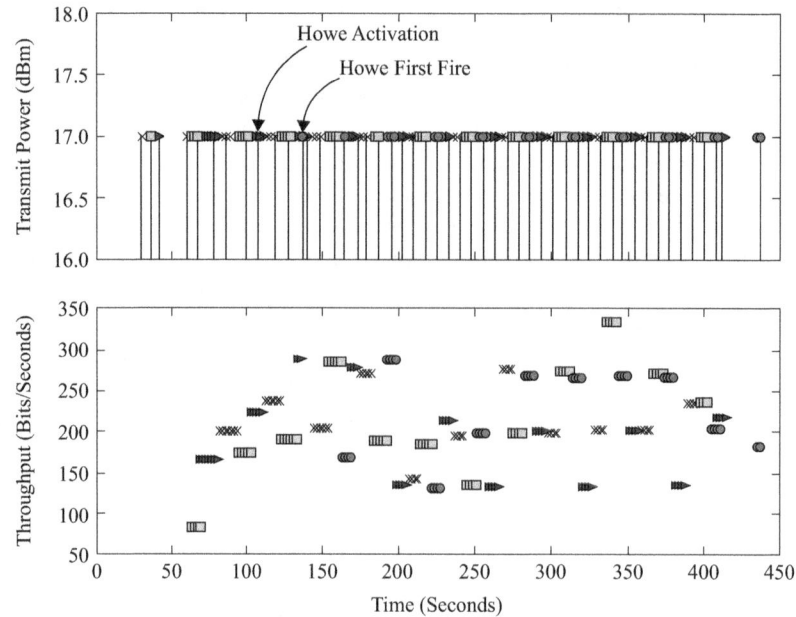

Figure 7.7 Entry scenario. © 2014 Nicholas J. Kaminski. Reprinted, with permission, from Reference 4.

Figure 7.7 presents a typical result, with the time organization of the radios at the top and the associated throughputs at the bottom. The time slots occupied by the radios are indicated by collections of each radio's individual symbol, and the firing of radios as a vertical line topped by a radio's symbol. The bottom portion also shows the slot length as a line of a radio's symbol and indicates the throughput averaged over that slot.

Figure 7.7 displays the start of fuhr, gillies, and joliat and the establishment of their slots prior to howe's entry. This entry occurs at roughly 110 seconds, as indicated. After its entry, howe listens to the beacons of its peers, gathering information about their firings. Our entrant then fires its first beacon at roughly 140 seconds, just before joliat fires its own beacon. As shown, this cuts joliat's slot short, as intended, and allows for all the radios to adjust to the new entrant. The radios then reorganize themselves. They do not necessarily reach their final convergence before shutdown begins. This test does not focus on convergence because its purpose is to examine and verify the entry sequence. It is clear that the radios are able to organize themselves in time when a new radio joins an existing network.

The mechanisms that allow for radio entry are the primary focus for the tests described in this section. Note that radio implementation's focus on maintaining the ability to receive information allowed joliat to receive howe's beacon and end its own slot to allow for necessary time position adjustments. In this situation, joliat ended its slot in recognition that its assumptions about the firing positions of its peers were incorrect. This recognition and response to an assumption error large enough to allow for a slot to overlap the firing of a neighbor gives the system a chance to recover from errors. While the beacons were not close enough together to trigger the time avoidance behavior, the system's focus on listening for social language messages and

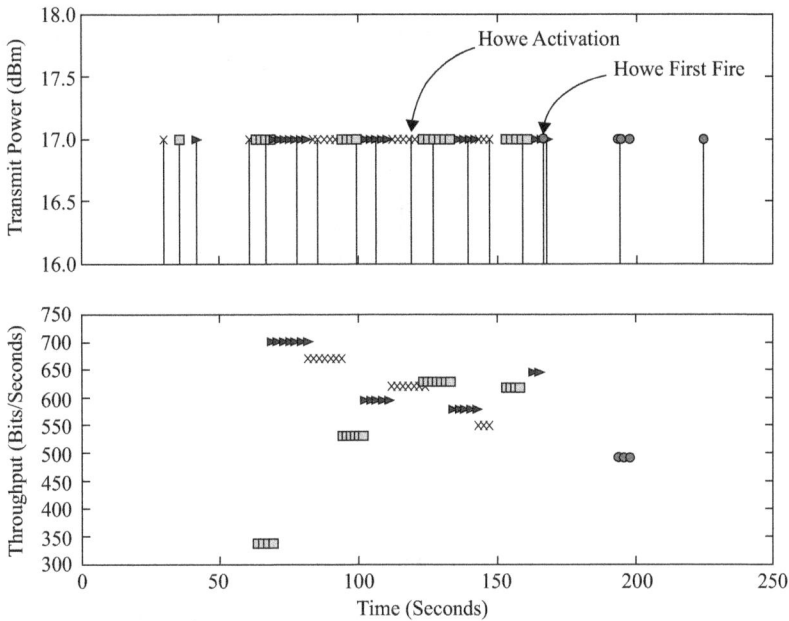

Figure 7.8   Lock out scenario. © 2014 Nicholas J. Kaminski. Reprinted, with permission, from Reference 4.

appropriately responding to those messages allows for new radios to join existing groups, as intended.

In fact, the listening aspect of time avoidance merely contributes to the mechanisms that allow new nodes to join the system. To examine this, the priorities of the system must be slightly altered. Recall that time avoidance mandates that radios listen to the channel transmitting, including application transmissions. In addition, to ensure reception, radios listen for acknowledgments after transmitting application messages. To accomplish this, a radio keeps its front end in receive mode until just before transmitting and then immediately returns to receive mode. The result of this action is that the radios prioritize listening for the entry of new peers or possible interference. This individual priority results in an emergent sharing that allow for new nodes to enter an existing group.

Altering the individual behavior of the radios, even slightly, changes the emergent properties of the overall system. For example, if the radios are not required to reenter receive mode for acknowledgments directly following each transmission, they send application messages more aggressively, no longer hampered by the delays of starting to receive, processing acknowledgment packets, and stopping to receive prior to transmission. Instead, the radios can stop receiving, and then transmit throughout their entire slot. The obvious benefit of higher throughput comes at the price of individual focus on listening and therefore the emergent time sharing.

Figure 7.8 displays the results of a brief test that exemplifies the effect on overall system behavior of altering individual radio focus. We conducted this test under the same conditions as for Figure 7.7, except for the alterations to application data sending policy discussed above.

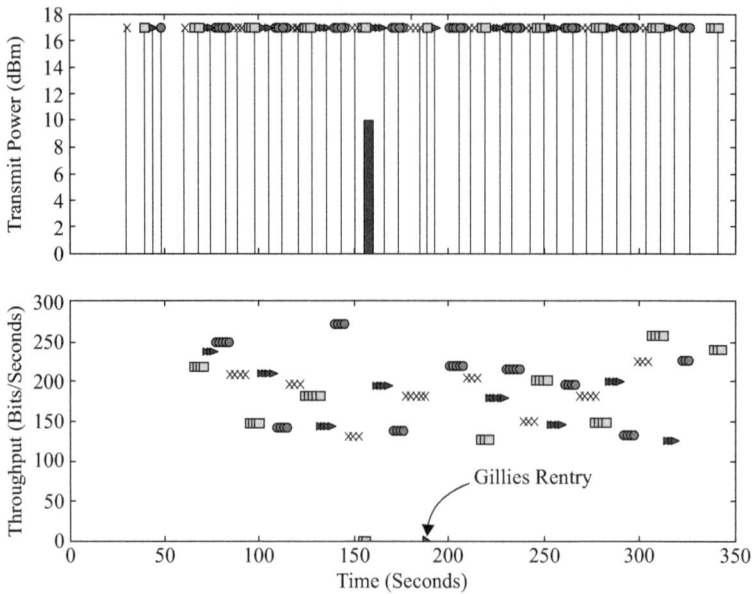

Figure 7.9   Test results with one beacon period of interference. © 2014 Nicholas J. Kaminski. Reprinted, with permission, from Reference 4.

Note that these do not affect the time avoidance behavior. While howe still enters the system at 120 seconds, it is prevented from firing until joliat fires its final beacon, just before joliat shuts down. The individual radio focus of maximally transmitting application data during its slot results in radios forming a defensive phalanx in time. Note that the radios that start earlier (fuhr, gillies, and joliat) still organize themselves in time. Its only when howe attempts to join that the difference is evident.

Focusing on the mechanics that lead to this time phalanx, the combination of the aggressive slot usage and the time avoidance behavior cause this particular emergence. Since the radios don't establish slots prior to determining their neighbor's positions, the initial time organization is not affected. Instead, once the radios form slots they aggressively use them. This means that a new radio attempting to join the system will be prevented from firing its own beacon because the channel is already busy with application data, as shown in Figure 7.8. Note that this alteration provides an indication of the emergence that results from the interactions between the behavior of various radios.

7.7.8 Social Learning

Interference affecting a single radio in the system demonstrates the system's capacity for social learning. Figure 7.9 depicts system behavior when one radio is interfered with after a time organization has been established. The interference depicted by the blue bar at about 150 seconds has prevented gillies from transmitting application data and its beacon after the group has established slots for each radio. This causes gillies to effectively exit this system.

Once this happens the other radios expand in time to absorb the empty space left from gillies leaving. The degree to which this space can be fully absorbed is controlled by the system factor α. This limiting is especially helpful when gillies fires its next beacon one round after it was forced to exit. Specifically, note that in one round's time only gillies' neighbors, fuhr and joliat, are affected by gillies' absence. Both of these radios are pulled into the void left by gillies; note their close spacing directly before 200 seconds. This motion in time would allow fuhr and joliat to expand their own slots by the time left by gillies. This additional time would later be shared with howe as the system reorganized itself. However, gillies reenters the system before such reorganization can take place. Specifically, gillies enters the system after joliat's beacon, as joliat moved toward fuhr to account for new space, but during joliat's slot. Thus, as discussed above, joliat ends its slot in recognition of the error in assuming that gillies had truly exited the group, and the system proceeds. Thus, the memory of gillies in the initial organization of the system maintains a space, albeit with different phase neighbors, for gillies even one round after gillies exits the system.

In this time flocking system, the time relative positions and slot lengths represent the knowledge that the system has gained over time about how to organize radios in time. This knowledge exists in the form of the spacing between radios, rather than within any individual radios. Additionally, note that the system is able to make use of this information even when a radio exits; i.e., the system remembers how to organize four radios in time even after gillies departs. But the system will not remember this information forever and immediately started taking action to forget information about four radios in favor of information about three radios once gillies was gone. Additionally, if all radios exit the system, no information is maintained, as there are no interactions in which the information may be stored, as would happen if the interference lasted for a full round.

7.7.9 Total System Behavior

The total system provides the means to develop self-organizing, decentralized networks through the coordination of radios in time and frequency. This coordination arises from the combinations of the behaviors discussed above. Both the interactions of radios and the interactions of behaviors within a single radio contribute to the final operation of the system.

Examination of the total system is simply a matter of reviewing system behaviors to ensure they contribute as expected. We first review the radios' ability to find one another in frequency and establish a time organization; then we task the radios with doing the same thing in the presence of interference.

Figure 7.10(a) displays the radios finding one another in frequency and establishing a time organization that allows for transmitting application data. This figure combines the information shown in the time flocking figures and that shown in the frequency flocking figures. Here, the slots of a radio are displayed by its symbol with no lines. The beacons of a radio are indicated by a vertical line below a radio's symbol and frequency changes are depicted by a vertical line above a radio's symbol. Throughput is determined and displayed in the same manner as above. Figure 7.10(b) shows the network gathering in frequency, avoiding interference. The interference is represented by the horizontal blue bar at the top. Each of these tests demonstrates the successful combination of the behaviors discussed above.

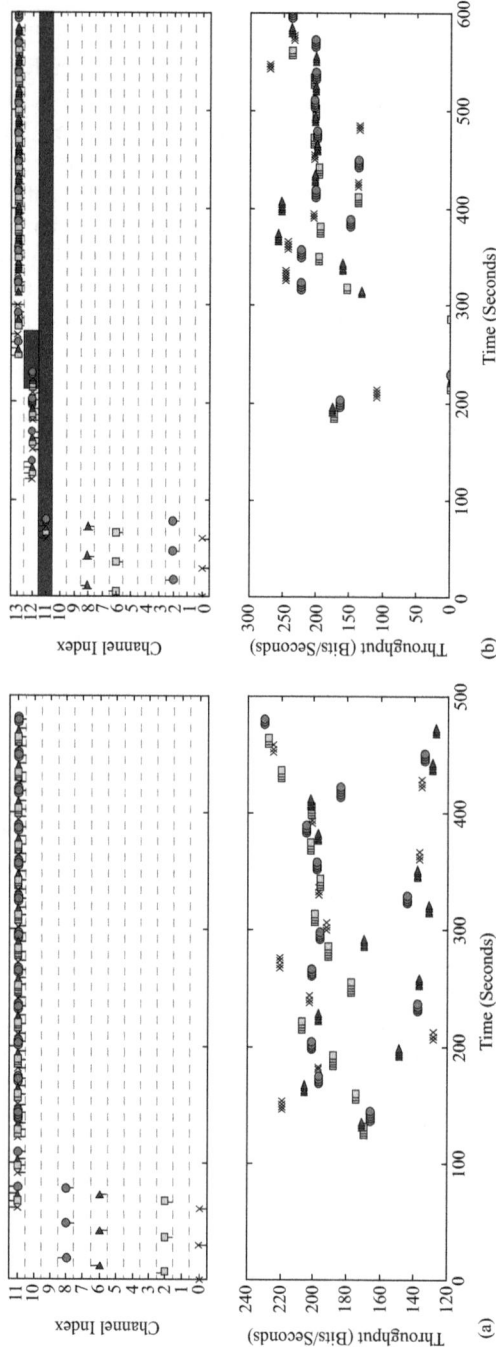

Figure 7.10 Total system testing. (a) Without interference. (b) With interference. © 2014 Nicholas J. Kaminski. Reprinted, with permission, from Reference 4.

Bibliography

[1] R. W. Thomas, L. A. DaSilva, and A. B. MacKenzie, "Cognitive Networks," in *First IEEE International Symposium on New Frontiers in Dynamic Spectrum Access Networks, 2005. DySPAN 2005.* Institute of Electrical & Electronics Engineers (IEEE), 2005. doi: org/10.1109/dyspan.2005.1542652

[2] 3GPP, "Telecommunication Management; Self-Organizing Networks (SON); Concepts and Requirements," 3rd Generation Partnership Project (3GPP), TS 32.500, December 2014. [Online]. Available: http://www.3gpp.org/DynaReport/32500.htm

[3] L. Doyle, J. Kibiłda, T. K. Forde, and L. DaSilva, "Spectrum Without Bounds, Networks Without Borders," *Proceedings of the IEEE*, vol. 102, no. 3, pp. 351–365, Mar 2014. [Online]. Available: http://dx.doi.org/10.1109/jproc.2014.2302743

[4] N. J. Kaminski, "Social Intelligence for Cognitive Radios," Ph.D. Dissertation, Virginia Polytechnic Institute and State University, Blacksburg, Virginia, 2014.

[5] (2013) Social Language Use (Pragmatics). American Speech-Language-Hearing Association. Accessed February 2014. [Online]. Available: http://www.asha.org/public/speech/development/pragmatics.htm

[6] J. Habermas. (2011) Metaphysics, Research Lab, Stanford University. Accessed Feb 2014. [Online]. Available: http://plato.standford.edu/entries/habermas

[7] T. D. Seeley, *Honeybee Democracy*. Princeton University Press, Princeton, NJ, 2010.

[8] E. Axtell, Roger, *Gestures: The Do's and Taboos of Body Language Around the World Revised and Expanded Edition*. John Wiley & Sons, Inc., Hoboken, NJ, 1997.

[9] J. Habermas, *The Theory of Communicative Action*. Beacon Press, 1985.

[10] T. D. Seeley, *Wisdom of the Hive: The Social Physiology of Honey Bee Colonies*. Harvard University Press, Cambridge, MA, 2009.

[11] M. McHenry, K. Steadman, A. E. Leu, and E. Melick, "XG DSA Radio System," in *2008 3rd IEEE Symposium on New Frontiers in Dynamic Spectrum Access Networks*. Institute of Electrical & Electronics Engineers (IEEE), 2008. doi: 10.1109/dyspan.2008.59

[12] F. Ge, R. Rangnekar, A. Radhakrishnan, S. Nair, Q. Chen, A. Fayez, Y. Wang, and C. W. Bostian, "A Cooperative Sensing Based Spectrum Broker for Dynamic Spectrum Access," in *MILCOM 2009 – 2009 IEEE Military Communications Conference*. Institute of Electrical & Electronics Engineers (IEEE), October 2009. doi: 10.1109/milcom.2009.5379708

[13] C. Doerr, D. C. Sicker, and D. Grunwald, "What a Cognitive Radio Network Could Learn From a School of Fish," in *Proceedings of the 3rd International ICST Conference on Wireless Internet*. Institute for Computer Sciences, Social Informatics and Telecommunications Engineering (ICST), 2007. doi: 10.4108/wicon.2007.2299

[14] E. O. Wilson and M. Pavan, "Glandular Sources and Specificity of Some Chemical Releasers of Social Behavior in Dolichoderine Ants," *Psyche: A Journal of Entomology*, vol. 66, no. 4, 1959, pp. 70–76. doi: 10.1155/1959/45675

[15] A. Strandburg-Peshkin, C. R. Twomey, N. W. F. Bode, A. B. Kao, Y. Katz, C. C. Ioannou, S. B. Rosenthal, C. J. Torney, H. S. Wu, S. A. Levin, and I. D. Couzin, "Visual sensory networks and effective information transfer in animal groups," *Current Biology*, vol. 23, no. 17, 2013, pp. R709–R711. doi: 10.1016/j.cub.2013.07.059

[16] J. R. Boyd, *Destruction and Creation*. U.S. Army Command and General Staff College, Leavenworth, KS, 1987.

[17] U. Greggers, G. Koch, V. Schmidt, A. Durr, A. Floriou-Servou, D. Piepenbrock, M. C. Gopfert, and R. Menzel, "Reception and Learning of Electric Fields in Bees," *Proceedings of the Royal Society B: Biological Sciences*, vol. 280, no. 1759, 2013, pp. 20 130 528–20 130 528. doi: 10.1098/rspb.2013.0528

[18] R. E. Learned, S. E. Johnston, and N. J. Kaminski, "Cognitive Coexistence: A Throughput Study of MUD-enhanced Opportunistic Spectrum Access," in *2013 Asilomar Conference on Signals, Systems and Computers*. Institute of Electrical & Electronics Engineers (IEEE), November 2013. doi: 10.1109/acssc.2013.6810537

[19] S. Haykin, *Neural Networks: A Comprehensive Foundation*. Prentice Hall, 1998.

[20] A. Leon-Garcia and I. Widjaja, *Communication Networks: Fundamental Concepts and Key Architectures*, 2nd ed. McGraw-Hill, Boston, MA, 2004.

[21] M. J. Mataric, "Interaction and Intelligent Behavior," Ph.D. Dissertation, Massachusetts Institute of Technology, Cambridge, MA, 1994.

[22] R. A. Brooks, "A Robust Layered Control System For a Mobile Robot," Massachusetts Institute of Technology, Cambridge, MA, USA, Tech. Rep., 1985.

[23] R. A. Brooks, "Intelligence without Reason," in *Computers and Thought, IJCAI-91*. Morgan Kaufmann, 1991, pp. 569–595.

[24] M. J. Mataric, "Minimizing Complexity in Controlling a Mobile Robot Population," in *IEEE International Conference on Robotics and Automation*, Nice, France, vol. 1, 1992, pp. 830–835.

[25] M. J. Mataric, "Designing Emergent Behaviors: From Local Interactions to Collective Intelligence," in *Proceedings of the Second International Conference on Simulation of Adaptive Behavior*, 1993, pp. 432–441.

[26] M. J. Mataric, "Kin Recognition, Similarity, and Group Behavior," in *Proceedings of the Fifteenth Annual Conference of the Cognitive Science Society*, 1993, pp. 705–710.

[27] (2012, Dec) BeagleBoard-xM. [Online]. Available: http://beagleboard.org/Products/Beagle Board-xM

[28] (2012, Jun) RFM22B FSK transceiver. [Online]. Available: http://www.hoperf.com/rf/fsk_module/ RFM22B.htm

[29] A. R. Young and C. W. Bostian, "Simple and Low-cost Platforms for Cognitive Radio Experiments [Application Notes]," *Microwave Magazine, IEEE*, vol. 14, no. 1, January 2013, pp. 146–157.

[30] A. R. Young, "Unified Multi-domain Decision Making: Cognitive Radio and Autonomous Vehicle Convergence," Ph.D. dissertation, Virginia Polytechnic Institute and State University, 2013.

[31] J. Degesys, I. Rose, A. Patel, and R. Nagpal, "DESYNC: Self-Organizing Desynchronization and TDMA on Wireless Sensor Networks," in *Information Processing in Sensor Networks, 2007. IPSN 2007. 6th International Symposium on*, 2007, pp. 11–20.

[32] R. E. Mirollo and S. H. Strogatz, "Synchronization of Pulse-Coupled Biological Oscillators," *SIAM Journal on Applied Mathematics*, vol. 50, no. 6, 1990, pp. 1645–1662.

CHAPTER 8

Cognitive Radio Applications

8.1 Introduction

In this chapter, we focus on representative cognitive radio applications where working proto-type hardware and software exist or where standards or regulations are available to guide the development of such prototypes. We have not attempted to cover the many proposed applications that have been simulated but not yet adopted or built or for which standards work is still in its early stages. Our examples cover the top level of the application-driven taxonomy proposed in Reference 1: dynamic spectrum access, wireless network management, user-aware wireless services, and homogeneous networking. See Reference 2 for an alternative treatment of many of the same applications.

8.2 Zoned Dynamic Spectrum Access

In Chapter 1, we discussed how DSA currently operates in the U.S. TV bands. There the governing regulations combined database lookup and spectral sensing techniques. In this chapter, we look at other aspects of DSA and extend them to communications and radar system coexistence and to frequency sharing by satellite-based systems. One spatial approach to DSA defines zones surrounding a primary user, and within these zones secondary users must employ different operating protocols. This is particularly applicable to radars and the earth station portion of some satellite systems, where there is a single primary user in a fixed location. This primary user is surrounded by a primary exclusive region [3] or an exclusion zone [4], within which cognitive secondary users cannot operate because the probability that a secondary user will cause harmful interference is sufficiently high. Surrounding the exclusion zone are one or more zones where secondary users may operate if they are able to ascertain that they will not cause harmful interference to the primary user. Allowed techniques may include temporal frequency sharing, secondary user transmitter power reduction, etc. There may also be a region whose secondary users are sufficiently remote from the primary user that they can operate freely without sensing the spectrum so long as they verify that they are in this zone.

8.3 Cognitive WiFi and LTE Operation in TV White Space Spectrum

8.3.1 WiFi Frequency Translators

8.3.1.1 Introduction

An attractive and relatively simple cognitive radio application is to use cognitive frequency translators (converters, repeaters, transceivers) to allow modified WiFi and LTE user equipment to operate in available TV white space (TVWS) spectrum. As a first example, consider the system presented in Figure 8.1 [5]. See Reference 6 for a prototype converter for tablet devices.

Here, an 802.11 b/g router's input (uplink or receiving) and output (downlink or transmitting) channels are translated into unoccupied spectrum to the 470–790 MHz band. The converter switches to transmit mode whenever it senses a high-level signal coming from the router port and reverts to receive mode when no such signal is present. A similar approach can be used for translating signals from LTE terminal equipment. In either case, the process of designing such a converter requires careful attention to the specifications for TV white space operation at one end and to those for LTE and IEEE 802.11 b/g at the other.

8.3.1.2 RF Considerations

Homodyne or Low-IF and SDR Difficulties   The RF design of an SDR capable of opportunistic WiFi or LTE operation in the 470 to 790 MHz band presents a number of challenges, particularly for SDR. Directly digitizing the RF input band with sufficient resolution and dynamic range is difficult and economically prohibitive. For a homodyne or low-IF architecture, the local oscillator frequency falls inconveniently close to the desired input frequency range. The image frequency range overlaps the input frequency range, so that the images within this overlap cannot be eliminated by filtering. The receiver must have sufficient blocking dynamic range to handle all strong signals within the 320 MHz input bandwidth without loss of sensitivity (de-sensing). These include primary users (TV transmitters) in occupied portions of the band as well as other secondary users, transmitter feedthrough, and local oscillator feedthrough and harmonics. At present, these problems can better be addressed through analog hardware rather than software.

The High-IF Architecture   An attractive hardware solution is a high-IF architecture where all the signals in the incoming TV white space bandwidth are first up converted using a mixer and tunable local oscillator that centers the wanted part of the up converted signal at a fixed intermediate frequency (IF). A SAW filter then sets the IF bandwidth, eliminating LO harmonics and attenuating blocking signals. In an example presented in Reference 7, TV white space signals in the band f_{RF} = 470–790 MHz are up converted to an IF center frequency of f_{IF} = 2.6 GHz using a local oscillator that tunes f_{LO} = 1.81 to 2.13 GHz and high-side injection $f_{LO} > f_{RF}$. An 80 MHz bandwidth SAW filter centered at 2.6 GHz follows the up conversion mixer and eliminates or at least attenuates blocking signals that fall outside of its passband.

In the mixer system used in this example, the wanted RF frequency f_{RF} will be up converted according to

$$f_{IF} = f_{LO} - f_{RF} \tag{8.1}$$

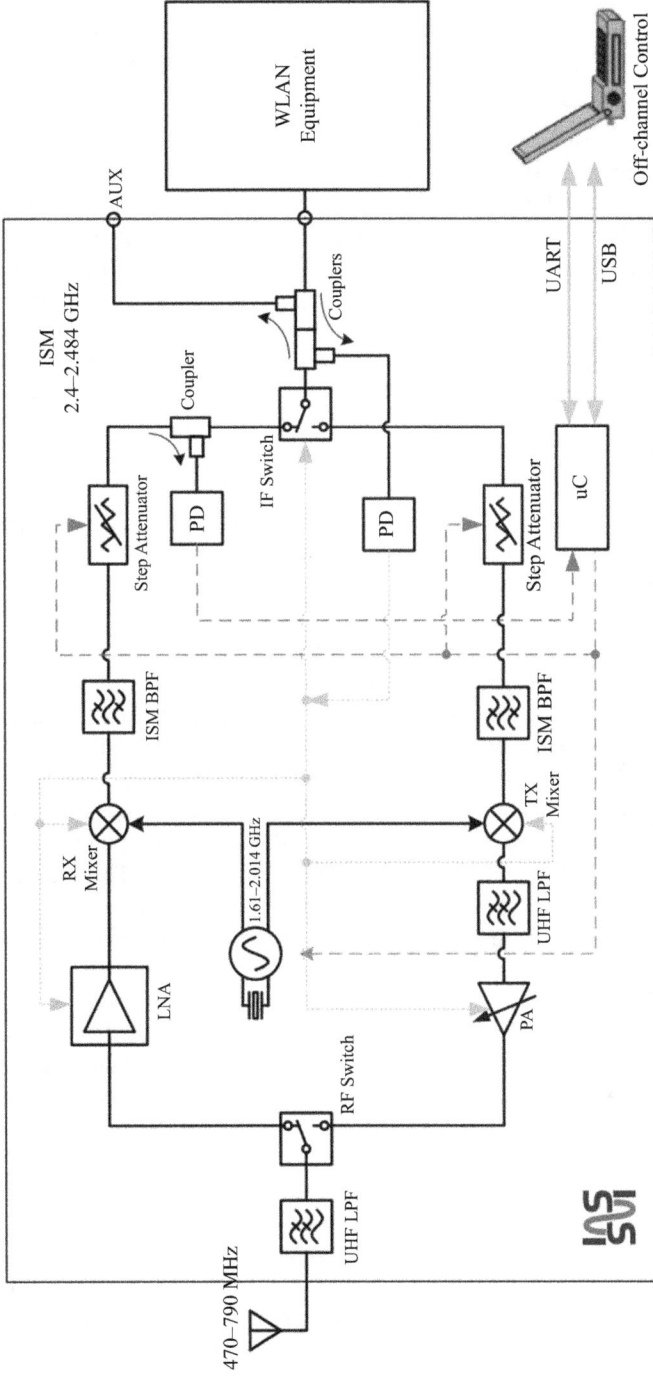

Figure 8.1 Prototype white space to WiFi converter. ©2013 IEEE. Reprinted, with permission, from Reference 5.

An input signal at the unwanted image frequency f_i will be down converted according to

$$f_{IF} = f_i - f_{LO} \tag{8.2}$$

and the image frequency f_i can be calculated from the RF and IF frequencies by

$$f_i = f_{RF} + f_{IF} \tag{8.3}$$

In this example, the images appear 5.2 GHz above the corresponding wanted signals and are easily eliminated by the UHF band pass filter following the antenna. LO harmonics are well outside of the IF passband, as are some of the out of band blocking signals.

If the complete system as shown in Figure 8.1 is to meet 802.11 b/g specifications, the designer must pay careful attention noise and strong signal behavior of the RF components and to the internal and external signal levels. RF specifications for a commercial 802.11 b/g router may not be readily available. In the notation of Chapter 3, we will use the generic values of $NF_{SR} = 5$ dB for noise figure and $IIP_3 = -12$ dBm for input third-order intercept point [5]. The 802.11 b/g standard does not directly specify NF_{SR} and IIP_3 values, but working from published 802.11 b/g values for sensitivity and adjacent channel selectivity, the authors of Reference 5 obtain $NF_{SR} = 11.7$ dB and $IIP_3 = -34.5$ dBm. To allow a margin for error, they use target values of $NF_{SR} = 8.5$ dB (maximum) and $IIP_3 = -15$ dBm (minimum) for the converter plus router combination.

Working with the component values and architecture of Figure 8.1 from Reference 5, we note that the authors picked components with good noise and strong signal performance and used an attenuator to adjust the converter performance. Here, we offer a slightly different analysis and our results differ in minor ways from theirs.

The attenuator has a small effect on the converter noise figure but a large effect on the converter gain. Reducing the converter gain increases the noise contribution of the router, and increasing the converter gain will lower the overall IIP_3. The gain must not be so large that strong signals within the anticipated 450–790 MHz frequency range drive the router into compression. To explore this, we developed a spread sheet incorporating the equations from Chapter 3 for calculating NF_{SR} and IIP_3 for the converter plus router. This indicates that there are multiple attenuator settings that will give us $NF_{SR} \leq 8.5$ dB and $IIP_3 \geq -15$ dBm. To choose an appropriate combination, we will follow the reference's suggestion and consider dynamic range.

Spurious Free Dynamic Range (SFDR) as derived in Chapter 3 is based on the input power that will raise third-order products at a two-port devices output to the level of the amplified input noise floor. Derived from a simple linear extrapolation of the devices intermodulation distortion, it is a standard but not necessarily realistic measure of the device's performance, widely used because it is easily calculated from the below equation:

$$SFDR = \frac{2}{3}(IIP_3 - MDS) = \frac{2}{3}(IIP_3 - NF_{SR} - 10 \log_{10}(B) + 174) \tag{8.4}$$

Here, MDS is the receiver noise floor or the so-called minimum detectable signal in dBm. The latter can be calculated from the noise figure NF_{SR}, the bandwidth B in Hz, and kT. For an antenna with a noise temperature of 290 K, $10 \log_{10}(kT) = -174$. A more useful measure would be the blocking dynamic range, BDR, which is the dB difference between the power level of an unwanted off-frequency interfering signal that would reduce a receiver's sensitivity to an on-frequency wanted signal by, say, 1 dB. Unfortunately, this cannot be calculated for a

Block No	Block ID	NF	G	IIP3	OIP3	Bk Te	Bk g	Bk nf	Ov T	Ov NF	Ov G	oip3 mW	Ov oip3 m	Ov OIP3	Ov IIP3	SFDR
1	Switch	0.8	-0.8	54	53.2	58.7	0.8	1.2	58.7	0.8	-0.8	2.09E+05	2.09E+05	53.2	54.0	84.1
2	UHF Filter	1.4	-1.4	1000	998.6	110.3	0.7	1.4	191.3	2.2	-2.2	7.24E+99	1.51E+05	51.8	54.0	83.2
3	LNA	0.5	22	12.5	34.5	35.4	158.5	1.1	250.0	2.7	19.8	2.82E+03	2.82E+03	34.5	14.7	56.7
4	Mixer	12.5	3.5	22	25.5	4867.0	2.2	17.8	301.0	3.1	23.3	3.55E+02	3.36E+02	25.3	2.0	47.9
5	SAW Filter	1.2	-1.2	60	58.8	92.3	0.8	1.3	301.4	3.1	22.1	7.59E+05	2.55E+02	24.1	2.0	47.9
6	Attenuator	20	-20	1000	980	28709.8	0.0	100.0	478.4	4.2	2.1	1.00E+98	2.55E+00	4.1	2.0	47.2
7	Coupler	1.1	-1.1	1000	998.9	83.6	0.8	1.3	530.0	4.5	1	1.76E+99	1.98E+00	3.0	2.0	47.0
8	Switch	1	-1	1000	999	75.1	0.8	1.3	589.6	4.8	0	7.94E+99	1.57E+00	2.0	2.0	46.8
9	Coupler	1.5	-1.5	1000	998.5	119.6	0.7	1.4	709.2	5.4	-1.5	7.08E+99	1.11E+00	0.5	2.0	46.4
10	Router	5	-5	-12	-17	627.1	0.3	3.2	1595.0	8.1	-6.5	2.00E-02	1.89E-02	-17.2	-10.7	36.1

Figure 8.2   Excel spreadsheet for calculating performance of converter and router. See text for explanation.

Attenuator NF	IIP3		SFDR
14	5	-16.6	34.3
15	5.4	-15.6	34.7
16	5.8	-14.6	35.1
17	6.3	-13.6	35.4
18	6.9	-12.7	35.7
19	7.5	-11.7	35.9
20	8.1	-10.7	36.1
21	8.8	-9.8	36.2
22	9.6	-8.9	36.4
23	10.4	-8	36.4

Figure 8.3   Results for selected attenuator settings.

general case and must be measured – see Reference 8 – so we will use SFDR as calculated by (8.4). Other definitions of SFDR exist and may quite appropriately be used. Figure 8.2 displays a spread sheet for the calculations yielding overall NF_{SR}, IIP_3, and SFDR for a 20-dB attenuator setting.

Each line in Figure 8.2 represents one stage in the block diagram of Figure 8.1. It gives the specifications for that stage and the overall values that would result if the system was truncated after that stage. Reading across, NF is the stage's standard noise figure in dB, G is its gain in dB, and IIP_3 is its input intercept point in dBm. Bk Te is the stage's equivalent input noise temperature in K, Bk g is its gain as a ratio, and Bk nf is its noise figure as a ratio. The next series of columns give the parameters that the overall system would have if it terminated at the end of the stage corresponding to that line. Thus Ov T is the equivalent input noise temperature in K, Ov NF is the overall noise figure in dB, and Ov G is the overall gain up to that point in the chain. The individual blocks individual output intercept point in mW appears next as oip3. This is followed by the overall output intercept point in mW, Ov oip3 mw, the overall IIP_3 in dBm, and the overall SFDR in dBm. The key input variable is the attenuator setting, lightly shaded and the overall NF, IIP_3, and SFDR, more darkly shaded.

Figure 8.3 presents a tabulation of the results for those attenuator settings which yield $NF_{SR} \leq 8.5$ dB and $IIP_3 \geq -15$ dBm, shaded. To optimize SF_{DR}, we choose a setting of 20 dB. This is the same result as obtained in Reference 5, although the two analyses use slightly different definitions of SFDR.

(a)

	Tx_out (UHF)	VLPF	Coupler	Power Amp	VBRF	VGA	V-BPF	MIX	ATT	BPF	TDD/FDD SW	TR SW	TDD/FDD SW	Band SW	Tx_in (2.6 GHz)
Gain (dB)		-2	-3	20	-3	8	-3	-8	-20	-3	-1	-1/-3	-1	-1	
Level (dBm) [TDD/FDD]	15	17	20	15	5	28/30	-20/-22	-17/-19	-9/-11	11/9	14/11	15/13	16	17	18

(b)

	Rx_in (UHF)	VLPF	Coupler	VBRF	LNA	BPF	Rx Amp	ATT	MIX	ATT	BPF	TDD/FDD SW	TR SW	TDD/FDD SW	Band SW	Rx_out (2.6 GHz)
Gain (dB)		-1	-3	-2	15.5	-1	15.5	-10	-8	-3	-1	-1	-1/-3	-1	-1	
Level (dBm) [TDD/FDD]	-50	-51	-54	-56	-40.5	-41.5	-26	-36	-44	-47	-48	-49/-51	-50/-52	-51/-53		

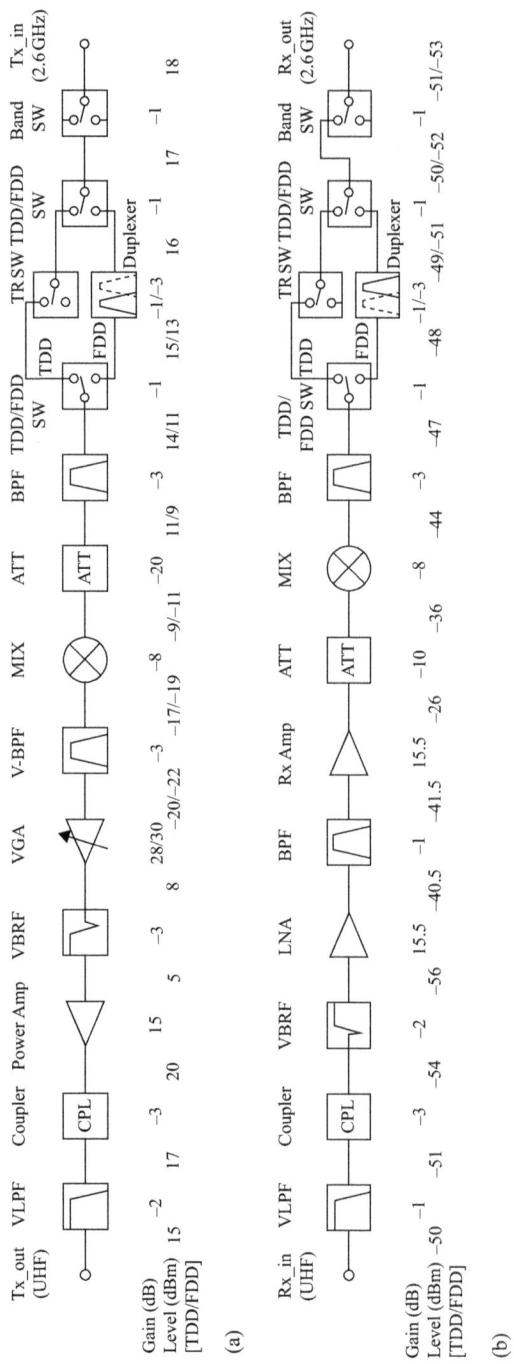

Figure 8.4 Block diagram and signal level tabulation for prototype LTE transmit/receive converter. ©2014 IEEE. Reprinted, with permission, from Reference 10. (a) Receiver. (b) Transmitter.

8.3.2 LTE Frequency Converters

Cognitive radio technology provides an opportunity for making TV white space spectrum available for LTE systems and potentially for new services carried by these systems. Applications presented in Reference 9 include

- Coverage extension – Base stations operating in TV white space using LTE provide Internet access to users via slave access points or directly to user devices.
- Backhaul – LTE eNBs use out-of-band backhaul in TV white space to increase capacity.
- Small cell deployment – Small LTE cells operate in TV white space to provide additional access to user equipment.
- Additional spectrum for capacity enhancement – LTE users needing additional capacity for high data rate applications can obtain supplemental spectrum in TV white space.

The authors of Reference 9 evaluate potential interference of white space LTE signals to TV users by *coverage loss*, the ratio of the number of users in an area who would lose TV coverage because of interference to the total number of users in the same area. For typical system parameters, they show that cognitive techniques reduce coverage loss to no more than 1.2%. Turning to interference by TV to LTE systems, this interference is much stronger on uplinks than on downlinks, given the relative antenna heights and transmitter powers involved in eNB and user equipment. Cognitive techniques reduce TV-to-LTE interference caused sector throughput loss and cell-edge throughput loss by about a factor of 10.

A prototype LTE transmit/receive converter similar in concept to the WiFi system discussed above appears in Reference 10. Figure 8.4 illustrates its block diagram and signal levels.

8.4 LTE Cognitive Repeaters for Indoor Applications

An alternative approach, presented by Wieruch *et al.* [11], employs cognitive repeaters to provide service inside buildings. A donor repeater using outdoor antennas or direct RF connections to a standard LTE-A eNB communicates with the eNB and relays uplink and downlink traffic for user terminals inside a building to a set of coverage repeaters inside the building. User terminals communicate with the coverage repeaters. See Figure 8.5. Frequencies used may be the same for each link, or each portion of the link may use a different set of frequencies. Isolation provided by the building walls provides opportunities for frequency re-use, both within the cellular bands or in TV white space. For example, the donor and coverage repeaters may communicate in available TV white space spectrum, and the user terminals may communicate with the coverage repeaters using cellular channels that are occupied outside the building, if the indoor/outdoor isolation is sufficient. Both the donor repeater and the coverage repeaters are able to monitor the spectrum in their environment. Since they are in fixed known locations, they collect real-time spectrum use data that can be shared by databases across the whole system.

An advantage of this system is the potential increase in capacity enjoyed by the indoor user units who are able to take advantage of the high capacity links to the donor repeater and by the use of the best available frequencies for MIMO techniques and coverage inside the building. Another advantage is that it puts all of the RF agile equipment and spectrum sensing capabilities in the repeaters and does not require any modifications to standard user equipment.

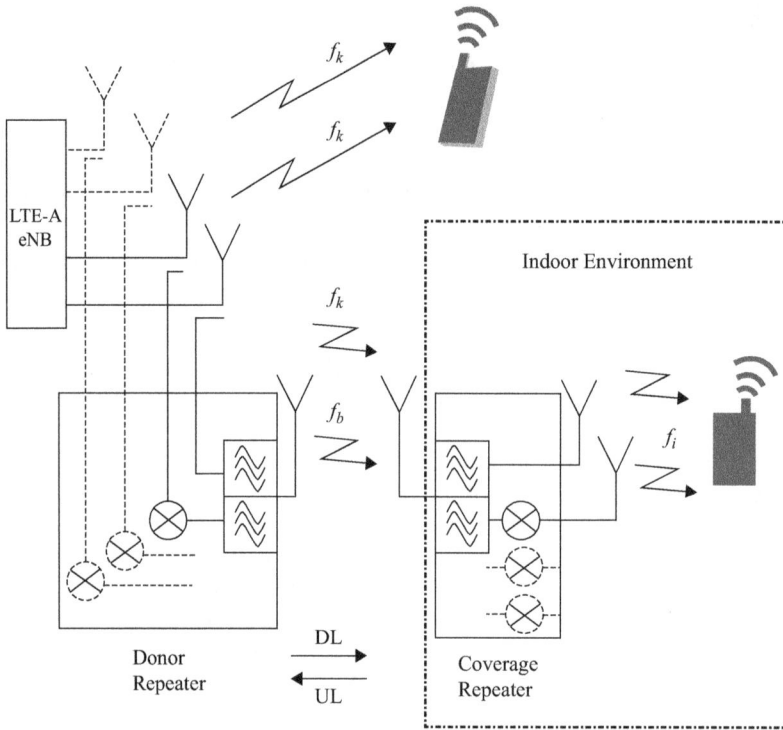

Figure 8.5 Cognitive repeater concept. ©2013 IEEE. Reprinted, with permission, from Reference 11.

The authors of Reference 11 distinguish between amplify-and-forward (AF) and decode-and-forward (DF) donor repeaters. While the DF design offers full decoding and re-encoding, it inherently suffers from the accompanying latency. The authors prototype multi-band-filter-based AF design offers latencies smaller than 200 nS.

8.5 Cognitive Radio and Cognitive Radar: Communications and Radar System Coexistence

8.5.1 Cognitive Radar

From the authors' perspective, a cognitive radar is a cognitive radio system employing a single collocated transmitter and receiver pair. It differs from a communications system where the received signal is generated by a distant transmitter operating under independent control. In a radar system, the received signal is generated when the transmitted signal is reflected back to the receiver from a distant target. The receiver has full knowledge of what was transmitted and extracts information about the target's range, velocity, trajectory, and identity from the characteristics of the received signal. In doing so, the receiver's signal processing systems must correct for propagation effects, reflections from scatterers other than the target, interfering and jamming signals, etc.

At first glance, cognitive radar design might appear to be a simpler subset of cognitive radio design, because a single cognitive engine can control both the transmitter and receiver, allowing both to be optimized simultaneously and avoiding the problems associated with keeping two or more geographically distributed radios in sync during an optimization process. In addition, radars are not usually part of networks, so networking issues like the *faculty meeting problem* do not arise. In principle, a cognitive radar could modify its transmitted waveform and antenna, receiver and signal processor characteristics as its mission, environment, and experience indicate. Adaptive radars routinely do some of these things now; the transition from adaptive to cognitive radars mirrors that from adaptive to cognitive radio.

With some exceptions, the cognitive radio and cognitive radar literature and research communities seems to be largely independent of each other. An exception is the work of Simon Haykin, whose early publications on both topics address important theoretical issues; see Reference 12. The cognitive radar literature focuses mainly on the intelligent adaptation process and on techniques like adding multi-input multi-output (MIMO) technology to the array of sensory inputs available to the radar's processor. While the cognitive radio literature includes intelligent adaptation to optimize terminal, link, and network performance, its primary focus is on dynamic spectrum access. The idea of cognitive radars cognitively exploiting vacant frequencies seems to have received relatively little attention in the radar community, with the exception of work by the authors of Reference 13. This results primarily from the practical difficulties in designing frequency agile radars.

Radars typically employ high-power non-linear transmitters whose output waveform time and frequency domain characteristics are carefully designed to extract wanted information about their intended targets. In contrast with most communications transmitters, which are designed for linear operation on set of channels within an assigned band, most radar transmitters operate on a fixed center frequency, generate as much output power as possible, and occupy a lot of bandwidth. Their power-added efficiency (the difference between the transmitter power amplifier's output and input drive power levels divided by the DC power supplied to the power amplifier) and spectral characteristics depend strongly on the load impedance that the power amplifier sees and on its operating frequency. This dependence is usually displayed on a *load-pull diagram,* where contours of constant power-added efficiency (PAE) and adjacent channel power ratio (ACPR) are plotted versus load reflection coefficient on Smith Chart coordinates. See Figure 8.6a from Reference 13.

Both the load reflection coefficient itself and the dependence of PAE and ACPR on the load reflection coefficient vary with frequency. When a cognitive radar changes frequency, its cognitive controller must obtain the appropriate load-pull diagram, determine acceptable values of PAE and ACPR for that frequency, and command the output matching network to deliver the required value of load reflection coefficient. The best values of PAE and ACPR will lie on a Pareto optimum contour, where one quantity cannot be further improved without degrading the other. See Figure 8.6 for an example and comparison of two algorithms. This process is considerably more complicated than simply commanding a communications transmitter to change frequency. For a discussion of the details and important references, see Reference 13.

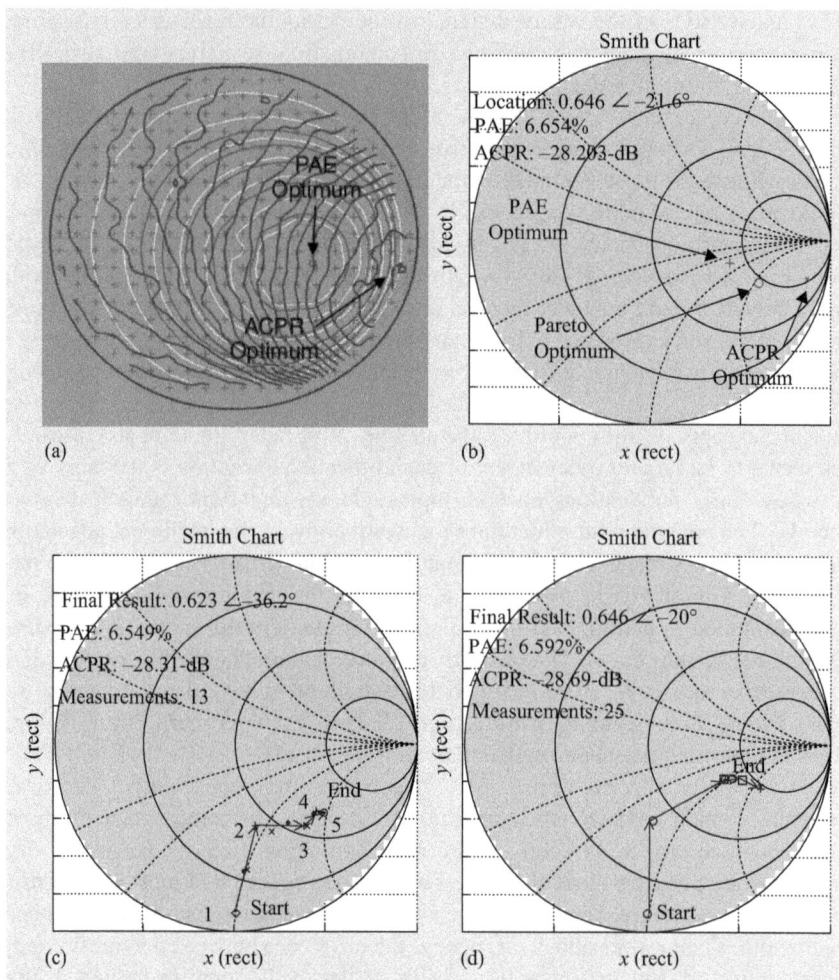

Figure 8.6 Load-pull contours and illustrations of the process of selecting a transmitter load impedance to optimize PAE and ACPR. (a) Full set of contours (b) Locations of PAE optimum, ACPR optimum, and Pareto optimum operating points, (c) and (d) Paths traced by two candidate algorithms for selecting transmitter operating point. ©2014 IEEE. Reprinted, with permission, from Reference 13.

8.5.2 Legacy Radar and Communications System Coexistence

8.5.2.1 Introduction

In this section, we focus on legacy radars which are not cognitive. They are assumed to operate as primary users on assigned frequencies and not to modify their operation in response to interference from secondary (communications) users. The radars may ignore the secondary users entirely, or they may broadcast some their transmitter parameters and observed interference levels for use by the secondary users. The radar parameters may also be available to the secondary users via a database. The degree to which information about the radar may be available obviously depends on the radar's mission, with weather radars and missile defense radars at opposite ends of a secrecy and security continuum. About 2 GHz of radar spectrum in L band (960–1400 MHz), S band (2.7–3.6 GHz), and C band (5.0–5.8 GHz) are potentially available for this kind of spectrum sharing [14].

Specifications and standards adopted at the time of writing require that the secondary users' radios be adaptive but not cognitive – i.e., they can sense the presence of radars and operate accordingly, but learning is not involved. More advanced techniques are under discussion, and we will return to these below.

Radar and communication systems coexistence raises both policy and technical issues that do not necessarily arise when communications system must share spectrum. Many radar systems are government owned while most communications systems belong to commercial enterprises; in the United States this means that they are regulated by different entities. Radars often perform critical safety or military functions and thus may require high levels of protection. Technical details of their operation may well be classified. Radar spectral leakage is a difficult problem. High-power non-linear radar transmitters have a much greater potential to interfere with signals in other parts of the spectrum than do communications transmitters. Their emissions must conform to specified spectral masks, but these are typically designed for a static environment, not for protecting a dynamically changing population of nearby communications signals. Other technical problems involve the spatial scanning inherent in most radar's and the time periods required for a radar receiver to listen for an echo from a target. A communications system may think a channel is open if it scans the radar's operating frequency while the radar is transmitting in another direction or while the radar is listening for echoes.

For a discussion of the policy issues involved in spectrum sharing between radar and other government systems and private users, see Reference 15. Reference 16 provides an excellent treatment of the spectrum engineering issues, including the trade-offs inherent in radar signal design and illustrations of the effects of interfering signals on radar performance.

8.5.2.2 Controlling Transmitter Power to Limit or Eliminate Interference

Looking now to some of the technical issues involved in spectrum sharing between cognitive communications systems and legacy radars, we will consider the European 5 GHz band where primary users include weather radars and other radar and satellite systems in the bands 5250–5350 and 5450–5725 MHz [17]. The susceptibility of a radar to interference is characterized by an interference threshold I_{thr} (dBm) which is the maximum interference level from a single interferer measured at the radar antenna output that the radar can incur without service degradation [14]. This is given by

$$I_{thr} = N + (I/N)\text{dBm} \tag{8.5}$$

where N is the radar receiver noise floor in dBm and I/N is the allowed interference-to-noise floor ratio in dB, normally set at -6 dB based on the assumption that interference of 6 or more dB below the noise floor will not degrade radar performance.

Assuming that the path loss between the radar and the secondary user is the same in both directions (this is a good assumption since both are on the same frequency), we can compute the minimum path loss between the secondary user and the radar that will guarantee that the interfering power received by the radar will be at or below I_{thr}. Assuming that all of the transmitter powers and antenna gains are known, we start with the link budget and then solve for the nominal path loss L in dB.

$$P_t^{SU} + G_t^{SU} + G_{radar} - L = I_{thr} \tag{8.6}$$

$$EIRP^{SU} + G_{radar} - L = I_{thr} \tag{8.7}$$

$$L = I_{thr} - EIRP^{SU} - G_{radar} \tag{8.8}$$

Here, G_{radar} is the gain in dB of the radar antenna. (We assume that the same antenna is used for transmitting and receiving.) P_t^{SU} and G_t^{SU} are the secondary user's transmit power in dBm and transmit antenna gain in dB, and they sum to $EIRP^{SU}$.

Usually, the radar and the secondary user signals will have different bandwidths. If the secondary user transmitter power is spread uniformly across its bandwidth and if the radar receiver bandwidth is the narrower of the two, then the received interfering power at the radar will be $10 \log_{10} \left(\frac{b_{radar}}{b_{SU}} \right)$ dB more than that calculated by the link budget, and the allowed path loss will be lower than the L value calculated above.

$$L_{allowed} = L + 10 \log_{10} \left(\frac{b_{radar}}{b_{SU}} \right) = L + B_{radar} - B_{SU} \tag{8.9}$$

Here, b_{radar} and b_{SU} are the bandwidths in Hz and B_{radar} and B_{SU} are the corresponding values in dBHz.

Thus,

$$L_{allowed} = I_{thr} - EIRP^{SU} - G_{radar} + B_{radar} - B_{SU} \, \text{dB} \tag{8.10}$$

Given the radar EIRP and the allowed path loss and assuming that the path loss is the same in both directions, we can calculate the maximum power that the secondary user can receive from the radar while guaranteeing that the secondary user's signal at the radar will be at or below the interference threshold I_{thr}. This quantity, P_{thr}, is called the detection threshold.

$$P_{thr} = EIRP^{radar} + G_r^{SU} - L_{allowed} \, \text{dBm} \tag{8.11}$$

Note that the value of P_{thr} calculated above depends on and will vary with $EIRP^{SU}$. Numerical values for typical radars and calculated results appear on p. 11 of Reference 17. See Table 8.1.

If a secondary user scans the radar frequencies and detects no radar signal at or above P_{thr}, then, subject to the assumptions made in this analysis (i.e. symmetric propagation paths, a single interferer, correct values for all of the parameters), the secondary user will not interfere harmfully with the radar. Note that this detection threshold is a rather conservative number, since it is calculated for the situation that the main lobes of the radar's antenna and the secondary user's antenna are pointed exactly at each other.

The authors of Reference 17 performed this calculation for a number of standard radars operating in Europe with an $EIRP^{SU}$ of 30 dBm (1 W). Disregarding one outlier, the minimum detection threshold for this dataset (rounded down to the nearest integer value) is -62 dBm for a secondary user with 30 dBm EIRP. This detection threshold can be adjusted for other EIRP values, for example, -55 dBm from a 0.2 W device and a -52 dBm for a 0.1 W device.

Table 8.1   Sample of typical values for radar and wireless access system interference parameters. © 2011 International Telecommunications Union. Reprinted, with permission, from Recommendation ITU-R M.1652-1 (05/2011) "Dynamic frequency selection in wireless access systems including radio local area networks for the purpose of protecting the radio determination service in the 5 GHz band."

	Characteristics	A	C	E
RADAR	Function	Meteo	Meteo	Meteo
	Platform type	Ground/ship	Ground	Ground
	Tx power into antenna, peak (kW)	250	250	250
	Receiver IF_{3dB} bandwidth (MHz)	0.5	20	0.91
	Antenna polarization	V	H	H
	Antenna main beam gain (dBi)	39	44	50
	Antenna height (m)	30	10	30
	EIRP radar (dBm)	123.0	128.0	134.0
	Receiver noise figure (dB)	7	4	2.3
	$N = kTBF$ (dBm)	−110.0	−97.0	−112.1
	$N − 6$ dB	−116.0	−103.0	−118.1
WAS	EIRP (dBm) outdoor	30		
	TPC (dB)	0		
	Bandwidth (MHz)	18		
	Antenna gain (onmi) (dBi)	0		
	10 log(Brad/BWAS)	−15.6	0.5	−13.0
	Nominal path loss (dB)	185.0	177.0	198.1
	Link budget for WAS signal received at radar receiver $N − 6$ dB	169.4	177.0	185.1
	Necessary detection threshold	−46.4	−49.0	−51.1

The values quoted above assume that only one secondary user is operating at a time. To allow for the cumulative interference that could be caused by multiple users, the ITU recommendation specifies a detection threshold of −64 dBm for devices with an EIRP between 200 mW and 1 W and of −62 dBm for an EIRP less than 200 mW. The EIRP values are normalized over a 1 microsecond period. The secondary user must listen on the radar channel for 60s. If it detects a radar it must not return to that channel for 30 minutes. If a secondary user begins operating on a radar channel and its in-service monitoring detects a radar signal, the secondary user must vacate the radar channel within 10 seconds.

8.5.2.3 Implementation

The procedure and protocol for communications systems sharing spectrum with legacy radars in the 5 GHz band and operating in Europe are specified in IEEE Standard 802.11h [18]. The basic operational procedure involves the following.

- Secondary users are part of a network associated with and controlled by an access point, as in other 802.11 systems.
- The access point can quiet the network prior to testing for the presence of a radar.
- Stations in the network test for the presence of a radar before using a channel and continue to test periodically while using the channel.

- The access point can request and receive reports on measurements by other stations both on the current and other channels.
- If a radar is detected in the current channel, the access point will select and advertise a new channel, and the network will vacate the occupied channel.

The standard specifies the waveform, frame structures, and bit values that enable this operation. It leaves specification of detection thresholds, listening periods, etc. to the cognizant regulatory authority and references ETSI EN 301 893 [19]. The latter provides detailed information on interference thresholds (basically the -63 or -64 dBm values mentioned above) and testing and measurement procedures.

8.5.2.4 Cooperative and Cognitive Approaches

The coexistence and dynamic frequency selection procedures described above are adaptive rather than cognitive and based on calculated worst case interference to the radar from a single secondary user. They do not consider the cumulative effects of multiple users or allow the secondary users to take advantage of temporal opportunities when a scanning radar antenna is pointed away from the secondary user. Considerable improvement could result from incorporating cognitive behavior by the secondary users and by enabling cooperation between the radar and the secondary users – for example, by equipping the radar with a beacon that identifies it and provides data about its current interference level and other operational conditions. See Reference 14 for further discussion of these ideas.

8.6 Ka Band Geostationary Satellite Applications

8.6.1 Introduction

A communication satellite typically serves as a repeater or base station in the sky, receiving signals in a specified uplink band and retransmitting them back to earth in a specified downlink band. In most applications, the uplink frequencies are higher than the downlink frequencies to maximize the receiving antenna gain that can be achieved subject to the satellite's volume and weight limitations. Depending on the satellite's mission and thus its orbital characteristics, a spacecraft may pass overhead rapidly (continuously scanning parts of the earths surface), remain stationary in the sky, or slowly rise and set.

Geostationary Satellite Orbit (GSO) satellites are assigned an orbital location (the orbital slot), above a specified point on the earth's equator nominally 22 242 km from the earth's center. Within the limits of their station-keeping systems, they appear at a nominal fixed location in the sky and do not usually require tracking by earth stations. The angular spacing between orbital slots for satellites in the same service (i.e., satellites using the same frequency allocations in the same way) is normally sufficient to prevent adjacent satellites from interfering with each other's uplinks and downlinks given the usual directional antennas. But it is possible for satellites in different services (Direct Broadcast Satellites and 17/24 GHz Broadcast Satellite Service spacecraft in the case treated here) to be within a few tenths of a degree of each other. The U.S. Federal Communications Commission (FCC) requires a minimum orbital separation of $0.2°$ between these kinds of satellites [20].

Non-geostationary orbit (NGSO) satellites are usually at a lower orbit (if operated for communications purposes) than GSOs to take advantage of the lower free space path losses

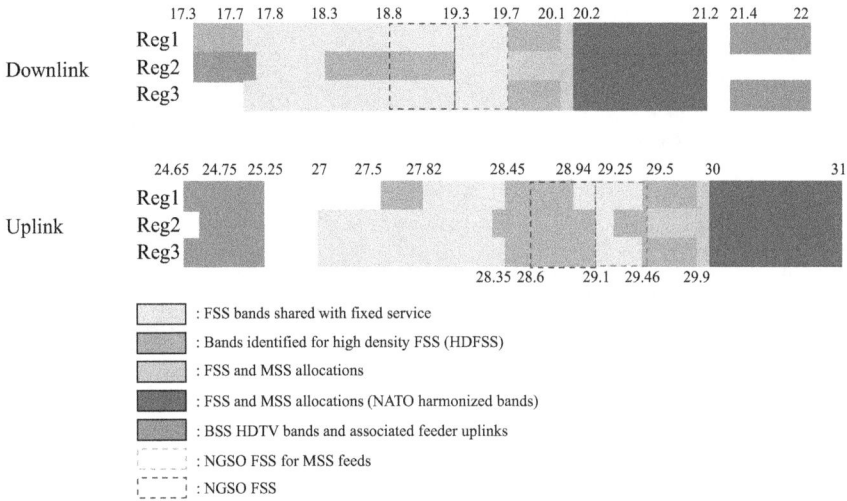

Figure 8.7   ITU Ka-band frequency allocations for satellite services. ©2014 ETSI. Reprinted, with permission, from Reference 24.

but mandating lower-gain (less directional) antennas or else precision tracking on the ground. NGSOs are particularly attractive for applications like satellite telephones and broadband Internet access in remote areas.

Regulatory bodies classify communications satellites in a variety of ways, using terminology that can seem confusing. Thus the Fixed Satellite Service (FSS) generally means satellites that provide communication links between earth stations at fixed locations, but FSS regulations may also apply to inter-satellite links (Inter-Satellite Service, ISS) or feeder links (uplinks) to satellites in the Broadcasting Satellite Service (BSS). BSS satellites provide downlink signals that are intended for reception by the general public, and the spacecrafts are often referred to in the U.S. regulations as Direct Broadcast Satellites (DBS). Those that provide audio rather than television may be further classified as belonging to the Satellite Digital Audio Radio Service (SDARS). There seems to be no one definitive glossary for satellite communications terms and acronyms; useful examples include References 21 and 22.

Our concern in this section is with geostationary satellite systems operating in Ka-band, which, following the terminology employed in Reference 23, we will take to be the frequency range 17.3–31.0 GHz. A portion of this band is allocated for High Throughput Satellites (HTS), FSS spacecraft providing high-speed Internet service (potentially 30 Mbps or higher) to large numbers of small earth terminals through hundreds of downlink beams. HTS operators include Eutelsat (www.eutelsat.com) and HughesNet (www.hughes.com). Worldwide, only 500 MHz of bandwidth is allocated to HTS primary user operation, 17.7–21.2 GHz downlink and 17.7–21.2 GHz uplink. But about 2.4 GHz bandwidth is potentially available in each direction using cognitive radio techniques to share spectrum with terrestrial systems and other satellite services [24]. Figure 8.7 from Reference 24 illustrates the International Telecommunications Union (ITU) Ka-band spectrum allocations. Here, Region 1 is Europe (including Russia), Africa, and the Arabian Peninsula. Region 2 is the Americas, and Region 3 is the Asia and the Pacific. In the figure, Fixed Service (FS) refers to terrestrial microwave links and MSS refers to Mobile Satellite Service.

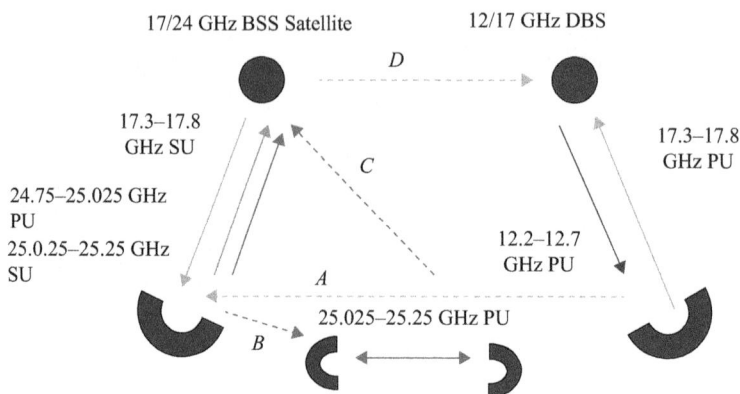

Figure 8.8   Intended paths and potential interference paths in US Ku-band satellite and terrestrial systems.

We will first consider applications of cognitive radio to U.S.-licensed BSS satellites that use 17.3–17.8 GHz for downlinks and 24.75–25.25 GHz for feeder uplinks. These are called 17/24 GHz BSS satellites. They share the 17 GHz spectrum with 12/17 GHz DBS satellites that downlink TV signals for home reception in the 12.2–12.7 GHz band and have their (uplink) feeder links in the 17.3–17.8 GHz band. See Reference 20. The 17/24 GHz BSS satellite uplinks must share 25.025–25.250 GHz with terrestrial fixed service (FS) systems, and the terrestrial systems are designated as primary users. See Figure 8.8 for a sketch of the systems involved.

Adopting the notation of Reference 25, we identify and label four potential interference scenarios. Scenario A is from a 12/17 GHz DBS uplink (PU) to a 17.24 GHZ BSS downlink (SU). Scenario B is from a 25.025–25.25 GHz uplink (SU) to a terrestrial link (PU) in the same band. Scenario C is from a terrestrial link (PU) in the same band to a 17/25 GHz BSS satellite uplink (SU) and Scenario D is from a 17/24 GHz BSS satellite 17.3–17.8 GHz downlink (SU) to a 12/17 GHz DBS satellite uplink (PU) Scenarios A and B are more likely than C and D because of the directional antennas involved on the downlinks and on the terrestrial microwave system.

The U.S. FCC regulations specify technical specifications and frequency coordination procedures designed to minimize these sources of interference without mentioning cognitive radio techniques. But in Europe, cognitive approaches are already proposed for an analogous situation involving a slightly different set of frequencies, as we will discuss below.

Earth stations receiving the 17.3–17.8 BSS downlinks must accept interference from the primary use DBS feeder link (Scenario A). They may avoid or minimize this interference by operating frequency choice and by appropriate modulation and encoding choices. Uplink interference at the DBS satellite from the 25.025–25.25 GHz terrestrial link (Scenario C) is probably negligible because of the directional antennas involved, but the spacecraft receiver and uplink transmitter must either accept it or use the same avoidance and minimization techniques. These cases are different from those usually considered in cognitive radio in that the primary user is the potential interferer rather than the station interfered with. In Scenario B, the 24.75–25.25 GHz uplink must avoid causing harmful interference to the terrestrial link. This situation closely resembles the problem of a secondary user interfering with a primary user radar receiver.

Cognitive radio operation in this part of the spectrum is also different from that found in the literature for lower frequencies (i.e., frequencies below 10 GHz) because of the propagation

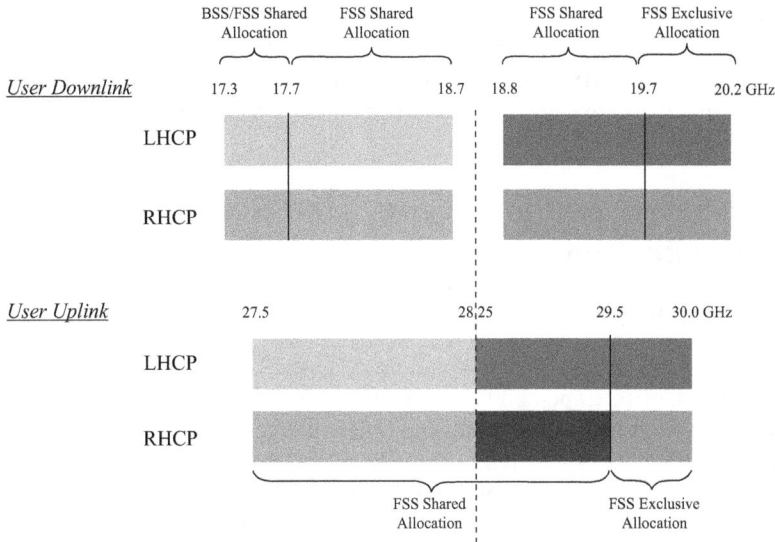

Figure 8.9   A proposed frequency reuse scheme for cognitive multibeam satellite networks. ©2015 IEEE. Reprinted, with permission, from Reference 25.

environment. Multipath is not normally a factor in satellite links, and the dominant causes of signal fading are rain attenuation and absorption by atmospheric water vapor. Such fades are distributed log-normally rather than following the Rayleigh or Rice distributions familiar to most readers. Rain also depolarizes Ka-band signals, causing cross-talk between polarizations that are nominally orthogonal. See Reference 26.

The European situation is slightly different. BSS feeder links operate in the 17.3–17.7 GHz band as primary users and share this spectrum with FSS downlinks in the same band. The number of BSS feeder link earth stations is restricted and small, but FSS downlink receivers can be deployed anywhere [25]. Operation with downlinks in the 17.7–19.7 GHz band is proposed to provide additional spectrum for High Throughput Satellites (HTS). These would uplink in the band 27.5–29.5 GHz. Both the downlinks and the uplinks would share spectrum as secondary users with terrestrial systems operating in the same bands. Figure 8.9 illustrates one possible frequency plan for such a system.

Here, the HTS downlinks to its users in the exclusive FSS satellite band 19.7–20.2 GHz and in part of the spectrum shared with BSS satellites (17.3–17.7 GHz). It shares spectrum with terrestrial microwave systems (FS) between 17.7 and 19.7 GHz. A guard band is provided between 18.7 and 18.8 GHz. Similarly, users uplink to the HTS both in the exclusive FSS band 29.5–30.0 GHz and in the 27.5–29.5 GHz band shared with FS systems. Both the uplink and the downlink allocations incorporate orthogonal polarization frequency sharing between Left-Hand Circularly Polarization signals and Right-Hand Circular Polarization signals.

See Reference 24 for a detailed discussion of the applicable cognitive radio techniques to such satellite systems. These range from defining exclusion zones where frequency sharing is not feasible, to establishing pre-coordinated areas where a small number of simple adjustments to transmitter and receiver parameters is sufficient. The ETSI document contains reference values for transmitter powers, bandwidths, antenna characteristics, interference thresholds, etc., for use in system design. Maleki *et al*. [27] provide an example where they calculate cognitive zones, geographic areas within which an FSS terminal must employ cognitive techniques to manage interference from BSS feeder links. They point out that rain attenuation may make service availability impossible beyond some threshold rain rate.

8.7 Public Safety and Emergency First Responder Communication

8.7.1 Introduction

The September 11, 2001, terrorist attacks and the August 29, 2005, Hurricane Katrina natural disaster exposed major interoperability problems between the radio systems used by United States local and state police, fire, rescue, and other emergency first responder (EFR) agencies, as well as radio incompatibility problems between local EFR organizations and federal disaster relief and military organizations. Even within a single jurisdiction, organizations like fire and police departments typically bought their communications systems from different vendors and on different budget cycles. There was little effective standardization of public safety systems. The result was a patchwork of networks operating on different frequencies with incompatible technical standards. Under normal circumstances, workarounds like mutual aid agreements, shared channels, and multiple radios in vehicles managed the resulting interoperability problems, but these solutions failed during large-scale incidents when help was summoned from remote locations and when repeaters, base stations, and other critical infrastructure were destroyed.

Emergency First Responder Communications historically were classified and regulated as land mobile radio (LMR) systems[1] where drivers in vehicles communicated via push-to-talk voice radios with dispatchers, or, less frequently and only in some systems, directly with each other. U.S. EFR radio systems began in 1928 with analog AM voice radios and soon evolved to FM analog and digital radios operating in the 150–174 MHz (VHF) and 421–512 MHz (UHF) bands using 25 kHz wideband channels. There was a gradual transition to 12.5 kHz and narrower channels (narrowband), but wideband operation was legal until January 1, 2013![2] In the late 1980s, the Association of Public Safety Communications Officers International (APCO) issued an evolving set of standards called APCO-25 (or simply P25) that, in a series of phases, specified analog and digital radios occupying 12.5 and 6.25 kHz channels and offered a number of operational features like talk groups (a private channel for a specified subset of radios to use) and trunking (dynamic channel assignment) [28]. P25 became the de facto standard for new radio system purchases in the United States, but its adoption was gradual and legacy (conventional) systems remained in use. Compatibility problems often existed between these radios and P25 systems, between new and old P-25 systems and between P-25 systems made by different manufacturers. But ideally, all interoperability problems would be solved

---

[1]The terms land mobile, maritime mobile, aeronautical mobile, fixed service, etc. predate modern transceivers which obviously can be carried by their operator and used anywhere.

[2]In this usage, "wideband" refers to the old 25 kHz analog radio channel, while "broadband" refers to a modern high data rate digital channel like that to be offered by FirstNet (see below).

when every first responder had a P25 radio. Practically, this would take too long, and, apparently in the opinion of many municipalities, would cost too much.

Cognitive radio offers a short-term solution to the interoperability problem by through a transceiver that could scan the public safety frequencies, identify to its operator the networks that are present, and configure itself to interoperate with whatever networks the operator selected. The radio could serve as an audio and data gateway between any two networks or operate as a repeater and establish its own network. See References 29 and 30 for an excellent FCC discussion of the pros and cons of these ideas. Public Safety is an inviting target for cognitive radio applications because the radio systems use only a small number of waveforms, operating frequencies fall in FCC specified channels that are licensed to particular users, and the licensee information is readily available. Hence the spectrum sweeping and waveform/user identification problems are well defined and bounded. The biggest problem is operational: what the FCC calls "Network Affiliation." This is who will be allowed to operate in the radio network. The FCC states "This is not a factor for negotiation by the radios themselves! It must be accomplished by the radio users who will establish the authentication and verification policies and rules for the radio set" [30].

The authors were part of a team that developed several working prototypes that we called "public safety cognitive radios." While for reasons to be discussed below, the market and the cognizant Federal agencies preferred adopting a nationwide standard, still (in 2015) under development, we will summarize that work below for its possible future value.

8.7.2 The Virginia Tech Public Safety Cognitive Radio

8.7.2.1 Description

The Virginia Tech Public Safety Cognitive Radio (PSCR) was a proof-of-concept prototype public safety cognitive radio that could interoperate with legacy voice radios in the 150, 450, and 700/800 MHz bands. The use case is for a police officer from a distant jurisdiction arriving in the aftermath of a large-scale disaster like Hurricane Katrina. The entire infrastructure is destroyed and arriving personnel have outrun their mutual aid agreements. The officer turns on the PSCR and operates it in scan mode. The radio scans all of the public safety bands and gives the officer a display of all the public safety signals that it sees, identifying networks by name if it knows them and otherwise identifying them by equipment type or other technical characteristics. With one click, the officer can configure the radio to interoperate with any selected network (talk mode). With two clicks, the officer can configure the radio to be in two incompatible networks at the same time. If desired, the officer can put the radio in gateway mode and provide a gateway between the two networks. Alternatively, the radio can form its own network and serve as a repeater. Development of the PSCR hardware and software solved problems in rapid signal recognition and radio reconfiguration, but it did not address the important command and control problems associated with officers arriving in a jurisdiction and being able to enter networks at will.

The VT PSCR evolved from a first prototype that ran on a high-end laptop computer to an embedded system prototype running on a Texas Instruments OMAP3530 board BeagleBoard computer built on the standard architecture used by the LMR industry. See References 31–33.

The reader will note that the radio was not fully cognitive. Operational concerns of EFR personnel required that the user select the operating mode and choose the network(s) to enter. These features could have been automated easily.

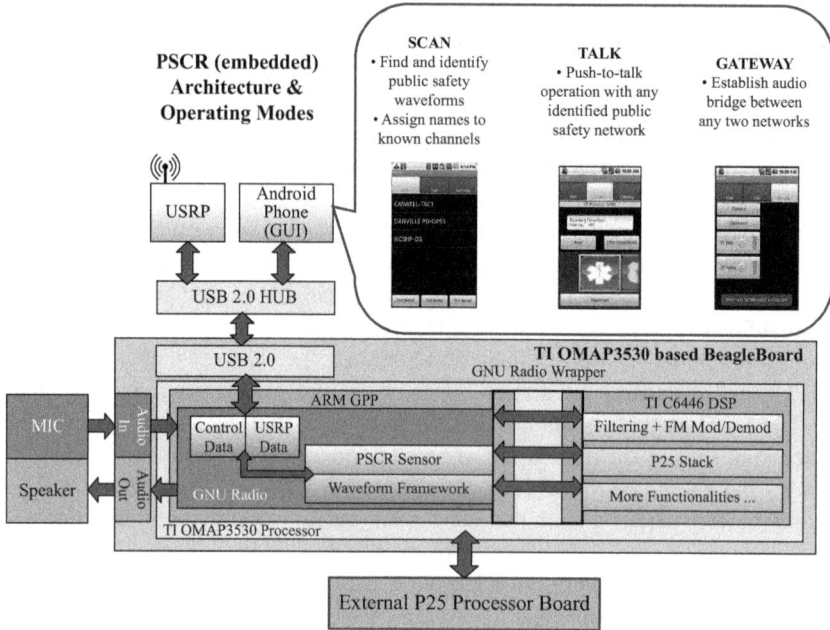

Figure 8.10   Virginia Tech Public Safety Cognitive Radio architecture.

8.7.2.2 Architecture and Major Components

Figure 8.10 illustrates the architecture and major components of the PSCR. A Google Nexus One Android smart phone serves as the user interface, providing a touch screen along with a microphone and a speaker. It offers access to the full capabilities offered by the PSCR on the BeagleBoard in a small hand-held form factor. The transmitter and receiver RF front-ends are provided by an Ettus Research USRP-1 motherboard with a WBX daughterboard. This architecture follows the three-board model (housekeeping and display, digital signal processing, and RF) that is standard in the Land Mobile Radio (LMR) industry. As delivered in June, 2011, the PSCR was able to interoperate with analog FM waveforms and lacked P25 capability. P25 (conventional) operation was to have been enabled by a proprietary daughter board that plugged into the BeagleBoard but which was not implemented. In the final version of the PSCR prototype, the BeagleBoard and USRP-1 were replaced with a USRP E100, where the E100 contains a Gumstix Overo module internally, which replaces the BeagleBoard functionality. Using the Ettus E100 provided a more integrated computing and RF system.

8.7.2.3 Operational Example

This section describes a standard operational demonstration of the VT PSCR that was presented to the FCC in 2011. It simulated a situation in which a major ice storm had destroyed the antenna towers and interoperability systems shared by all law enforcement agencies in the Caswell County, NC, and Danville, VA, adjoining areas, providing interoperable communications between users simulating three agencies (city police, county sheriff, and state highway patrol). These agencies were represented by a Motorola RDX Series on-site two-way business radio in the 150 MHz band, a Family Radio Service (FRS) 460 MHz band radio, and a

700/800 MHz E. F. Johnson 51SL ES handheld radio. The PSCR scanned one of these bands (say FRS) and listed the signals that it found. The operator selected the desired FRS signal and configured the PSCR to communicate with it. The PSCR then scanned another band (say 800 MHz), found the E.F. Johnson radio, and the operator configured the PSCR to communicate with that. The PSCR operator then set up a gateway between the two radios and their operators talked to each other.

8.7.2.4 Design Perspectives

The embedded PSCR system used a GPP and DSP processors (and an FPGA on the Ettus platform which was not used for computation purposes). CR applications spanning across multiple computational devices need to take into account certain architectural design needs. For example, if using a fixed point DSP (meaning it does not have a floating point unit) designers need to account for quantization effects in processing, i.e., overflow and underflow, this can be done in the form of filter coefficient calculation and scaling of data at various points in the radio chain, e.g., changing the decimal point location.

The DSP contained a single program with multiple functional blocks, e.g., FIR filters and modulator/demodulator. When switching between transmit and receive operations the radio requests different functional blocks to be executed to modify the radio operation. This was done to reduce the time overhead of switching between transmit and receive functionality; however, this added overhead and complexity to perform runtime data stream routing to the proper functional blocks. Alternatively, if an application is more tolerant of initial configuration time overhead, the designer can have separate DSP programs for receive and transmit operations and switch between them when radio operations are changed. This means that after the DSP is loaded, the overhead of determining what radio functional blocks are executed is eliminated. Using this approach, the overall DSP program size can be smaller since it would only contain the functional blocks necessary to execute the radio and it does not have the extra logic necessary to route data to the functional blocks.

The PSCR does not actively monitor and modify its configuration to overcome and/or improve quantization effects and/or latency caused by data routing to control blocks. However, CRs can monitor such effects and attempt to compensate for them by changing filter parameters, scaling factors, and switching between different implementation of radio functionality to choose settings that would be the most appropriate to application needs. This can be extended to other factors such as throughput and power consumption.

8.7.3 Current (2016) Situation

At the time of writing, public safety and emergency first responder communications technology is being driven by a number of somewhat conflicting social, regulatory, and technical considerations. We provide a brief and, we hope, unbiased summary here. See Reference 34 for a valuable tutorial.

First, EFR communication systems are different from most others because they involve matters of life and death. Emergency personnel require extremely reliable and robust systems and historically they have felt that these are best provided if they own their systems and have sole access to their frequencies. The ideas of obtaining shared services from a commercial provider or sharing spectrum with other users conflict sharply with this position. Second, EFR

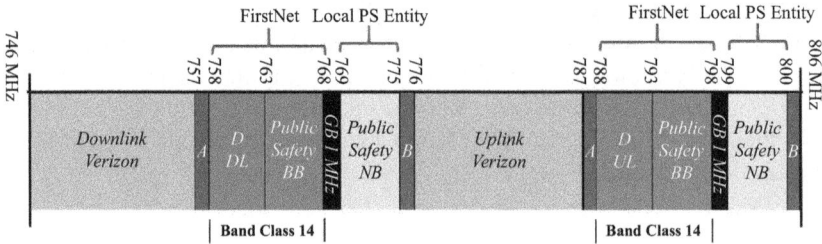

Figure 8.11   Partial band plan for 746–806 MHz. Bands A, B, and GB are all 1 MHz wide
guard bands. ©2015 IEEE. Reprinted, with permission, from Reference 34.

systems arose within the LMR community, predating both Internet Protocol (IP) and cellular
telephone technology. Thus, the architectures and networking protocols of P25, TCP/IP, and
LTE systems are different. Each has its own proponents; see, for example, Reference 35. EFR
users wish for radio systems that combine the best features of all three technologies, but they
do not exist.

The February 22, 2012, "Middle Class Tax Relief and Job Creation Act" created an authority (in the government sense of a "Port Authority") called FirstNet (First Responder Network Authority) [36] whose mission is "to build, operate and maintain the first high-speed
nationwide wireless broadband network dedicated to public safety." This would provide a
national network that all EFR radios could access, guaranteeing universal interoperability.
The FCC provided spectrum for FirstNet in the 700 MHz band. See Figure 8.11. Designing and building the FirstNet network is currently in progress. It is required by law to
be based on the minimum technical requirements of the commercial standards for Long
Term Evolution (LTE) service. For views of how this will relate to P25, see the APCO
International website [37]. For current information on frequency allocations for EFR communications and the U.S. regulatory environment in general, see the FCC website [38],
for example.

Cognitive radio offers a way to allow both new and legacy EFR radio systems to interoperate
until the day that all first responders have nationally compatible radio systems – if and when
that day comes.

8.8 Cognitive Radio and Autonomous Vehicles³

Often overlooked in the literatures of cognitive radio and autonomous vehicles is the great
similarity in the tasks that these two systems perform. In the words of Young [39] both analyze
their environment, make and execute a decision, evaluate the result (learn from experience),
and repeat as required. A single intelligent agent (a cognitive engine operating in multiple

---

³The material in this section is largely taken from Young's dissertation [39]. Details of the radio system also appeared
in Reference 40.

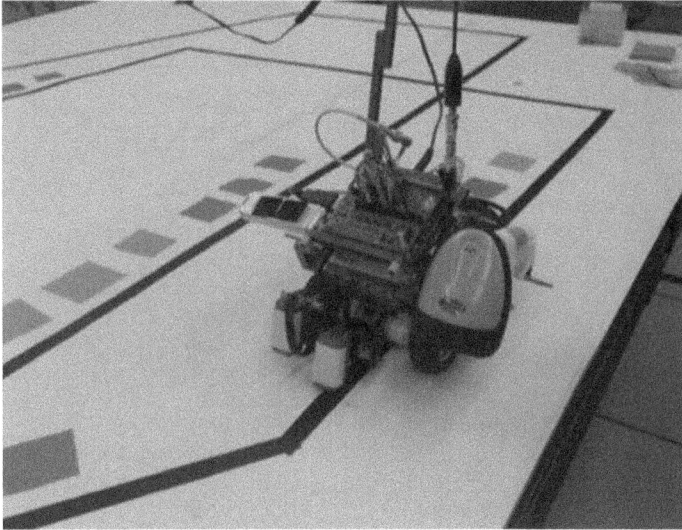

Figure 8.12   Prototype autonomous vehicle. ©2013 Alexander R. Young. Reprinted, with permission, from Reference 39.

dimensions) can control both an autonomous vehicle and its communication system, simultaneously observing the geographical and the radio frequency environment and controlling and optimizing multiple aspects of the vehicles mission. In this section, we briefly describe a prototype software and hardware package designed to address the following simplified use case.

A small UAV (Un-piloted Autonomous Aerial Vehicle) is traversing a nominally cyclic or repeating flight path (an orbit) seeking to observe targets and where possible avoid hostile agents. As the UAV traverses the path, it experiences varying RF effects, including multipath propagation and terrain shadowing. The goal is to provide the capability for the UAV to learn the flight path with respect both to motion and RF characteristics and modify radio parameters and flight characteristics proactively to optimize performance. Using sensor fusion techniques to develop situational awareness, the UAV should be able to adapt its motion or communication based on knowledge of (but not limited to) physical location, radio performance, and channel conditions. Using sensor information from RF and motion (MOT) domains, the UAV uses the mission objectives and its knowledge of the world, to decide on a course of action. The UAV develops and executes a multi-domain action; action that crosses domains, such as changing the RF power and increasing its speed.

The proof of concept prototype for this use case shown in Figure 8.12 is a low-cost (less than $250) software and hardware package called SKIRL, based on the BeagleBoard-xM computer and the Hope RF RFM22B RF integrated circuit that is suitable for installation in small experimental UAVs. Here, SKIRL is integrated with a set of target, navigational, and environmental sensors mounted on a LEGO-wheeled vehicle that executes a hypothetical

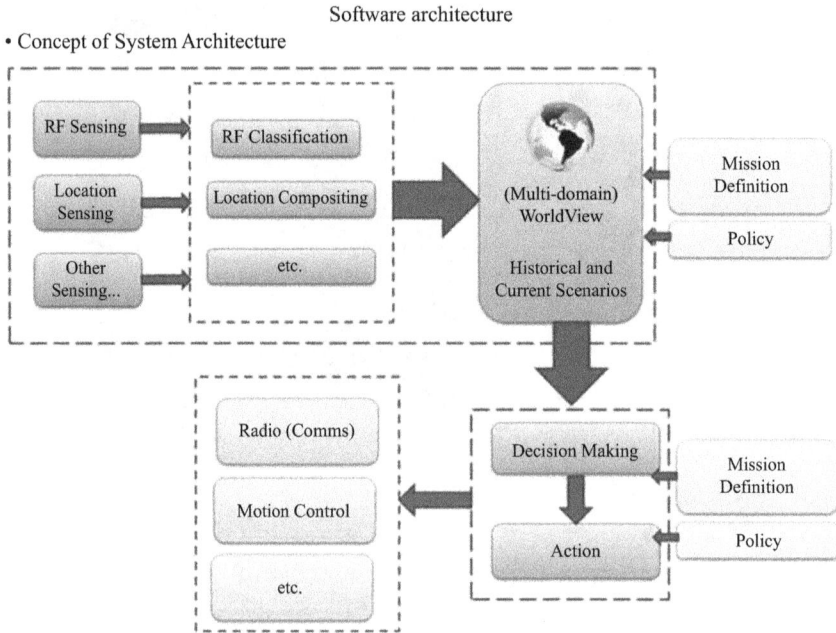

Figure 8.13   Software architecture of the prototype system. ©2013 Alexander R. Young. Reprinted, with permission, from Reference 39.

two-dimensional mission based on the UAV use case while avoiding the costs and potential security problems associated with a flight test. Experiments with the system demonstrated its ability to explore and learn a multidimensional environment that combines changing RF, location, and mission data and to optimize its mission performance intelligently.

The robot vehicle used an optical sensor to follow paths defined by black lines on a horizontal surface. It could choose which sequence of lines to follow and would periodically return to its starting point, traversing a closed path called an orbit. The vehicle determined its position by reading barcodes attached to the surface and looking up the corresponding locations in a table. It exchanged radio packets with a fixed base station; path loss, interference, and channel noise varied with position. Its multi-objective mission was to discover and traverse a route that would maximize targets (yellow patches), minimize anti-targets (orange patches), minimize orbital time, and maximize the number of packets delivered successfully in one orbit. Quantified, these were its cognitive engines meters. The radio knobs that the cognitive engine controlled were symbol rate and EIRP, and the vehicle knobs were speed and route choice. For each orbit the engine determined a score based on weighted objective functions of the meter readings. The vehicle was programmed to explore each of three possible paths (A,B, and C). It would then follow the best path for a predetermined number of orbits and then explore again, in case conditions had changed.

A sketch of the underlying system architecture appears in Figures 8.13–8.15.

A more ambitious system to be deployed in an aircraft is described in Reference 41. At the time that reference was published the prototype was not yet finished.

Software Architecture

• System module organization

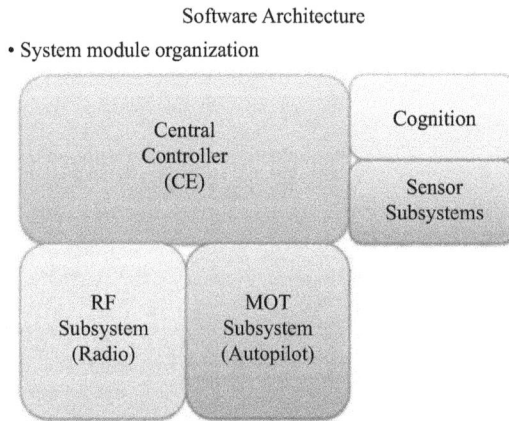

Figure 8.14   System module organization. ©2013 Alexander R. Young. Reprinted, with permission, from Reference 39.

Software Architecture

• Stack: each layer provides functionality to the layer above, and builds on the layer below

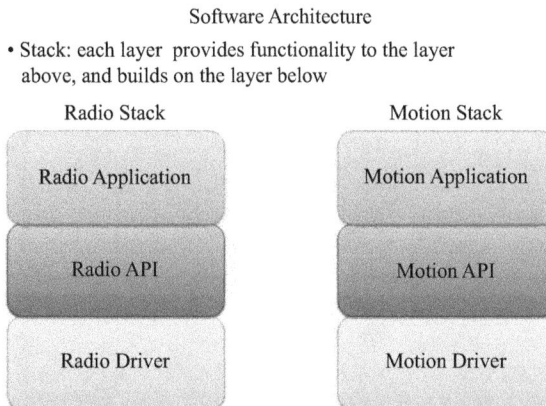

Figure 8.15   Radio and motion stacks. ©2013 Alexander R. Young. Reprinted, with permission, from Reference 39.

8.9 Smart Grids

The electrical grid traditionally depends on centralized sources and on manual recovery when needed as it lacks a two-way communication which would allow autonomy, sensing, and control. Historically grids operate based on load forecasting models, where when consumer consumption exceeds a particular threshold, power companies would enable secondary power generators knows as *peaker plants* that counteract excess demands [42]. Large-scale energy storage is expensive [43], which implies that when power generation exceeds demand it must be consumed and wasted which is wasteful and costly. When energy is generated via non-renewable energy sources, this behavior is not environmentally conscious. However, incorporating renewable resources in the electric grid presents its own challenges in terms of predictability, e.g., power output from wind turbines and solar arrays are dependent on nature elements which aren't necessarily predictable [43].

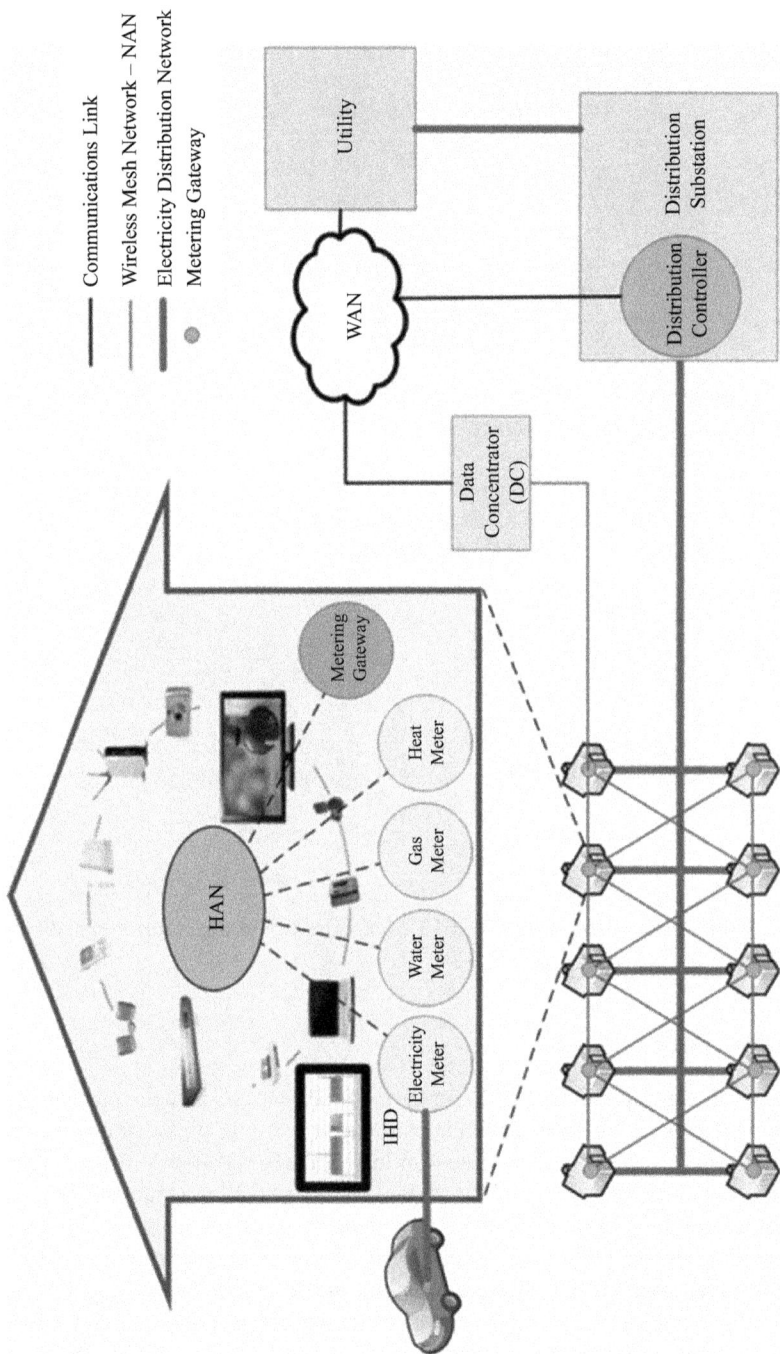

Figure 8.16 Typical smart metering architecture. ©2013 IEEE. Reprinted, with permission, from Reference 42.

Legend:
— Communications Link
--- Wireless Mesh Network – NAN
≡ Electricity Distribution Network
• Metering Gateway

Utility

WAN

Distribution Controller

Distribution Substation

Data Concentrator (DC)

Metering Gateway

Heat Meter

Gas Meter

Water Meter

Electricity Meter

HAN

IHD

Smart grid refers to electrical distribution networks which leverage communication and computing for improved collection, efficiency, reliability, and safety [44]. Such a grid allows the matching between consumer demand and power generation, in addition it also provide pricing incentives for deferring power use during peak hours. For example, a two-way communication between the customer and grid can result in the customer delaying turning on their AC/Heat [44]. Some of the functionality needed from smart grids include [45]:

- Detecting any abnormalities, e.g., a natural disaster causing a breakdown, and self healing where it is able to recover without causing service disruption by re-balancing itself.
- Offering a high level of reliability and quality.
- Resistant to cyber attacks.
- Incorporating a variety of energy distribution and storage options.
- Optimizing utilization of grid resources.
- Minimizing expenses for maintenance and operation.

Typical smart grid communication is three tiered [42, 44], which is also shown in Figure 8.16:

- Home Area Network (HAN): The network formed inside a house between appliances, devices, and home smart meter (which enables two communication between the house and the grid).
- Neighborhood Area Network (NAN): Forms the network between multiples HANs and a central data collector.
- Wide Area Network (WAN): The network between NANs and the control centers which monitors energy consumption.

Communication in smart grid spans multiple devices and networks. For example, HAN communication might use Zigbee and WiFi while the WAN might use LTE or fiber-based backbone; the 802.15.4g working group aims to specify a global standard for large-scale and network diverse Smart Utility Networks (SUN) communication [46, 47]. Cognitive radio (CR) can be leveraged to provide interoperability between the various standards. In addition, the operation of many radio nodes in the HAN can lead to significant interference especially if they are all operation in the ISM band, and the use of a CR can mitigate such interference. Smart grids will generate data on the order of tens of terabytes [48] which might be challenging for the current communication infrastructure to handle, the use of CRs and Dynamic Spectrum Access (DSA) would allow NAN and WAN communication to increase overall bandwidth by opportunistically leveraging unused spectrum.

More details can be found in the works cited in this section.

Bibliography

[1] A. Medeisis, O. Holland, and L. De Nardis, "Taxonomy of Cognitive Radio Applications," in *Dynamic Spectrum Access Networks (DYSPAN), 2012 IEEE International Symposium on.* IEEE, 2012, pp. 166–170.

[2] J. Wang, M. Ghosh, and K. Challapali, "Emerging Cognitive Radio Applications: A Survey," *Communications Magazine, IEEE*, vol. 49, no. 3, 2011, pp. 74–81.

[3] M. Vu, N. Devroye, and V. Tarokh, "On the Primary Exclusive Region of Cognitive Networks," *Transactions on Wireless Communication*, vol. 8, no. 7, 2009, pp. 3380–3385. doi: 10.1109/TWC.2009.080454

[4] C. de Souza Lima Jr, F. Paisana, J. Ferreira de Rezende, L. DaSilva, "A Cooperative Approach for Dynamic Spectrum Access in Radar Bands," in *Telecommunications Symposium (ITS), 2014 International*. IEEE, 2014, pp. 1–5.

[5] I. Subbiah, M. Schrey, A. Ashok, *et al.*, "Design of a TV White Space Converter Prototype Towards Cognitive Radio for WLAN Routers," in *Cognitive Radio Oriented Wireless Networks (CROWN-COM), 2013 8th International Conference on*. IEEE, 2013, pp. 208–213. doi: 10.1109/CROWN-Com.2013.6636819

[6] T. Matsumura and H. Harada, "Prototype of Tablet-type TV Band Portable Device with UHF Converter for TV White-spaces Utilization," in *Personal Indoor and Mobile Radio Communications (PIMRC), 2013 IEEE 24th International Symposium on*. IEEE, 2013, pp. 2748–2752.

[7] I. Subbiah, A. Ashok, G. Varga, M. Schrey, and S. Heinen, "System Design of a High-IF to UHF Converter Enabling Cognitive Radio," in *Cognitive Cellular Systems (CCS), 2014 1st International Workshop on*. IEEE, 2014, pp. 1–5.

[8] R. E. Watson, "Receiver Dynamic Range: Part 1," retrieved October 4, 2015. [Online]. Available: http:www/rfcafe.com/references/articles/wj-tech-notes/Rec dyn range1.pdf

[9] J. Xiao, R. Hu, Y. Qian, L. Gong, and B. Wang, "Expanding LTE Network Spectrum with Cognitive Radios: From Concept to Implementation," *Wireless Communications, IEEE*, vol. 20, no. 2, 2013, pp. 12–19.

[10] T. Matsumura, K. Ibuka, K. Ishizu, H. Murakami, and H. Harada, "Prototype of FDD/TDD Dual Mode LTE Base Station and Terminal Adaptor Utilizing TV White-spaces," in *Cognitive Radio Oriented Wireless Networks and Communications (CROWNCOM), 2014 9th International Conference on*. IEEE, 2014, pp. 317–322. doi: 10.4108/icst.crowncom.2014.255388

[11] D. Wieruch, T. Wirth, O. Braz, A. Dußmann, M. Mederle, and M. Muller, "Cognitive Repeaters for Flexible Mobile Data Traffic Offloading," in *Cognitive Radio Oriented Wireless Networks (CROWN-COM), 2013 8th International Conference on*. IEEE, 2013, pp. 214–219. doi: 10.1109/CROWN-Com.2013.6636820

[12] S. Haykin, "Cognitive Radar: A Way of the Future," *IEEE Signal Processing Magazine*, vol. 23, no. 1, January 2006, pp. 30–40. doi: 10.1109/msp.2006.1593335

[13] C. Baylis, M. Fellows, L. Cohen, and R. J. Marks II, "Solving the Spectrum Crisis: Intelligent, Reconfigurable Microwave Transmitter Amplifiers for Cognitive Radar," *IEEE Microwave Magazine*, vol. 15, no. 5, pp. 94–107, July 2014. [Online]. Available: http://dx.doi.org/10.1109/mmm.2014.2321253

[14] C. de Souza Lima Jr, F. Paisana, J. F. de Rezende, and L. A. DaSilva, "A Cooperative Approach for Dynamic Spectrum Access in Radar Bands," in *2014 International Telecommunications Symposium (ITS)*. IEEE, August 2014. doi: 10.1109/its.2014.6947955

[15] M. Marcus, "Sharing Government Spectrum with Private Users: Opportunities and Challenges," *IEEE Wireless Communication*, vol. 16, no. 3, June 2009, pp. 4–5. doi: 10.1109/mwc.2009.5109457

[16] H. Griffiths, L. Cohen, S. Watts, *et al.*, "Radar Spectrum Engineering and Management: Technical and Regulatory Issues," *Proceedings of the IEEE*, vol. 103, no. 1, January 2015, pp. 85–102. doi: 10.1109/jproc.2014.2365517

[17] R. S. of ITU (ITU-R), "Dynamic Frequency Selection in Wireless Access Systems (Recommendation ITU-R M.1652-1 (05/2011))," International Telecommunications Union, Geneva, Switzerland, Tech. Rep., 2011.

[18] *802.11h-2003 – IEEE Standard for Information Technology – Local and Metropolitan Area Networks – Specific Requirements – Part 11: Wireless LAN Medium Access Control (MAC) and Physical Layer (PHY) Specifications – Spectrum and Transmit Power Management Extensions in the 5 GHz Band in Europe*, IEEE Std., 2003.

[19] Draft ETSI EN 301 893 V1.7.2 (2014-07). ETSI. Accessed August 12, 2015. [Online]. Available: http://www.etsi.org/deliver/etsi_en/301800_301899/301893/01.07.02_20/en_301893v010702a.pdf

[20] Federal Communications Commission (US). "Ka-band Geostationary Satellite Orbit (GS) and Non-Geostarionary Orbit (NGSO) Fixed-Satellite Servicew (FSS). [Online]. Available: http://www2.fcc.gov/ib/sd/ssr/ka_band.html

[21] "Glossary," Via Satellite. (n.d.). Retrieved August 28, 2015. [Online]. Available: http://www.satellitetoday.com/glossary-glossary1-html/#A

[22] Alliance for Telecommunications Glossary, "Alliance for Telecommunications Industry Solutions." *Retrieved from the Internet on August*, vol. 28, 2015, p. 2006. [Online]. Available: http://www.atis.org/glossary/

[23] J. Christensen, "ITU Regulations for Ka-band Satellite Networks," September 25, 2015. [Online]. Available: http://www.itu.int/md/dologin_md.asp?id=R12-ITURKA.BAND-C-0001!!MSW-E

[24] E. T. S. Institute, Cognitive Radio Techniques for Satellite Communications Operating in Ka Band (ETSI TR 103 263 V1.1.1 (2014-07).

[25] S. Maleki, S. Chatzinotas, B. Evans, *et al.*, "Cognitive Spectrum Utilization in Ka Band Multibeam Satellite Communications," *Communications Magazine, IEEE*, vol. 53, no. 3, 2015, pp. 24–29. doi: 10.1109/MCOM.2015.7060478

[26] T. Pratt, C. W. Bostian and J. Allnutt, *Satellite Communications, 2nd Edition*. Hoboken, NJ: John Wiley & Sons, 2003.

[27] S. Maleki, S. Chatzinotas, J. Krause, K. Liolis, and B. Ottersten, "Cognitive Zone for Broadband Satellite Communication in 17.3–17.7 GHz Band," 2015.

[28] A. Paulson and T. Schwengler, "A Review of Public Safety Communications, from LMR to Voice Over LTE (VoLT E)," in *Personal Indoor and Mobile Radio Communications (PIMRC), 2013 IEEE 24th International Symposium on*. IEEE, 2013, pp. 3513–3517.

[29] Federal Communications Commission (US) Public Safety and Homeland Security Bureau 8. Techtopics Topic 8: Cognitive Radio for Public Safety. [Online]. Available: https://transition.fcc.gov/pshs/techtopics/techtopic8.html

[30] Federal Communications Commission (US) Public Safety and Homeland Security Bureau 9. Topics Topic 9: Cognitive Radio Potential for Public Safety. [Online]. Available: https://transition.fcc.gov/pshs/techtopics/techtopic9.html

[31] A. S. Fayez, "Designing a Software Defined Radio to Run on a Heterogeneous Processor," Ph.D. dissertation, Virginia Polytechnic Institute and State University, 2011.

[32] B. Le, F. A. Rodriguez, Q. Chen, *et al.*, "A Public Safety Cognitive Radio Node," in *Proceedings of the 2007 SDR Forum Technical Conference*, SDRF, 2007.

[33] B. Le, "Building a Cognitive Radio: From Architecture Definition to Prototype Implementation," *Doctor of Philosopy Dissertation*, Virginia Polytechnic Institute and State University, 2007.

[34] A. Kumbhar and I. Güvenç, "A Comparative Study of Land Mobile Radio and LTE-Based Public Safety Communications," in *SoutheastCon 2015*. Ft. Lauderdale, FL: Institute of Electrical & Electronics Engineers (IEEE), April 2015. doi: 10.1109/secon.2015.7132951

[35] R. Ferrus, O. Sallent, G. Baldini, and L. Goratti, "LTE: The Technology Driver for Future Public Safety Communications," *Communications Magazine, IEEE*, vol. 51, no. 10, 2013, pp. 154–161.

[36] FirstNet. First Responder Network Authority. [Online]. Available: http://www.firstnet.gov/

[37] APCO International. Project 25. [Online]. Available: https://www.apcointl.org/spectrum-management/resources/interoperability/p25.html

[38] F. C. Commission. Federal Communications Commission (US). [Online]. Available: https://apps.fcc.gov/edocs_public/attachmatch/FCC-11-93A1.pdf

[39] A. R. Young, Unified Multi-domain Decision Making: Cognitive Radio and Autonomous Vehicle Convergence, *Ph.D. Dissertation*, Electrical and Computer Engineering, Virginia Tech, Blacksburg, 2013.

[40] A. R. Young and C. W. Bostian, "Simple and Low-cost Platforms for Cognitive Radio Experiments [Application Notes]," *Microwave Magazine, IEEE*, vol. 14, no. 1, 2013, pp. 146–157.

[41] A. Anderson, E. W. Frew, and D. Grunwald, "Cognitive Radio Development for UAS Applications," in *Unmanned Aircraft Systems (ICUAS), 2015 International Conference on*. IEEE, 2015, pp. 695–703.

[42] Z. Fan, P. Kulkarni, S. Gormus, *et al.*, "Smart Grid Communications: Overview of Research Challenges, Solutions, and Standardization Activities," *Communications Surveys & Tutorials, IEEE*, vol. 15, no. 1, 2013, pp. 21–38.

[43] D. Lindley, "Smart Grids: The Energy Storage Problem," *Nature News*, vol. 463, no. 7277, 2010, pp. 18–20.

[44] R. Ma, H.-H. Chen, Y.-R. Huang, and W. Meng, "Smart Grid Communication: Its Challenges and Opportunities," *Smart Grid, IEEE Transactions on*, vol. 4, no. 1, 2013, pp. 36–46.

[45] R. E. Brown, "Impact of Smart Grid on Distribution System Design," in *Power and Energy Society General Meeting-Conversion and Delivery of Electrical Energy in the 21st Century, 2008 IEEE*. IEEE, 2008, pp. 1–4.

[46] C.-S. Sum, F. Kojima, and H. Harada, "Coexistence of Homogeneous and Heterogeneous Systems for IEEE 802.15. 4g Smart Utility Networks," in *New Frontiers in Dynamic Spectrum Access Networks (DySPAN), 2011 IEEE Symposium on*. IEEE, 2011, pp. 510–520.

[47] I. C. S. L. M. S. Committee *et al.*, "Wireless Medium Access Control (MAC) and Physical Layer (PHY) Specifications for Low-Rate Wireless Personal Area Networks (LR-WPANs), IEEE Std 802.15. 4," 2003.

[48] R. Yu, Y. Zhang, S. Gjessing, C. Yuen, S. Xie, and M. Guizani, "Cognitive Radio Based Hierarchical Communications Infrastructure for Smart Grid," *Network, IEEE*, vol. 25, no. 5, 2011, pp. 6–14.

Index

additive white Gaussian noise (AWGN) 27, 57,
    137, 159, 191
adjacent channel power ratio (ACPR) 87,
    223–4
Altera Cyclone V SoC series 109
amplify-and-forward (AF) 222
analog front end (AFE) 73, 75–9, 81–2, 88
analog hardware radio frequency (RF)
    platforms 1
analog-to-digital converter (ADC) 65, 73,
    75–7, 82–3, 85, 88–9, 94–5, 97, 127
ant action-based social language 193
applications, cognitive radio
    autonomous vehicles, cognitive radio
      and 236–9
    cognitive radar 222–4
    Ka band geostationary satellite applications
      228–32
    legacy radar and communications system
      coexistence 224–5
      controlling transmitter power to limit/eliminate
        interference 225–7
      cooperative and cognitive approaches 228
      implementation 227–8
    LTE cognitive repeaters for indoor
      applications 221–2
    LTE frequency converters 221
    public safety and emergency first responder
      communication 232–3
      current (2016) situation 235–6
      Virginia Tech Public Safety Cognitive
        Radio 233
    smart grids 239–41
    WiFi frequency translators 216–20
      RF considerations 216
    zoned dynamic spectrum access 215
application-specific integrated
    circuit (ASIC) 92, 109

Arithmetic Logic Unit (ALU) 104
ARM-based processors 109
artificial intelligence (AI) technique 18, 23
artificial neural networks (ANNs) 32,
    158–9, 195
Association of Public Safety Communications
    Officers International (APCO) 232, 236
automatic gain control (AGC) stage 76
automatic neighbor relation (ANR)
    actions 181
autonomous vehicles, cognitive radio
    and 236–9

Baseband Scale (BS) 135
Basic Service Set (BSS) 15
BeagleBoard 132, 234
BeagleBoard-xM 198
bee waggle dance 185
behavior-based design 197
Berkley Design Technology, Inc. (BDTI) 130
blocker 84–5
blocking signal 84, 216
book-keeping 31
Boolean-controlled Data flow (BDF) 116–19
broadband 232
Broadcasting Satellite Service (BSS) 229–32

cache memory 105
Calculus of Communicating Systems (CCS)
    122–3
case-based decision theory (CBDT) 34, 52
case-based learning 31, 34
Center for Wireless Telecommunications
    (CWT) 132
CE-Radio Interface 9, 52
Channel Availability Query (CAQ) 16
Channel Filter (CF) 135
Channel Schedule Management (CSM)
    procedure 16

choosing a platform 128
  benchmarks 130
  between RF alternatives 128–9
  processor choices 129–30
  processor interconnect 130–1
classification 32–3
cognitive engine (CE) 3–4, 23, 186
  architecture 42
    CSERE architecture 53–5
    European Telecommunications Standards
      Institute (ETSI) 46
    flow of control 42–4
    monolithic versus distributed 44
    original CE architecture 49–52
    Wireless Innovation Forum (WINNF) 45–6
  basic function of 24–5
  components, in the cognition loop 26
  estimation 38–9
  general CE architecture 45
  information flow in 55
    example use of uncertainty coefficient
      57–60
  machine learning 30
    case-based learning 34
    classification 32–3
    Markov models 33–4
    reinforcement learning 31–2
  optimizers 35
    evolutionary algorithms 36–8
    expert systems 38
    mathematical approaches 36
  organization 25
    controller 30
    knowledge base 28
    objective analyzer 27
    optimizer 27
    radio interface 28
    ranker 28
    sensor 28–9
    user interface 29–30
  original CE functional architecture 51
  original CE logical architecture 50
  sensing 39
    cooperative sensing 41
    energy detection 39–40
    feature detection 40–1
    Radio Environment Map (REM)
      sensing 42
cognitive network
  building, with social language 196
    behavior-based design and social language
      197
    entry scenario 207–10

hardware considerations and
  implementation 198–200
implementing behaviors in software and
  hardware 200–6
MAC layer considerations 196–7
network evaluation 207
social learning 210–11
tasks and behaviors 197–8
total system behavior 211–12
  goals 180–2
  versus networks of cognitive radios 179–80
cognitive radar 222–4
cognitive repeater concept 222
Cognitive System Enabling Radio Evolution
  (CSERE) architecture 9, 29, 52–5
cognitive versus adaptive quandary 18
collective cognition 180–1
Communicating Sequential Processes (CSP)
  119–22, 135
communications and radar system coexistence
  222–3
  cognitive radar 222–4
  legacy radar 224–5
    controlling transmitter power 225–7
    cooperative and cognitive approaches 228
    implementation 227–8
complementary metal-oxide-semiconductor (CMOS)
  technology 4
compression region 78, 84
computational devices 107
  alternative computational devices 109
    Application-specific Integrated Circuits
      (ASICs) 109
    computational heterogeneity 109–10
    digital signal processors (DSPs) 107–8
    field-programmable gate arrays (FPGAs)
      108–9
    general purpose processors (GPPs) 108
computational platform 3, 44, 54, 65, 92, 95,
  127
computing hardware and cognitive radio
  architecture 103
consensual coordination 185
Contact Verification Signal (CVS) 15
continuous wave modulation (CW) 75
control flow computing 103
  control and data paths 104–5
  memory hierarchy 105–6
  multiprocessors 106
controller 9, 43, 48–9, 54–5, 104
  of cognitive engine 30
  fixed FEC cognitive radio 172–4
  free FEC cognitive radio 167–8

control ports 116
cooperative sensing 41–2
Cyclo-dynamic Data Flow (CDDF) 119
Cyclo-static Data flow (CSDF) 118–19

data flow computing 103, 106–7, 113
Data Flow Interchange Format (DIF) language
    123
data flow models of computation 112
    Boolean-controlled Data flow
        (BDF) 116–18
    Cyclo-static Data flow (CSDF) 118–19
    Kahn Process Networks (KPN) 113
    Parametrized Synchronous Data flow (PSDF)
        119
    Synchronous Data Flow (SDF) 113–16
Data Source (DS) 135
Data Source block 139–40
data transmission 200
DBPSK Modulator (DM) 135
Decision Maker 9, 52
decode-and-forward (DF) donor receptors 222
decoding messages 189
Defense Advanced Research Project Agency
        (DARPA) 3–4, 94
desensing 84
de-sensitizing 84
DESYNC algorithm 201
    modifying 204–5
desynchronization 201
Differential Binary Phase-Shift Keying (DBPSK)
        radios 135, 137–9
digital signal processor (DSP) system 65, 92,
        103, 107–10, 128, 130, 132–4, 235
digital-to-analog converter (DAC) 65, 94–5,
        97, 127
diplexers 88
Direct Broadcast Satellites (DBS) 228–30
direct-launch frequency converter 68
direct launch transmitter 68, 71
Direct RF DAC 68
distributed CEs 44
domain-mapping tables 135–6
DSM Trigger Entity 47
duplexers 88
Dynamic Self-Organizing Network Planning
        and Management (DSONPM) block
        46, 48–9
dynamic spectrum access (DSA) 1, 11–12, 40,
        42, 46, 66, 88, 159, 182, 215, 241
    in broadcast television bands 11
        cognitive radio design for U.S. TV White
            Space 17–18

FCC rule implementation by the IEEE
        802.11af Standard 16
FCC rules and commercial standards for
        unlicensed Television Band Devices
        13–16
    Standard ECMA-392 16–17
    TV channel occupancy in the United States
        12–13
    TV White Space Databases 16
Dynamic Spectrum Management (DSM) block
        46–9
DySPAN 2007 4, 7

Electronic Code of Federal Regulations 14
Embedded Microprocessor Benchmark
        Consortium (EEMBC) 130
emergency first responder (EFR) agencies 232
encoding phase 155–7
    continued performance heuristics 161–3
    high-level group heuristics 163
    policy group heuristics 163
end-to-end functionality 179
energy conservation 2
energy detection 39–40
entropy 55–7
equipment cost 1
error end state 155
error vector (EV) 85–6
error vector magnitude (EVM) 87
estimation 38–9
Ettus products 95
Ettus Research Universal Software Radio Periph-
        eral (USRP) 4
Ettus RF daughter boards 97
European Computer Manufacturers
        Association (ECMA-International)
        ECMA-392 16–17
European Telecommunications Standards
        Institute (ETSI) Reconfigurable
        Radio Systems (RRS) 46
    Dynamic Self-organizing Network Planning
        and Management Block (DSONPM)
        48–9
    Dynamic Spectrum Management (DSM)
        block 46–7
Eutelsat 229
evaluation, of cognitive radio 147
    example code 167
        fixed FEC cognitive radio 172–6
        free FEC cognitive radio 167–72
        interpolation code 176–8
    example evaluation
        encoding phases 161–3

interpolation phase 163–6
logging phase 160–1
setup phase 159–60
metrics and factors for 149–50
actions 154
language 151–4
purpose 150–1
performance evaluation principles 148
practical evaluation methods 154–5
alternative approaches to 159
encoding phase 155–7
interpolation phase 157–9
logging phase 155
setup phase 155
event-driven CE 43–4
evolutionary algorithms 36–8
evolved node B (eNB) 181
expert systems 38, 47, 57, 59

faculty meeting problem 3, 223
fast Fourier transform (FFT) 75, 82–3, 89
feature detection 40–1
Federal Communications Commission
(FCC) 3
FCC rules and commercial standards for
unlicensed Television Band Devices
13–16
field programmable gate array (FPGA) 43, 95,
108–9, 128, 198
finite state machines (FSMs) 33, 123
firing vector 115–16
first in first out (FIFO) buffer 92, 112, 115
FirstNet 236
5.8-GHz Proxim *Tsunami* radio 4
fixed devices 15
Fixed Satellite Service (FSS) 229
forward error correction (FEC) CR
fixed 172
controller 172–4
knowledge base 176–7
optimizer 174
radio 175–6
ranker 174–5
free 167
controller 167–8
knowledge base 171–2
optimizer 168–9
radio 170–1
ranker 169
frequencies identified for cognitive radio
operation 12–13
frequency division duplexing (FDD) 88
frequency flocking 198, 205–6, 211

gain compression 78, 84
Gaussian noise blocks 139–40
general-purpose input/output (GPIO) 92
general purpose processors (GPPs) 103,
108–10, 128, 130, 132–3
generic metrics, definitions of 156
Genetic Algorithm (GA)-based optimizer
component 29
genetic algorithm (GA) CE 192
Geostationary Satellite Orbit (GSO) satellites
228–32
global positioning system (GPS) 42
GNU Radio 94–5, 97, 108, 123, 132–3, 135, 138
stream-to-vector block 139
vector-to-stream block 139
group learning 194–6

Hardware Description Languages (HDLs) 109, 134
Hardware-in-the-Loop (HIL) simulation 134
high-IF architecture 216
High Throughput Satellites (HTS) 229, 231
hive mind 180
Home Area Network (HAN) 241
homodyne/low-IF and SDR difficulties 216
HughesNet 229

IEEE 802.11af Standard 14–16
information flow in cognitive engines 55
example use of uncertainty
coefficient 57–60
Information Process Architecture (IPA) 45
information theory analysis 194
Integer-controlled Data flow (IDF) MoC 118
Integrated Development Environments (IDEs)
128, 142
integrated radio front end 71
integrated SDR platform 127
interaction methods for cognitive radios 182
components of interaction 187
observability 187–9
understanding 189–90
social language 183–7
interactions, analyzing 190
analysis results 192–4
example analysis 191–2
intermediate frequency (IF) 216
intermodulation products 17, 78
International Telecommunications
Union (ITU) 229
Internet Engineering Task Force Protocol to
Access White Space (IETF PAWS)
database 14
Internet Protocol (IP) 236

interpolation phase 157–9
Inter-Satellite Service (ISS) 229

Ka band geostationary satellite applications
        228–32
Kahn Process Networks (KPN) model 113
knowledge base 28
    of cognitive engine 28
    fixed FEC cognitive radio 176–7
    free FEC cognitive radio 171–2

Land Mobile Radio (LMR) 232, 234
language 151–2
    external language 154
    human language 152–3
    internal language 153–4
legacy radar 224–5
    controlling transmitter power 225–7
    cooperative and cognitive approaches 228
    implementation 227–8
logging phase 155–61, 163
Long Term Evolution (LTE) service 123, 236
    LTE cognitive repeaters for indoor
        applications 221–2
    LTE frequency converters 221
Low Power Auxiliary Devices (LPAD) 13

machine learning 30
    case-based learning 34
    classification 32–3
    Markov models 33–4
    reinforcement learning 31–2
Mango FMC-RF-2X245 99
Markov decision process (MDP) 33
Markov modeling 33–4, 38
mathematical optimization 36
Mathwork's Simulink 109, 128, 132, 134
Maxim MAX5688 RF DAC 72
maximum delay value 119
maximum token transfer function 119
Measurement Collecting Entity 47
medium access control (MAC) 196
    and performance considerations 88–90
memory hierarchy 105–6
mental language lexical items 152
meta-data 135–6
metrics and factors for cognitive radio
        evaluation 149–50
    actions 154
    language 151–4
        external language 154
        human language 152–3
        internal language 153–4
    purpose 150

environment 151
    mission 150–1
microprocessors 103
minimum detectable signal 75
Mixed Integer Program (MIP) 136
Mobile Satellite Service (MSS) 229
Model-Based Design (MBD) 128, 134
models-of-computation 134
    computational knobs and meters 141
    core framework 135–6
    first stage 136–7
    mapping of, for radio applications 135
    second stage 137–41
models of computation (MoCs) 110, 132
    application of 134
        computational knobs and meters 141
        core framework 135–6
        first stage 136–7
        mapping for radio applications 135
        second stage 137–41
    data flow models of computation 112
        Boolean-controlled Data flow
            (BDF) 116–18
        Cyclo-static Data flow (CSDF) 118–19
        Kahn Process Networks (KPN) 113
        Parametrized Synchronous Data flow
            (PSDF) 119
        Synchronous Data Flow (SDF) 113–16
    Process Algebra 119
        Calculus of Communicating Systems (CCS)
            122–3
        Communicating Sequential Processes (CSP)
            119–22
    reactive and real-time systems 111–12
    use 123
modular cognitive radio architecture 6
Modulation Coding Scheme (MCS) 24–5
monolithic design 44
multi-agent systems 197, 207
multi-armed bandit (MAB) approach 33
multi-carrier receiver 68–9
multi-input multi-output (MIMO) technology
        223
Multiple Instruction streams Multiple Data
        streams (MIMD) 106
Multiple Instruction streams Single Data stream
        (MISD) 106
multiprocessors 106

National Instrument's LabVIEW 109, 134
National Science Foundation (NSF) Network
        Technology and Systems (NeTS)
        program 4

natural social language 183–5
need for cognitive radio 1–2
Neighborhood Area Network (NAN) 241
nested cognition loop 52–3
  Virginia Tech implementation of 8
Network Affiliation 233
network resource managers 23
network self-organization 2
networks of cognitive radios
  versus cognitive networks 179–80
neural network interpolator (NNI) 158
neural networks (NNs) 32
noise, noise performance, and weak
    signal behavior 72–7
non-determinism 206
non-geostationary orbit (NGSO)
    satellites 228–9
Non-recurring Engineering (NRE) cost 109
NVIDIA Graphic Processor Units (GPUs) 123

objective analyzer, of cognitive engine 27
observe, orient, decide, and act (OODA)
    loop 186–7
Occam radio flowgraph definitions 135
one-armed bandits 33
OpenEmbedded (OE) project 108
operation, of cognitive radio 5–11
operations and management (OAM)
    system 181
optimization of network operation 182
optimizer 5, 35
  of cognitive engine 27
  evolutionary algorithms 36–8
  expert systems 38
  fixed FEC cognitive radio 174
  free FEC cognitive radio 168–9
  mathematical approaches 36
original CE architecture 49–52
original cognition loop concept 8
Original Virginia Tech cognitive engine
    concept 5
origin of cognitive radio 3–5
overdrive region 78

packaged RF front ends and all-in-one
    platforms 94
  RF daughter boards 97–8
  USRP family of products and GNU radio
    94–7
  Wireless Open-Access Research Platform
    (WARP) products 98–9
Packet Encoder (PE) 135
Parameter Generator (PG) 135

Parametrized Synchronous Data flow
    (PSDF) 119
Pareto optimality 35
partially observable Markov decision processes
    (POMDPs) 33
peaker plants 239
peak-to-average power ratio (PAPR) 76
Performance API 9
performance evaluation principles 148
Periodic Admissible Parallel Scheduler
    (PAPS) 114
Periodic Admissible Sequential Schedule
    (PASS) 114
Π-calculus 123
polling control flow approach 43
power-added efficiency (PAE) 223
power amplifier (PA) 68
practical evaluation methods 154–5
  alternative approaches to 159
  encoding phase 155–7
  interpolation phase 157–9
  logging phase 155
  setup phase 155
probably approximately correct (PAC)
    learning 32
Process Algebra 119
  Calculus of Communicating Systems (CCS)
    122–3
  Communicating Sequential Processes (CSP)
    119–22
processor choices 129–30
processor interconnect 130–1
Program Making Special Event (PMSE)
    devices 13
prototype LTE transmit/receive
    converter 220–1
prototype white space to WiFi converter 217
public safety and emergency first responder
    communication 232–3
  current (2016) situation 235–6
  Virginia Tech Public Safety Cognitive
    Radio 233
    architecture and major components 234
    design perspectives 235
    operational example 234–5
Public Safety Cognitive Radio (PSCR) 132,
    233
PyBrain 165

Q-learning 32
quiescent points 119

Radio Environment Map (REM)
    sensing 42

radio frequency (RF) architectures 67
    receivers 67–8
    transmitter 68
radio frequency (RF) daughter boards 97–8
radio frequency (RF) platforms 65, 127
    medium access control (MAC) and
        performance considerations 88–90
    preliminary considerations in
        choosing 66–7
    receiver RF specifications 71
        noise, noise performance, and weak signal
            behavior 72–7
        strong signal behavior 77–85
    software defined radio, platforms for 94
        RF daughter boards 97–8
        USRP family of products and GNU radio
            94–7
        Wireless Open-Access Research Platform
            (WARP) products 98–9
    transmitter RF specifications 85–8
radio frequency integrated circuit (RFIC) 1, 65,
    127, 198
    computational support for 92
    RFM69CW 90–2
radio interface, of cognitive engine 28
Radio Performance Monitor 9, 52
Radio Resource Monitor 9, 52
ranker 28, 48
    of cognitive engine 28
    fixed FEC cognitive radio 174–5
    free FEC cognitive radio 169
Raspberry Pi 92
reactive and real-time systems 111–12
Real-time Operating Systems (RTOSs) 43
real-time spectrum analyzer, operation
    of 89–90
received signal strength indicator (RSSI)
    measures 199–200
receiver RF specifications 71
    noise, noise performance, and weak signal
        behavior 72–7
    strong signal behavior 77
        blocking 84–5
receivers 67–8
Registered Location Secure Server (RLSS) 16
reinforcement learning 31–4
repetition vectors 115, 118, 136
RFM69HCW 91–2
RF Out (RO) 135
rule-based social languages 194
rule-based system 38, 57
rules 38

Satellite Digital Audio Radio Service (SDARS)
    229
saturation 78
scheduler 114
self-configuration 181–2
self-healing 180–2
self-optimization 180–2
self-organizing networks (SONs) 181
sensing 15, 39
    cooperative sensing 41
    energy detection 39–40
    feature detection 40–1
    Radio Environment Map (REM)
        sensing 42
sensing only devices 15–16
sensor component, of cognitive engine 28–9
serial peripheral interface (SPI) bus 92
setup phase 155
    cognitive radio 160
    mission 159–60
signal-to-noise ratio (SNR) 75, 162
SINAD derivation 75–7
single-carrier receiver 67–8
Single Instruction stream Multiple Data streams
    (SIMD) 106
Single Instruction stream Single Data stream
    (SISD) 106
SKIRL platform 198–200, 205–6, 237
smart grids 239–41
smart metering architecture 240
Smart Utility Networks (SUN) communication
    241
social insects 180
social language 183–7
    building cognitive network with 196
        behavior-based design and social language
            197
        entry scenario 207–10
        hardware considerations and
            implementation 198–200
        implementing behaviors in software and
            hardware 200–6
        MAC layer considerations 196–7
        network evaluation 207
        social learning 210–11
        tasks and behaviors 197–8
        total system behavior 211–12
    cognitive radio 185
    natural 183
social learning 210–11
social transmission of information 185
societal learning 194–6
societal rules 183–6, 189–91, 193

Software Defined Radio (SDR) platforms 1,
    3–4, 67, 94, 127–8
  choosing 128
    benchmarks 130
    between RF alternatives 128–9
    processor choices 129–30
    processor interconnect 130–1
  computational platform 127
  integrated SDR platform 127
  packaged RF front ends and all-in-one
      platforms 94
  RF daughter boards 97–8
  USRP family of products and GNU radio
      94–7
  Wireless Open-Access Research Platform
      (WARP) products 98–9
  programming 131–2
    classic approach 132
    Model-Based Design (MBD) 134
    models-of-computation, application
        of 134–41
  RF platform 127
  software application 128
  supporting hardware 127–8
spectrum shortage 1
Spurious Free Dynamic Range (SFDR) 81,
    218
Standard ECMA-392 16–17
Standard Performance Evaluation Corporation
    (SPEC) 130
strong signal behavior 77, 218
  blocking 84–5
superheterodyne 67
super-organism learning 195
surface acoustic wave (SAW) filters 85
swept-tuned spectrum analyzer 88–9
symbol error rate (SER) 75
Synchronous Data Flow (SDF) 113–16,
    118, 135

Television Band Devices (TVBDs), unlicensed
    13–16
third-order input intercept point
    (IIP3) 79, 84
third-order intercept point 79–82, 84, 218
third-order output intercept point (OIP3) 79
time division duplexing (TDD) 88
time division multiple access (TDMA) 200
topology matrix 114–15, 118, 136–9
transmitter 12, 40–2, 67–8, 87, 194, 223–6
transmitter RF specifications 85–8

TV white space (TVWS) spectrum
  LTE frequency converters 221
  WiFi frequency translators 216–20
    RF considerations 216
TV White Space Databases 16

UAV (Un-piloted Autonomous Aerial Vehicle)
    237–8
Ubuntu Linux operating system 198
Universal Grammar 152
Universal Software Radio Peripheral (USRP)
    94–7
Universal Software Radio Platform (USRP)
    series 128–9
user interface, of cognitive engine 29–30
USRP 4, 9, 94–5, 98, 128, 131–2, 135, 234
U.S. TV White Space, cognitive radio design for
    17–18

variable gain amplifier (VGA) 68
Virginia Tech (VT) cognitive engine 3, 5, 8–10
Virginia Tech implementation of nested
    cognition loop 8
Virginia Tech Public Safety Cognitive
    Radio 233
  architecture and major components 234
  design perspectives 235
  operational example 234–5
visitation probability 206
Von Neumann architecture 104–6

Wide Area Network (WAN) 241
wideband 232
WiFi frequency translators 216–20
  RF considerations 216
WiFi transmission 24–5
Wireless Innovation Forum (WINNF) 45–6
Wireless Modeling System 9, 52
Wireless Monitoring System 9
Wireless Open-Access Research Platform (WARP)
    products 94, 98–9
  WARP v3 hardware architecture 98
Wireless System Genetic Algorithm (WSGA)
    9, 52

Xilinx Zynq-7000 109

Yocto project 108

zero-IF (ZIF) architecture 68, 79
zoned dynamic spectrum access 215